AF625478

Functional Integration and Semiclassical Expansions

Mathematics and Its Applications



Volume 10

F. Langouche, D. Roekaerts
*Institute for Theoretical Physics, Catholic University of Leuven, Belgium*

and

E. Tirapegui
*Facultad de Ciencias Fisicas y Matematicas, Universidad de Chile, Santiago, Chile*

# Functional Integration and Semiclassical Expansions

SPRINGER-SCIENCE+BUSINESS MEDIA, B.V.

**Library of Congress Cataloging in Publication Data**

CIP

Langouche, F. (Flor), 1955–
Functional integration and semiclassical expansions.

(Mathematics and its applications ; v. 10)
Includes bibliographical references and index.
1. Integration, Functional. 2. Integrals, Path. 3. Mathematical physics. I. Roekaerts, D. (Dirk) II. Tirapegui, Enrique. III. Title. IV. Title: Semiclassical expansions. V. Series: Mathematics and its applications (D. Reidel Publishing Company) ; v. 10.
QC20.7.F85L36 1982 530.1'55 82-13165
ISBN 978-90-481-8377-7 ISBN 978-94-017-1634-5 (eBook)
DOI 10.1007/978-94-017-1634-5

FOREWORD

This book is intended as a fairly complete presentation of what we call the discretization approach to functional integrals, i.e. path integrals defined as limits of discretized axpressions. In its main parts it is based on the original work of the authors. We hope to have provided the readers with a rather complete and up-to-date bibliography, and we apologize to authors whose work has not been cited through ignorance on our part. Our main concern has been to present a formalism that is practical and which can be adapted to make computations in the numerous areas where path integrals are being increasingly used. For these reasons applications, illustrative examples, and detailed calculations are included.

The book is partially based on lectures given by one of us (E.T.) at the Institut de Physique Théorique of the U.C.L. (Louvain-la-Neuve). We thank Dr. M.E. Brachet (University of Paris) for his help in the redaction of chapter 8. We are indebted to many of our colleagues and especially to the members of the Instituut voor Theoretische Fysica, K.U. Leuven for their interest and encouragement. We also thank Professor Claudio Anguita, Dean of the Faculty of Physics and Mathematics of the University of Chile, for his constant support. Special thanks are due to Christine Detroije and Lutgarde Dubois for their very fine and hard work in typing the manuscript.

F. Langouche is Research Assistant Inter-University Institute for Nuclear Research, Belgium, and D. Roekaerts is Research Assistant National Fund for Scientific Research, Belgium.

## TABLE OF CONTENTS

# INTRODUCTION

The present work aims to treat of functional integrals. In our research it became evident to us that since the use of functional integrals had become so widespread in various branches of modern physics, it would be impossible to give a full account of all its applications. This work therefore does not attempt to explore all the practical ramifications of functional integral methods but confines itself to a more general treatment. We are concerned mostly with path integrations in a finite dimensional space, a d-dimensional configuration or a 2d-dimensional phase space and our primary interest lies in the ways of representing propagators of Fokker-Planck and Schrödinger equations, and the way of computing these. Although we are aware of the mathematical difficulties associated with a rigorous definition of path integrals these latter problems will be outside the scope of the work, and we shall treat both equations on the same footing. Our greatest concern is to be practical and to show how one has to handle the objects that we call functional integrals. The computation of the semiclassical expansion is also presented from the practical point of view.

We define functional integrals as limits of discretized expressions and we notice at once one of the essential features: several discretized expressions give the same limit and lead to formally distinct functional integral representations. Instead of rejecting this multiplicity by searching for a unique or privileged representation, we have been willing to work with it and have explored and exploited the consequences of such a multiplicity.

A first part of this programme is carried out in chapters 2 to 5, where we treat the notion of discretization and the relation of the multiplicity of representations to the ordering problem of operators (correspondence rules) and the stochastic properties of the paths. As a first application we study the perturbation expansion by means of phase space path integrals. We also consider nonlinear point transformations and introduce the geometric interpretation of Riemannian manifold that can be given to the configuration space.

In chapter 6 we treat covariant definitions of functional integrals : i.e. definitions which allow the use of the usual rules of calculus in the continuum expression.

In chapter 7 the previous techniques are applied to the case of Fokker-Planck dynamics and to the generalization of the Onsager-Machlup theory to general diffusion processes. There we also approach the multiplicity problem from the point of view of stochastic differential equations, which illustrates the stochastic origin of the different representations.
Product integrals and their relations to path integrals are treated in chapter 8.
In chapters 9 and 10 we study the semiclassical expansion. We first give the general method and go into more detail when we give the covariant and gauge invariant calculation of the correction terms to the WKB approximation on a Riemannian manifold. At the same time we analyze systematically how the calculation can be done in the configuration space integral.
Chapter 11 is devoted to the work of other authors on the subject, and we present in some detail the stochastic methods that have been used to obtain the WKB-approximation. We relate this to our approach by using the familiar discretized integrals
Finally in chapter 12, an exact calculation of the propagator on the manifold of SU(2) is presented which is a nice illustration of discretization techniques and differs from earlier treatments in that it succeeds in the computation using only functional integrals.

Appendices are placed immediately after the chapter to which they refer. Some of them are of a general character and present technical notions used in the text, while others are concerned with an alternative way of obtaining a result or give a generalization.

# CHAPTER I : FUNCTIONAL INTEGRALS DEFINED AS LIMITS OF DISCRETIZED EXPRESSIONS

## 1.1. Introduction

The fundamental solution or propagator $K(q,t;q_o,t_o)$ of a first order (in time t) partial differential equation

$$[\frac{\partial}{\partial t} - \mathcal{L}(q,\partial,t)] \quad \Phi(q,t) = 0 \tag{1.1}$$

is defined as its solution satisfying the initial condition

$$K(q,t_o;q_o,t_o) = \delta(q - q_o) \tag{1.2}$$

Boundary conditions in q-space can be specified and $\mathcal{L}(q,\partial,t)$ is a linear, not necessarily hermitean and possibly time dependent operator. ($\partial \equiv \frac{\partial}{\partial q}$) In the following we shall often restrict it to be of second order in $\partial$. It can then always be written in the form

$$\mathcal{L}(q,\partial,t) = \frac{1}{2} c \; \partial_\mu \partial_\nu \; G^{\mu\nu}(q,t) + \partial_\mu A^\mu(q,t) + \frac{1}{c} V(q,t) \tag{1.3}$$

where the range of the Greek indices is from 1 to d, d being the dimension of q-space. The number c in (1.1) is in general complex with positive real part. For c = i eq. (1.1) is a Schrödinger equation, for c = 1 it is a Fokker-Planck or diffusion type equation. The fundamental property of K leading to its path integral representation is the semigroup law

$$K(q,t;q_o;t_o) = \int dq' \; K(q,t;q',t') \; K(q',t';q_o,t_o) \tag{1.4}$$

In Fokker-Planck dynamics, where K has the interpretation of a transition probability density, this is called the Chapman-Kolmogorov equation. It expresses the Markov property of the stochastic process q (see Chapter 7).

A functional integral representation of K can be obtained in the following way. Divide the time interval $[t_o,t]$ into N + 1 pieces of length $\varepsilon = (t - t_o)/(N+1)$, denote the intermediate times by

$$t_j = t_o + j\varepsilon\ ,\ j = 0,\ 1,\ \ldots\ldots\ ,\ N+1 \tag{1.5}$$
$$t_{N+1} = t$$

and iterate (1.4). One obtains ($q_{N+1} \equiv q$)

$$K(q,t;\ q_o,t_o) = \int dq_N\ dq_{N-1}\ldots dq_1\ \ K(q,t;q_N,t_N)\ K(q_N,t_N;q_{N-1},t_{N-1})$$
$$\ldots.\ K(q_1,t_1;q_o,t_o)$$
$$= \int \prod_{i=1}^{N} dq_i \prod_{j=1}^{N+1} K(q_j,t_j;\ q_{j-1},t_{j-1}) \tag{1.6}$$

We are interested in the limit of (1.6) for $N \to \infty$ . In this limit $\varepsilon = t_j - t_{j-1}$ becomes arbitrarily small and $K(q_j,t_j;q_{j-1},t_{j-1})$ is called the <u>short time propagator</u>. The useful aspect of the procedure is that (if the limit exists) every short time propagator contributes to the final result $K(q,t;q_o,t_o)$ with terms up to order $\varepsilon$ only. Once this approximation is determined, a functional integral representation can be defined.
It was assumed by Feynman[F3], following Dirac[D19] that in quantum mechanics it is proportional to $\exp\ [\frac{i}{\hbar}\ A\ (q_j,t_j;q_{j-1},t_{j-1})]$ with $A$ the classical action of the classical path between $(q_j,t_j)$ and $(q_{j-1},t_{j-1})$.
Also approximations of this function were used.[F3,F5]

Let us consider the example of the Schrödinger equation for a particle in a one dimensional harmonic oscillator potential :

$$i\hbar\ \frac{\partial}{\partial t}\ \ \psi(q,t) = [-\ \frac{\hbar^2}{2\ m}\ \ \frac{\partial^2}{\partial q^2}\ +\ \frac{1}{2}\ m\,\omega^2 q^2]\ \psi(q,t) \tag{1.7}$$

The propagator is function of $t - t_o$ and for $t-t_o < \frac{\pi}{\omega}$ , it is given by (*)

$$K(q,t;q_o,0) = [\frac{m\omega}{2\pi i\hbar \sin\ \omega t}]^{1/2}\ \exp\ \{\frac{im\omega}{2\hbar \sin\omega t}\ [(q^2+q_o^2)\ \cos\ \omega t\ -2qq_o]\} \tag{1.8}$$

The argument of the exponential in (1.8) is $\frac{i}{\hbar}\ A(q,q_o,t)$ with $A$ the classical action for the harmonic oscillator Lagrangian

$$L(\dot{q},q) = \frac{1}{2}\ m\ \dot{q}^2\ -\ \frac{1}{2}\ m\ \omega^2 q^2 \tag{1.9}$$

This is always so for systems with Lagrangian quadratic in $\dot{q}$ and q.
By developing sin $\omega t$ and cos $\omega t$ in (1.8) we obtain

---

(*) For a discussion of the phase factor present for $t-t_o > \frac{\pi}{\omega}$ ,see e.g. [C10]

$$K(q_j,\varepsilon;q_{j-1},0) = [\frac{m}{2\pi i\hbar\,\varepsilon}]^{1/2} \exp\{\frac{i\varepsilon}{\hbar}\,[\frac{1}{2}m(\frac{q_j-q_{j-1}}{\varepsilon})^2 - \frac{1}{2}m\omega^2(\frac{q_j^2+q_{j-1}^2}{2}) + \frac{1}{12}\omega^2(q_j-q_{j-1})^2] + O(\varepsilon^{3/2})\} \quad (1.10)$$

If $(q_j-q_{j-1})$ is appreciably different from zero the dominating exponential $\exp\{\frac{i}{\hbar}\frac{1}{2}m\frac{(q_j-q_{j-1})^2}{\varepsilon}\}$ oscillates very rapidly and the integral of (1.10) over $q_j$ gives a very small value. Only if $q_j-q_{j-1}$ is of order $\sqrt{\varepsilon}$ we get important contributions ([F 5] p.77). Even stronger : we can consider in the functional integral $q_j-q_{j-1}$ as being of order $\sqrt{\varepsilon}$ . Consequently the third term in the exponential in (1.10) which is proportional to $\varepsilon(q_j-q_{j-1})^2$ is of order $\varepsilon^2$ and can be neglected since it will not contribute when the limit $N\to\infty$ $(\varepsilon\to 0)$ is taken. For the same reason we can replace in the second term $\frac{1}{2}(q_j^2+q_{j-1}^2)$ by other possibilities such as $[\frac{1}{2}(q_j+q_{j-1})]^2$. Then we arrive by replacing (1.10) in (1.6) and taking the limit at the following path integral representation for the propagator of the harmonic oscillator :

$$K(q,t;q_o,t_o) = \lim_{N\to\infty}\int(\prod_{i=1}^{N} dq_i \prod_{j=1}^{N+1}[\frac{m}{2\pi i\hbar\varepsilon}]^{1/2})\exp\{\frac{i}{\hbar}\varepsilon\sum_{j=1}^{N+1} L(\frac{q_j-q_{j-1}}{\varepsilon},\frac{q_j+q_{j-1}}{2})\} \quad (1.11)$$

with L the harmonic oscillator Lagrangian (1.9).

Although in principle the variables $q_i, i=1,\ldots,N$ are independent, in the limit $N\to\infty$ , $q_j-q_{j-1}$ goes to 0 and we can speak about an integral over continuous paths $q(\tau)$. The paths however are nondifferentiable since $\dot{q}(\tau)\sim\frac{q_j-q_{j-1}}{\varepsilon}\sim\varepsilon^{-1/2}$ diverges. An instructive reference on the nature of these quantum mechanical paths as self similar fractal curves is [A1] .

The short hand notation for (1.11) is

$$K(q,t;q_o,t_o) = \int_{q_o}^{q}\mathcal{D}q\,\exp[\frac{i}{\hbar}\int_{t_o}^{t}d\tau\,L(\dot{q}(\tau),q(\tau))] \quad (1.12)$$

or also

$$K(q,t;q_o,t_o) = \int\mathcal{D}q\,\exp[\frac{i}{\hbar}\int_{t_o}^{t}d\tau\,L(\dot{q}(\tau),q(\tau)]\,\delta(q(t)-q)\delta(q(t_o)-q_o) \quad (1.13)$$

and we call this a <u>functional integral in configuration space</u>.

A second type of functional integral can be obtained by using in (1.11) (A an invertible and symmetric nxn matrix, b and x n-component vectors)

$$[\frac{1}{\det A}]^{1/2}\exp\{\tfrac{1}{2}b_jb_k(A^{-1})_{jk}\} = \int\prod_{j=1}^{n}\frac{dx_j}{\sqrt{2\pi}}\exp[-\frac{1}{2}x_kx_\ell A_{k\ell}+x_kb_k] \quad (1.14)$$

for $n = N+1$, $A = \text{diag.}\left(\frac{i\varepsilon}{\hbar m}\right)$, $b_j = \frac{i}{\hbar}(q_j - q_{j-1})$. One obtains

$$K(q,t;q_o,t_o) = \lim_{N\to\infty} \int \left(\prod_{i=1}^{N} dq_i \prod_{j=1}^{N+1} \frac{dp_j}{2\pi\hbar}\right) \exp\left\{\frac{i}{\hbar}\,\varepsilon \sum_{j=1}^{N+1} \left[p_j \frac{q_j - q_{j-1}}{\varepsilon} - H\left(p_j, \frac{q_j + q_{j-1}}{2}\right)\right]\right\} \tag{1.15}$$

where $H(p,q)$ is the harmonic oscillator Hamiltonian $\frac{1}{2m}p^2 + \frac{1}{2}m\omega^2 q^2$.

The short hand notation for (1.15) is

$$K(q,t;q_o,t_o) = \int \mathcal{D}q\, \mathcal{D}_\hbar p \exp\left\{\frac{i}{\hbar}\int_{t_o}^{t} d\tau\, [p\dot{q} - H(p,q)]\right\} \tag{1.16}$$

or also

$$K(q,t;q_o,t_o) = \int \mathcal{D}q\, \mathcal{D}_\hbar p \exp\left\{\frac{i}{\hbar}\int_{t_o}^{t} d\tau\, [p\dot{q} - H(p,q)]\right\}.\delta(q(\tau)-q)\delta(q(t_o)-q_o) \tag{1.17}$$

We call this a <u>functional integral in phase space</u>.

The expressions (1.12 - 1.13) and (1.16 - 1.17) are defined by their discretized versions (1.11) and (1.15). In both cases the prescription was that in the interval $[t_{j-1}, t_j]$, $\dot{q}(\tau)$ stands for $\frac{q_j - q_{j-1}}{\varepsilon}$ and $q(\tau)$ for $\frac{1}{2}(q_j + q_{j-1})$.

In the following it will be the procedure of discretization that will be taken as the basic definition of a functional integral. Discretization implies a limiting procedure and all manipulations will be done on the condition that the limit exists or can be made to exist by considerations as analytic continuation. We shall introduce a common notation for the path integrals that appear in the study of stochastic processes (Wiener integrals and their generalizations) and for the ones that appear in quantum mechanics (Feynman integrals) because the formalism that will be developed applies to both. But we never forget that mathematically they are essentially different.

Many interesting contributions to the mathematical theory of Feynman path integrals can be found in [A4]. One can see that the approach by limiting procedure appears in various forms : "jump processes" (Chebotarev and Maslov), "polygonal path" (Truman), "families of projections" (Tarski)... . Our formulation however is most closely related to the work of Cohen [C8] , Berezin [B4] , Dowker [D27] and Leschke and Schmutz [L32].

It will not be before chapter 6 that we pay attention in a systematic way to other definitions of functional integrals than the one given in the next section.

In section 1.3 we compute explicitly a functional integral as the limit of a multidimensional integral and show that its value depends on the discretization. In section 1.4 we derive functional integral representations from the operator formalism associated to equation (1.1) and show the relation of the discretization dependence of functional integrals to the problem of ordering non-commuting operators.

## 1.2. Definitions

### Functional integrals in phase space

A functional integral in the 2d-dimensional phase space $(p,q) \equiv (p_1,\dots,p_d, q^1,\dots,q^d)$ is defined as follows:

$$I = \int_\gamma \mathcal{D}q\mathcal{D}_\eta p \exp\{\frac{i}{\eta}\int_{t_o}^{t} d\tau[p.\dot{q} - H(p(\tau),q(\tau),\tau)]\} \quad \{\delta\} \quad . \tag{1.18}$$

In this expression $\gamma$ stands for the discretization which means that one is given a function $h^\gamma(p,q',q,\tau)$ such that

$$h^\gamma(p,q,q,\tau) = H(p,q,\tau) \tag{1.19}$$

in terms of which I is defined as

$$I = \lim_{N\to\infty} I_N, \tag{1.20}$$

$$I_N = \int(\prod_{i=o}^{N+1} dq_i \prod_{j=1}^{N+1} \frac{dp_j}{(2\pi\eta)^d}) \exp\{\frac{i\varepsilon}{\eta}\sum_{j=1}^{N+1}[p_j.\frac{\Delta q_j}{\varepsilon} - h^\gamma(p_j,q_j,q_{j-1},\bar{t}_j]\}\{\delta\} \tag{1.21}$$

where

$$\varepsilon = (t-t_o)/(N+1) \ , \tag{1.22}$$

$$t_j = t_o+j\varepsilon \ , j=0,1,\dots,N+1; t_{N+1}=t, \tag{1.23}$$

$$\bar{t}_j \in [t_{j-1},t_j] \ , \tag{1.24}$$

$$\Delta q_j = q_j - q_{j-1} \tag{1.25}$$

and

$\eta$ is a parameter corresponding with Planck's constant $\hbar$ . In general also $h^\gamma$ and H can depend on $\eta$ . In the following we will set $\eta = 1$ and note $\mathcal{D}_1 p \equiv \mathcal{D}p$ in

(1.18).
We remark that the number of integrals over p in (1.21) is by one less than the number of integrals over q. This follows from the fact that $p_j$ is the momentum variable associated with the interval $[t_{j-1}, t_j]$ while $q_j$ is the coordinate at time $t_j$.
In (1.18) $\{\delta\}$ stands for a set of $\delta$-functions specifying the initial and/or final values of the paths in q-space. In the case of a propagator (section(1.1)) $\{\delta\} = \delta(q(t)-Q)\ \delta(q(t_\circ)-Q_\circ)$ which in the discrete version becomes $\delta(q_{N+1}-Q)$ $\delta(q_\circ-Q_\circ)$ and we say that we integrate over all paths $(p(\tau),q(\tau)), t_\circ<\tau<t$, with $q(t)=Q$ and $q(t_\circ)=Q_\circ$ and no restrictions on $p(\tau)$.(*)
The independence of the limit (1.20) of the choice of $\bar{t}_j$ in $[t_{j-1};t_j]$ will be shown in section 2.3.
In (1.31) all integrations go from $-\infty$ to $+\infty$.
New features arise when the range of the variables q is different from $(-\infty,+\infty)$. This can happen for example after a change from cartesian to polar coordinates. Some of the problems concerning nonlinear transformations of coordinates in path integrals are studied in chapters 6 and 12.
Phase space path integrals for systems subject to $p > 0$ have been treated by Klauder in [K4].

Here we restrict ourselves to variables q and p with range $(-\infty,+\infty)$. In chapter 3 the definition (1.18) will be generalised to the functional integral

$$\langle F[p(\tau),q(\tau),\tau]\rangle = \int_\gamma \mathcal{D}q\mathcal{D}p\ F[p(\tau),q(\tau),\tau]\exp\{i\int_{t_\circ}^{t} d\tau[p.\dot{q}-H(p(\tau),q(\tau),\tau)]\}\ \{\delta\} \tag{1.26}$$

for a certain class of functionals $F[p(\tau),q(\tau),\tau]$.

In section 1.1 we have already seen that sometimes it is possible to change the discrete approximation without changing the value of the functional integral. Therefore we can introduce the notion of <u>equivalence of discretizations</u>:

two discretizations $\gamma'$ and $\gamma''$ defined by $h^{\gamma'}(p,q',q,\tau)$ and $h^{\gamma''}(p,q',q,\tau)$ satisfying (1.19) are called equivalent when the limit of $I_N$, $N\to\infty$ is the same for both. In fact, it is useful to introduce this equivalence because (as we shall see) the value of the functional integral can depend on the discretization. It will also become clear that some discretizations are equivalent only for a restricted set of functions $H(p,q,\tau)$.

(*)In this case the integrals over $q_o$ and $q_{N+1}$ can be done and in the integrand one must put $q_o=Q_o$, $q_{N+1}=Q$.

Examples

To illustrate the definition, let us introduce two discretizations depending on a parameter $\alpha$ , namely $\gamma_1(\alpha)$ and $\gamma_2(\alpha)$.

Set

$$q^{(\alpha)} = q + \alpha(q'-q) \tag{1.27}$$

then $\gamma_1(\alpha)$ is defined by

$$h^{\gamma_1(\alpha)}(p,q',q,\tau) = H(p,q^{(\alpha)},\tau) \; . \tag{1.28}$$

On the other hand $\gamma_2(\alpha)$ is defined by

$$h^{\gamma_2(\alpha)}(p,q',q,\tau) = (1-\alpha)H(p,q,\tau) + \alpha H(p,q',\tau) \tag{1.29}$$

Both definitions (1.28) and (1.29) satisfy the condition (1.19) for any $\alpha$.

Special names are frequently given to the discretizations $\gamma_1(0), \gamma_1(\frac{1}{2})$ and $\gamma_1(1)$. They are called prepoint-discretization midpoint-discretization and postpoint-discretization respectively. This way of saying originates from the fact that in the lattice approximation $q_{j-1}^{(\alpha)} = q_{j-1} + \alpha(q_j - q_{j-1})$ and $q_{j-1}^{(o)} = q_{j-1}$ is the earlier point, $q_{j-1}^{(1)} = q_j$ the later point. $q_{j-1}^{(1/2)} = \frac{1}{2}(q_{j-1} + q_j)$ is the "midpoint".

The discretizations $\gamma_1(\alpha)$ and $\gamma_2(\alpha)$ were introduced first in [L32] and [L4] respectively.

Functional integrals in configuration space

A functional integral in the d-dimensional configuration space $(q) \equiv (q^1,\ldots,q^d)$ is defined as follows:

$$I = \int_\gamma \mathcal{D}\left(\frac{\sqrt{G}}{(\eta c)^{d/2}} q\right) \exp\left\{-\frac{1}{\eta c}\int_{t_o}^{t} d\tau\, L(\dot q,q,\tau)\right\}\{\delta\} \tag{1.30}$$

where $L(\dot q,q,\tau)$ is of the form

$$L(\dot q,q,\tau) = \frac{1}{2} G_{\mu\nu}(q,\tau)\dot q^\mu \dot q^\nu + A_\mu(q,\tau)\dot q^\mu - V(q,\tau) \tag{1.31}$$

and

$$G(q,\tau) = \det G_{\mu\nu}(q,\tau) \; . \tag{1.32}$$

Here the discretization $\gamma$ means that one is given functions $g_{\mu\nu}^{[\gamma]}(q',q,\tau)$, $a_\mu^{[\gamma]}(q',q,\tau)$ , $v^{[\gamma]}(q',q,\tau)$ and $g^{[\gamma]}(q',q,\tau)$ such that

$$g_{\mu\nu}^{[\gamma]}(q,q,\tau) = G_{\mu\nu}(q,\tau) \tag{1.33}$$

$$a_\mu^{[\gamma]}(q,q,\tau) = A_\mu\ (q,\tau) \tag{1.34}$$

$$v^{[\gamma]}(q,q,\tau) = V(q,\tau) \tag{1.35}$$

$$g^{[\gamma]}(q,q,\tau) = G(q,\tau) \tag{1.36}$$

in terms of which I is defined as $I = \lim_{N\to\infty} I_N$, with

$$I_N = \int (\prod_{i=o}^{N+1} dq_i \prod_{j=1}^{N+1} \quad [(2\pi\eta c\varepsilon)^{-d}\ g^{[\gamma]}(q_j,q_{j-1},\bar{t}_j)]^{1/2}).$$
$$\cdot \exp\{-\frac{\varepsilon}{\eta c}\sum_{j=1}^{N+1}[\frac{1}{2}g_{\mu\nu}^{[\gamma]}\ (q_j,q_{j-1},\bar{t}_j)\ \frac{\Delta q_j^\mu}{\varepsilon}\ \frac{\Delta q_j^\nu}{\varepsilon} +$$
$$+ a_\mu^{[\gamma]}(q_j,q_{j-1},\bar{t}_j)\ \frac{\Delta q_i^\mu}{\varepsilon} - v^{[\gamma]}(q_j,q_{j-1},\bar{t}_j)]\ \}\ \{\delta\}\ . \tag{1.37}$$

In section (2.3) it will be shown that when $G_{\mu\nu}$ depends explicitly on time we have to impose the condition $g^{[\gamma]} = \det g_{\mu\nu}^{[\gamma]}$ in order that the limit of $I_N$ for $N \to \infty$ is independent of $\bar{t}_j \in [t_{j-1}, t_j\ ]$.
In the following we set $\eta = 1$ and shall often write simply $\mathcal{D}(\sqrt{G}\ q)$ in (1.30).
Two discretizations $\gamma'$ and $\gamma''$ defined by $\{g_{\mu\nu}^{[\gamma']},\ a_\mu^{[\gamma']},\ v^{[\gamma']},\ g^{[\gamma']}\}$ and $\{g_{\mu\nu}^{[\gamma'']},\ a_\mu^{[\gamma'']},\ v^{[\gamma'']},\ g^{[\gamma'']}\}$ satisfying (1.33-36) are called <u>equivalent</u> when the limit of $I_N$, $N \to \infty$ is the same for both.

In section 1.1. we have met the following equivalence: all discretizations of the harmonic oscillator potential $V(q,\tau)$ are equivalent to the discrétization $V(q_{j-1},\bar{t}_j)$.

In section 2.3. the equivalence classes for arbitrary $L(\dot{q},q,\tau)$ of the form (1.31) will be determined.

<u>Examples</u>
The discretizations $\gamma_1(\alpha)$ and $\gamma_2(\alpha)$ are defined in analogy to their counterparts for the functional integral in phase space: let $F(q,\tau)$ be any of the functions $G_{\mu\nu}(q,\tau)$, $A_\mu(q,\tau)$, $V(q,\tau)$ or $G(q,\tau)$ and $f^{[\gamma]}(q',q,\tau)$ the corresponding discretized quantity, then

$$f^{[\gamma_1(\alpha)]}(q',q,\tau) = F(q^{(\alpha)},\tau) \ , \tag{1.38}$$

and

$$f^{[\gamma_2(\alpha)]}(q',q,\tau) = (1-\alpha)F(q,\tau) + \alpha F(q',\tau) \ . \tag{1.39}$$

Relation between functional integrals in phase space and in configuration space

When $H(p,q,\tau)$ in (1.18) is quadratic in the momenta p we can perform, using the formula (1.14) , the integration over the variables $p_j, j=1,\dots,N+1$ in (1.21) in order to obtain the relation between the two types of functional integrals. (See section 2.2.4 ).

When $H(p,q,\tau)$ is more than quadratic in the momenta (this corresponds to a partial differential equation (1.1) of more than second order in $\partial$ ) only the functional integral in phase space can easily be defined.

We will restrict ourselves to a detailed study of the case that $\mathcal{L}$ in (1.1) is quadratic in $\partial$ . This comprises the most important physical applications (e.g. Schrödinger and Fokker-Planck equation) and is also the situation in field theory where the limit $d \to \infty$ is taken. (In Appendix 9.3 an operator $\mathcal{L}$ more than quadratic in $\partial$ is considered).

## 1.3. An explicit calculation showing the discretization dependence of the functional integral.

Consider in a one dimensional configuration space the example

$$L(\dot q,q) = \frac{1}{2}\dot q^2 + \dot q q + \frac{1}{2}q^2. \tag{1.40}$$

We will show by explicit calculation of the functional integral (1.30) in the $\gamma_2(\alpha)$ discretization that its value depends on $\alpha$ , i.e. depends on the discretization . This illustrates the necessity of introducing this notion.

According to (1.37),(1.39) and $\{\delta\} \equiv \delta(q(t)-q)\delta(q(t_\circ)-q_\circ)$ $I_N$ is given by (c=1)

$$I_N = \int\Big(\prod_{i=1}^{N} dq_i \prod_{j=1}^{N+1} (2\pi\varepsilon)^{-1/2}\Big) \exp\Big\{-\varepsilon \sum_{j=1}^{N+1} \Big[\frac{1}{2}\frac{\Delta q_j}{\varepsilon}\frac{\Delta q_j}{\varepsilon} + \\ + ((1-\alpha)q_{j-1} + \alpha q_j)\frac{\Delta q_j}{\varepsilon} + \frac{1}{2}((1-\alpha)q_{j-1}^2 + \alpha q_j^2)\Big]\Big\} \tag{1.41}$$

where $q_{N+1} = q$. One has that owing to the simple form of $L(\dot q,q)$ $\gamma_1(\alpha)$ and $\gamma_2(\alpha)$ are equivalent here. Both discretize q in $\dot q q$ as $(1-\alpha)q_{j-1} + \alpha q_j$ and both discretize $\frac{1}{2}q^2$ in a way equivalent to $\frac{1}{2}q_{j-1}^2$ . (cf. section 2.3). The argument of the exponential in (1.41) is of the form

$$-\frac{1}{2\varepsilon}\left[q_{N+1}^2(1+\varepsilon\alpha)^2 + q_\circ^2\,(\varepsilon(1-\alpha)-1)^2\right] - \frac{1}{2}\sum_{i,j=1}^{N} B_{ij}^{(N)}\, q_i q_j + \sum_{i=1}^{N} c_i q_i \tag{1.42}$$

where the NxN - matrix $\underline{B}^{(N)}$ is

$$\underline{B}^{(N)} = \frac{1}{\varepsilon}\,\underline{\tilde{B}}^{(N)}$$

$$\underline{\tilde{B}}^{(N)} = \begin{pmatrix} b(\varepsilon,\alpha) & a(\varepsilon,a) & 0 & 0 & \cdots & & 0 \\ a(\varepsilon,\alpha) & b(\varepsilon,a) & a(\varepsilon,\alpha) & 0 & \cdots & & 0 \\ 0 & a(\varepsilon,a) & b(\varepsilon,a) & 0 & \cdots & & \\ \vdots & & & & & & \\ 0 & \cdots & & & a(\varepsilon,\alpha) & b(\varepsilon,\alpha) & a(\varepsilon,\alpha) \\ 0 & \cdots & & & 0 & a(\varepsilon,\alpha) & b(\varepsilon,\alpha) \end{pmatrix} \tag{1.44}$$

with

$$a(\varepsilon,\alpha) = (1 + \varepsilon\alpha)(\varepsilon(1-\alpha)-1) \tag{1.45}$$

$$b(\varepsilon,\alpha) = (1 + \varepsilon\alpha)^2 + (\varepsilon(1-\alpha)-1)^2 \tag{1.46}$$

and the vector $\vec{c}$ has components

$$\begin{aligned} c_1 &= -\frac{1}{\varepsilon}\, q_\circ\; a(\varepsilon,\alpha) \\ c_2 &= c_3 \ldots = c_{N-1} = 0, \\ c_N &= -\frac{1}{\varepsilon}\, q\; a(\varepsilon,\alpha). \end{aligned} \tag{1.47}$$

Formula (1.14) gives us the result of the integration over $q_i, i=1,\ldots,N$ in (1.41). Using $\det \underline{B}^{(N)} = \varepsilon^{-N} \det \underline{\tilde{B}}^{(N)}$ one obtains

$$I_N = [2\pi\varepsilon \det \underline{\tilde{B}}^{(N)}]^{-1/2} \exp\{-\frac{1}{2\varepsilon}[q^2(1+\varepsilon\alpha)^2 + q_\circ^2(\varepsilon(1-\alpha)-1)^2] + \frac{1}{2}\sum_{i,j} c_i c_j (B^{(N)})^{-1}_{ij}\}. \tag{1.48}$$

Now we have to compute $\det \underline{\tilde{B}}^{(N)}$ and $(B^{(N)})^{-1}$. From (1.44) one has

$$\det \underline{\tilde{B}}^{(N)} = b(\varepsilon,\alpha)^N \det \underline{D}^{(N)} \tag{1.49}$$

where $\underline{D}^{(N)}$ is the matrix of the same form as $\underline{\tilde{B}}^{(N)}$ but with $b(\varepsilon,\alpha)$ replaced by 1 and $a(\varepsilon,\alpha)$ by

$$d(\varepsilon,\alpha) = a(\varepsilon,\alpha)/b(\varepsilon,\alpha). \tag{1.50}$$

Its determinant $D^{(N)}$ (without bar underneath) satisfies the recursion relation

$$D^{(N)} = D^{(N-1)} - d^2\, D^{(N-2)}, \quad D^{(1)} = 1, \; D^{(2)} = 1 - d^2$$

which implies that $D^{(N)}/D^{(N-1)}$ is the $(N-1)^{th}$ approximant of a continued fraction [W1]

$$\frac{D^{(N)}}{D^{(N-1)}} = 1 - \cfrac{d^2}{1 - \cfrac{d^2}{1 - \ddots - \cfrac{d^2}{1-d^2}}}$$

and has the value ( [W1],p.23)

$$\frac{D^{(N)}}{D^{(N-1)}} = 1 - \cfrac{v}{\frac{v}{u} + \cfrac{1}{\sum\limits_{p=o}^{N-2} (\frac{v}{u})^p}} = u\,\frac{\sum\limits_{p=o}^{N} (\frac{v}{u})^p}{\sum\limits_{p=o}^{N-1} (\frac{v}{u})^p} \tag{1.51}$$

with

$$u = \frac{1}{2}\left(1 + \sqrt{1-4d^2}\,\right), \quad v = \frac{1}{2}\left(1 - \sqrt{1-4d^2}\right). \tag{1.52}$$

From (1.51) it follows that

$$D^N = \prod_{k=1}^{N} D^{(k+1)}/D^{(k)} = u^N \sum_{p=o}^{N} (\frac{v}{u})^p = u^N(1 - (\frac{v}{u})^{N+1})/(1 - \frac{v}{u}). \tag{1.53}$$

Replacing (1.49-50,53) in the first factor of (1.48) one obtains

$$[2\pi\varepsilon \;\det \underline{\tilde{B}}^{(N)}]^{-1/2} = [2\pi\varepsilon\; b(\varepsilon,\alpha)^N\, u^N(1 - (\frac{v}{u})^{N+1})/(1 - \frac{v}{u})]^{-1/2}. \tag{1.54}$$

From (1.45-46) and (1.52) one has that

$$\begin{aligned} b(\varepsilon,\alpha) &= 2 + 2\varepsilon(2\alpha-1) + \mathcal{O}(\varepsilon^2) \\ u &= \frac{1}{2}\left(1 + \varepsilon + \mathcal{O}(\varepsilon^2)\right) \\ v &= \frac{1}{2}\left(1 - \varepsilon + \mathcal{O}(\varepsilon^2)\right) \\ \frac{u}{v} &= 1 - 2\varepsilon + \mathcal{O}(\varepsilon^2) \end{aligned} \tag{1.55}$$

and therefore

$$[2\pi\varepsilon \det \underline{\tilde{B}}^{(N)}]^{-1/2} = [2\pi\varepsilon\ (1+\varepsilon(2\alpha-1) + \mathcal{O}(\varepsilon^2))^N (1+\varepsilon+\mathcal{O}(\varepsilon^2))^N \,.$$

$$.(1-(1-2\varepsilon+\mathcal{O}(\varepsilon^2))^{N+1})/(2\varepsilon+\mathcal{O}(\varepsilon^2))]^{-1/2} . \qquad (1.56)$$

In the limit $N \to \infty$ , $\varepsilon = (t-t_\circ)/(N+1) \to 0$ this gives

$$\lim_{N\to\infty} [2\pi\varepsilon \ \det \underline{\tilde{B}}^{(N)}]^{-1/2} = [\pi \exp\{2\alpha(t-t_\circ)\}\ (1-\exp\{-2(t-t_\circ)\}]^{-1/2} . \qquad (1.57)$$

This result gives an explicit and simple illustration of the general feature that terms of order $\varepsilon^2$ don't contribute in the limit $N \to \infty$ .

Let us now compute $(\underline{B}^{(N)})^{-1}$ . Because of the special form of $\vec{c}$ (see (1.47)) we only need the elements 11,1N,N1 and NN. Using (1.44) , (1.49) and (1.53) one obtains

$$(\underline{\tilde{B}}^{(N)})^{-1}_{11} = (\underline{\tilde{B}}^{(N)})^{-1}_{NN} = \frac{\det \underline{\tilde{B}}^{(N-1)}}{\det \underline{\tilde{B}}^{(N)}} = \frac{1}{ub}\ \frac{1-(\frac{v}{u})^N}{1-(\frac{v}{u})^{N+1}} , \qquad (1.58)$$

and

$$(\underline{\tilde{B}}^{(N)})^{-1}_{1N} = (\underline{\tilde{B}}^{(N)})^{-1}_{N1} = \frac{(-1)^{N+1}a^{N-1}}{\det \underline{\tilde{B}}^{(N)}} = \frac{(-1)^{N+1}a^{N-1}(1-\frac{v}{u})}{b^N u^N (1-(\frac{v}{u})^{N+1})} . \qquad (1.59)$$

From $(\underline{B}^{(N)})^{-1} = \varepsilon(\underline{\tilde{B}}^{(N)})^{-1}$ ,(1.45-46) and (1.55) one then derives that

$$\frac{1}{2}\sum_{i,j=1}^{N} c_i c_j (\underline{B}^{(N)})^{-1}_{ij} = \frac{1}{2}\,(q_\circ^2 + q^2)\,\frac{a^2}{\varepsilon}\,(\underline{\tilde{B}}^{(N)})^{-1}_{11} + 2\,qq_\circ\,\frac{a^2}{\varepsilon}\,(\underline{\tilde{B}}^{(N)})^{-1}_{1N}$$

$$= (q_\circ^2 + q^2)\ (\frac{1}{2\varepsilon} + \alpha - \frac{1}{1-\exp\{-2(t-t_\circ)\}} + \mathcal{O}(\varepsilon)\ ) \ +$$

$$+ \ 2qq_\circ\ \frac{2\exp\{-(t-t_\circ)\}}{1-\exp\{-2(t-t_\circ)\}} + \mathcal{O}(\varepsilon) \ . \qquad (1.60)$$

Inserting this in (1.48) one obtains finally, in the limit $N \to \infty$

$$I = [\pi \exp\{2\alpha(t-t_\circ)\}(1-\exp\{-2(t-t_\circ)\}]^{-1/2}$$

$$.\ \exp\{-\frac{(q-q_\circ \exp\{-(t-t_\circ)\})^2}{1-\exp\{-2(t-t_\circ)\}}\} \ . \qquad (1.61)$$

We see that through the factor $\exp\{-\alpha(t-t_\circ)\}$ the value of I depends explicitly on $\alpha$ . For $\alpha' \neq \alpha''$ the discretizations $\gamma_1(\alpha')$ and $\gamma_1(\alpha'')$ are therefore inequivalent. I is the propagator for the $\alpha$-dependent equation

$$\frac{\partial}{\partial t}\phi(q,t) = [\frac{1}{2}\frac{\partial^2}{\partial q^2} + \frac{\partial}{\partial q}q - \alpha]\phi(q,t). \tag{1.62}$$

For $\alpha = 0$ this is the Fokker-Planck equation for Brownian motion with friction. (The linear Ornstein-Uhlenbeck Process) (see Chapter 7).

## 1.4. Path integral from the operator formalism

Let us introduce the operators $\hat{q} = (\hat{q}^1,\ldots,\hat{q}^d)$ and $\hat{p} = (\hat{p}_1,\ldots,\hat{p}_d)$ with commutation relations

$$[\hat{q}^\mu,\hat{p}_\nu] = i\,\delta^\mu{}_\nu \tag{1.63}$$

$$[\hat{q}^\mu,\hat{q}^\nu] = [\hat{p}_\mu,\hat{p}_\nu] = 0$$

and representation $\hat{q} \to q$, $\hat{p} \to -i\partial$ .
They are hermitean with respect to the scalar product

$$(\phi,\psi) = \int dq\phi^*(q,t)\psi(q,t) \tag{1.64}$$

and we use the quantummechanical ket and bra notations

$$\hat{q}^\mu|q\rangle = q^\mu|q\rangle \tag{1.65}$$

$$\hat{p}_\mu|\hat{p}\rangle = p_\mu|p\rangle$$

with normalization

$$\langle q'|q\rangle = \delta(q'-q) \tag{1.66}$$

$$\langle q|p\rangle = (2\pi)^{-d/2}\exp[i\,p.q]\ . \tag{1.67}$$

We can identify from (1.1) an operator

$$\hat{H}(\hat{p},\hat{q},t) = i\,\mathcal{L}(\hat{q},i\hat{p},t) \tag{1.68}$$

that we will call "Hamiltonian (operator)". Then the propagator is the coordinate representation of the evolution operator $U(t,t_\circ)$ :

$$K(q,t;q_\circ,t_\circ) = \langle q|U(t,t_\circ)|q\rangle \tag{1.69}$$

with $U(t,t_\circ)$ satisfying the operator equivalent of (1.1) :

$$[\,i\frac{\partial}{\partial t} - H(\hat{p},\hat{q},t)]U(t,t_\circ) = 0,\quad U(t_\circ,t_\circ) = 1\ . \tag{1.70}$$

By repeated use of the semigroup property $U(t_1,t_2) = U(t_1,t')U(t',t_2)$ and the completeness relation $\int dq|q\rangle\langle q| = 1$ one obtains from (1.69) (compare with (1.6), $q_{N+1} = q$)

$$K(q,t;q_o,t_o) = \int \prod_{i=1}^{N} dq_i \prod_{j=1}^{N+1} \langle q_j|U(t_j,t_{j-1})|q_{j-1}\rangle . \tag{1.71}$$

Since we are interested in the limit for $N \to \infty$ of (1.71) we only need the short time propagator $\langle q_j|U(t_j,t_{j-1})|q_{j-1}\rangle$ up to $O(\varepsilon)$. The usual procedure is to approximate $U(t_j,t_{j-1})$ up to $O(\varepsilon)$. An explicit proof that this is allowed is presented in chapter 5. (See also [B4]). Let us use then that

$$U(t_j,t_{j-1}) = 1 - i\varepsilon\, \hat{H}(\hat{p},\hat{q},\bar{t}_j) + O(\varepsilon^2) \tag{1.72}$$

where

$$t_{j-1} \leq \bar{t}_j \leq t_j \tag{1.73}$$

and consider in detail the example of a time independent $\hat{H}$ and a two dimensional phase space:

$$\hat{H}(\hat{p},\hat{q}) = -\frac{1}{2}\, ic\, \hat{p}^2\, G(\hat{q}) - \hat{p}\, A(\hat{q}) + \frac{i}{c}\, V(\hat{q}) . \tag{1.74}$$

Introducing in (1.71) N+1 times the completeness relation $\int dp\, |p\rangle\langle p| = 1$ one obtains

$$K(q,t;q_o,t_o) = \int \prod_{i=1}^{N} dq_i \prod_{j=1}^{N+1} (dp_j\, \langle q_j|p_j\rangle\langle p_j|(1 - i\,\varepsilon\,\hat{H}(\hat{p},\hat{q}) + O(\varepsilon^2))|q_{j-1}\rangle ). \tag{1.75}$$

$\hat{H}(\hat{p},\hat{q})$ is given by (1.74) in so-called anti-standard order (all operators $\hat{p}$ at the left hand side of all operators $\hat{q}$). Therefore ,(1.75) leads (using also 1.67) to an approximation of the short time propagator

$$\langle q_j|U(t_j,t_{j-1})|q_{j-1}\rangle = \int \frac{dp_j}{(2\pi)^d} \exp\{ip_j\cdot(q_j - q_{j-1}) - i\varepsilon H(p_j,q_{j-1})\} . \tag{1.76}$$

Comparing this with the definition (1.19-25) of functional integral in phase space and the definitions (1.28-29) of $\gamma_1(\alpha)$ and $\gamma_2(\alpha)$ discretization we see that we will arrive at a functional integral representation in the $\gamma_1(0)$ , $\gamma_2(0)$ or prepoint discretization.

We will now show in a simple and intuitive way how different discretizations are related to different possible orderings of non-commuting factors in the operator $\hat{H}(\hat{p},\hat{q})$.

For any value of the parameter $\alpha$ , $\hat{H}(\hat{p},\hat{q})$ can be written as

$$\begin{aligned}\hat{H}(\hat{p},\hat{q}) &= (1-\alpha)\ \{-\tfrac{1}{2}\ ic\ \hat{p}^2 G(\hat{q}) - \hat{p}\ A(\hat{q})\} \\ &+ \alpha\ \{-\tfrac{1}{2}\ icG(\hat{q})\hat{p}^2 - A(\hat{q})\hat{p}\ \} \\ &+ \alpha\ \{-\tfrac{1}{2}\ ic\ [\hat{p}^2,G(\hat{q})] - [\hat{p},A(\hat{q})]\} + \tfrac{i}{c}\ V(\hat{q}).\end{aligned} \tag{1.77}$$

From the commutation relation $[\hat{q},\hat{p}] = i$ it follows that for any operator $F(\hat{q})$ ($F' = \frac{\partial F}{\partial q}$ , etc... )

$$[\hat{p},F(\hat{q})] = -\ i\ F'(\hat{q}) \tag{1.78}$$

$$\begin{aligned}[\hat{p}^2,F(\hat{q})] &= -\ 2i\ \hat{p}\ F'(\hat{q}) + F''(\hat{q}) \ . \\ &= -\ 2i\{(1-\alpha)\hat{p}\ F'(\hat{q}) + \alpha F'(\hat{q})\hat{p}\ \} - 2(\alpha - \tfrac{1}{2})F''(\hat{q}).\end{aligned} \tag{1.79}$$

Using (1.78-79) in (1.77) we obtain

$$\begin{aligned}\hat{H}(\hat{p},\hat{q}) &= -\ \tfrac{1}{2}\ ic\{\ (1-\alpha)\hat{p}^2 G(\hat{q}) + \alpha\ G(\hat{q})\hat{p}^2\ \} \\ &- \{(1-\alpha)\hat{p}(A(\hat{q}) + \alpha cG'(\hat{q})) + \alpha(A(\hat{q}) + \alpha cG'(\hat{q}))\hat{p}\ \} \\ &+ \tfrac{i}{c}\ V(\hat{q}) - i\alpha\ A'(\hat{q}) + ic\alpha(\alpha - \tfrac{1}{2})\ G''(\hat{q}) .\end{aligned} \tag{1.80}$$

Repeating the procedure that led to (1.76) but using $\hat{H}(\hat{p},\hat{q})$ written as in (1.80) one arrives at

$$\langle q_j|U(t_j,t_{j-1})|q_{j-1}\rangle = \int \frac{dp_j}{2\pi}\ \exp\ \{ip_j(q_j-q_{j-1}) - i\ \varepsilon h^\alpha(p_j,q_j,q_{j-1})\} \tag{1.81}$$

with

$$\begin{aligned}h^\alpha(p_j,q_j,q_{j-1}) &= -\ \tfrac{1}{2}\ icp_j^2\ \{(1-\alpha)G(q_{j-1}) + \alpha G(q_j)\} \\ &- \ p_j\ \{(1-\alpha)(A(q_{j-1}) + \alpha cG'(q_{j-1})) + \alpha(A(q_j) + \alpha cG'(q_j))\} \\ &+ \tfrac{i}{c}\ V(q_{j-1}) - i\alpha A'(q_{j-1}) + ic\alpha(\alpha - \tfrac{1}{2})\ G''(q_{j-1}).\end{aligned} \tag{1.82}$$

Recalling now the definition of $\gamma_2(\alpha)$-discretization (formula (1.29)) we see that we have constructed the path integral representations

$$K(q,t,;q_\circ,t_\circ) = \int_{\gamma_2(\alpha)} \mathcal{D}q\mathcal{D}p\ \ \exp\ \{i\int_{t_\circ}^{t} d\tau[p\dot{q} - H^\alpha(p,q)]\}\delta(q(t)-q)\ \delta(q(t_\circ)-q_\circ) \tag{1.83}$$

where

$$H^{\alpha}(p,q) = h^{\alpha}(p,q,q) = -\frac{1}{2}\, ic\, p^{2} G(q) - p(A(q) + \alpha c G'(q)) +$$
$$+ \frac{i}{c} V(q) - i\alpha A'(q) + ic\alpha(\alpha - \frac{1}{2})G''(q) \,. \tag{1.84}$$

The fact that the p-independent terms can be discretized in an arbitrary way is here a clear consequence of the construction $(\langle q_j|V(\hat{q})|q_{j-1}\rangle = V(q_j)\langle q_j|q_{j-1}\rangle = V(q_{j-1})\langle q_j|q_{j-1}\rangle)$ . An essential feature is that by doing the commutators we have introduced $\alpha$-dependent terms in the phase space function appearing in the functional integral: $H^{\alpha}(p,q)$. I.e. the integrand in (1.83) depends on the definition of the integral. The interpretation of this fact is clear: since we are studying representations of a fixed quantity $K(q,t;q_0,t_0)$ which is independent of any discretization, in particular independent of $\alpha$ , the $\alpha$-dependence of the integrand $\exp\{i\int_{t_0}^{t} d\tau[p\dot{q} - H^{\alpha}(p,q)]\}$ is there to cancel the $\alpha$-dependence of the definition of the functional integral .

In the next chapter we will show in a systematic way how all possible ordering of an operator $\hat{H}(\hat{p},\hat{q},t)$ lead to functional integral representations of the propagator of the form

$$K(q,t;q_0,t_0) = \int_{\gamma} \mathcal{D}q\mathcal{D}p \;\exp\{i\int_{t_0}^{t} d\tau[p.\dot{q} - H^{\gamma}(p,q,\tau)]\}\;\delta(q(t)-q)\,\delta(q(t_0)-q_0) \tag{1.85}$$

where $H^{\gamma}$ depends on the equivalence classes of discretizations $\gamma$ .

# CHAPTER II: CORRESPONDENCE RULES AND FUNCTIONAL INTEGRAL REPRESENTATIONS.

In section 1.3. we have seen that the value of the functional integral can depend on the discretization or limiting procedure. If so it represents a different quantity for every equivalence class of discretizations.
On the other hand in section 1.4. we have seen that a fixed propagator can be represented by functional integrals in different discretizations. A clear link has been made with the non-commutativity of operators in the associated operator formalism.
In this chapter we present a systematic treatment of these problems, based on the formalism of correspondence rules.

The problem of obtaining a quantum mechanical Hamiltonian from its classical counterpart via Feynman path integrals was first treated in a systematic way by Cohen [C8], Testa [T3] and Mayes and Dowker [M9 -M10]. The first work presenting different functional integral representations of one quantummechanical operator according to the way it is ordered is the work of Berezin [B4]. More recently Leschke and Schmutz [L32] have studied in a unified fashion operator ordering schemes and functional formulations of quantum and stochastic dynamics. The treatment of the present chapter is based on our work [L8] which also covers applications in quantum mechanics and stochastic dynamics. A study of correspondence rules and path integrals in the frame work of an approach without limiting procedure has been done by Mizrahi [M16]. Extensions to functional integrals for Fermi systems and coherent states have been made by Suzuki [S10-S11].

Some other works on the subject of path integral representations of the propagator where ordering is taken into account are [C9,D27,E5,H1,M14,K2].

## 2.1. Correspondence between phase space functions and operators

### 2.1.1. Correspondence rules

A correspondence rule is a linear mapping from phase space functions $B(p,q)$ into operators $\hat{B}(\hat{p},\hat{q})$. It is completely characterized by the way it acts on the phase space function $\exp[iu.q + iv.p]$ which serves as a generating function for the monomials

$$M(n,m) \equiv q^{\mu_1} \dots q^{\mu_n} \, p_{\nu_1} \dots p_{\nu_m} \,. \tag{2.1}$$

All usual correspondence rules are of the following form

$$\exp[iu.q + iv.p] \rightarrow D_\Omega(u,v;\hat{q},\hat{p}) \equiv \Omega(u,v)\ \exp\,[iu.\hat{q} + iv.\hat{p}] \qquad (2.2)$$

where $\Omega(u,v)$ is an analytic function of the components $u_\lambda, \lambda = 1,\ldots,d$ of u and $v^\lambda$, $\lambda = 1,\ldots,d$ of v.
From (2.2) it follows that the monomials M(n,m) are mapped as

$$M(n,m) \rightarrow \frac{1}{i^{n+m}} \left[ \frac{\partial^{n+m} D_\Omega(u,v;\hat{q},\hat{p})}{\partial u_{\mu_1}\ldots\partial u_{\mu_n}\ \partial v^{\nu_1}\ldots\partial v^{\nu_m}} \right]\Bigg|_{u=v=o} \quad . \qquad (2.3)$$

The notation $\Omega(u,v)$ was introduced by Agarwal and Wolf [A3]. The idea of using a function of two variables to define correspondence rules goes back to the work of Cohen [C7] on generalized phase space distribution functions.

In table 2.1. we have listed the better known correspondence rules with their $\Omega$ -function. ($u.v \equiv u_\mu v^\mu$, summation over repeated indices)

Table 2.1 : Correspondence rules and $\Omega$-functions

| correspondence rule | $\Omega(u,v)$ |
|---|---|
| Weyl | $1$ |
| Symmetric | $\cos\,(\frac{1}{2}u.v)$ |
| Standard | $\exp\,(-\frac{1}{2}\,iu.v)$ |
| Anti-Standard | $\exp\,(+\frac{1}{2}\,iu.v)$ |
| Born-Jordan | $\sin\,(\frac{1}{2}\,u.v)/(\frac{1}{2}\,u.v)$ |

Using the formula $\exp\,(\hat{A}+\hat{B}) = \exp \hat{A} \exp \hat{B} \exp\,(-\frac{1}{2}[\hat{A},\hat{B}])$ , valid when the operators $\hat{A}$ and $\hat{B}$ both commute with the commutator $[\hat{A},\hat{B}]$ , and the form of $\Omega(u,v)$ for standard, anti-standard and symmetric correspondence one obtains immediately that by these rules $\exp\,[iu.q + iv.p]$ is mapped into $\exp\,(iu.\hat{q})$ $\exp\,(iv.\hat{p})$, $\exp\,(iv.\hat{p})\exp\,(iu.\hat{q})$ and half the sum of these two results respectively. Therefore we have that the monomials M(n,m) are mapped into $\hat{q}^{\mu_1}\ldots\hat{q}^{\mu_n}$ $\hat{p}_{\nu_1}..\hat{p}_{\nu_m}$, $\hat{p}_{\nu_1}\ldots\hat{p}_{\nu_m}\hat{q}^{\mu_1}\ldots\hat{q}^{\mu_n}$ and half the sum of these respectively.
When $\Omega(u,v)$ is nonvanishing the inverse mapping of (2.2) is defined by

$$\exp\,[iu.\hat{q} + iv.\hat{p}] \rightarrow [\Omega(u,v)]^{-1}\ \exp[iu.q + iv.p] \ . \qquad (2.4)$$

An inverse mapping for monomials $\hat{p}_{\nu_1} \ldots \hat{p}_{\nu_m} \hat{q}^{\mu_1} \ldots \hat{q}^{\mu_n}$ follows in analogy to (2.3) from

$$\exp(iv.\hat{p}) \exp(iu.\hat{q}) \rightarrow [\Omega(u,v)]^{-1} \exp[\tfrac{i}{2}u.v] \exp[iu.q + iv.p] \tag{2.5}$$

Remarks

In the literature on quantization this inverse mapping has received no attention, since from the point of view of quantization procedure there is one c-number function to which correspond by the correspondence rules many operators. From all these operators the original c-number function is recuperated by putting Planck's constant $\hbar$ equal to zero. There is no contradiction to what we do here since if we would specify to quantum mechanics and introduce $\hbar$ we would see that a fixed operator is mapped into many but in general $\hbar$ -dependent c-number functions which all reduce to one $\hbar$-independent (and therefore "classical") function when we let $\hbar \rightarrow 0$.

In the literature [A3] one also finds the terminology "operator written in... ordered way". We shall use it only for operator written in anti-standard way. By this we mean that, by using the commutation relations we have written it such that all operators $\hat{p}$ stand at the left hand side of all operators $\hat{q}$. Replacing then $\hat{p}$ by p and $\hat{q}$ by q one obtains immediately the phase space function that is mapped into the operator by the anti-standard rule of correspondence.

Examples for a two dimensional phase space

The correspondence rules listed in table 2.1 map the monomials $q^n p^m$ in the following operators:

Weyl [C5] : $2^{-n} \sum_{\ell=o}^{n} \binom{n}{\ell} \hat{q}^{n-\ell} \hat{p}^m \hat{q}^\ell$ (2.6)

Symmetric : $\frac{1}{2} (\hat{q}^n \hat{p}^m + \hat{p}^m \hat{q}^n)$ (2.7)

Standard: : $\hat{q}^n \hat{p}^m$ (2.8)

Anti-Standard : $\hat{p}^m \hat{q}^n$ (2.9)

Born-Jordan [C5] : $\frac{1}{m+1} \sum_{\ell=o}^{n} \hat{p}^{m-\ell} \hat{q}^n \hat{p}^\ell$ . (2.10)'

The operator (2.6) equals the sum of all operators which are a product of n $\hat{q}$'s and m $\hat{p}$'s, divided by the number of these operators : $\binom{n+m}{n}$.

### 2.1.2. Closed form expressions

We shall now derive closed form expressions for the operator associated to a given phase space function and for the phase space function associated to a given operator.

Let B(p,q) be an arbitrary c-number function on phase space, representable by the Fourier integral

$$B(p,q) = \int \tilde{B}(u,v) \exp (iu.q + iv.p)dudv \tag{2.10}$$

with

$$\tilde{B}(u,v) = (2\pi)^{-2d} \int B(p,q) \exp(-iu.q - iv.p) \, dp \, dq \; . \tag{2.11}$$

From (2.2) and the linearity of the correspondence rule it follows that B(p,q) is mapped by the correspondence rule $\Omega$ into the operator $\hat{B}^{\Omega}(\hat{p},\hat{q})$ given by

$$\hat{B}^{\Omega}(\hat{p},\hat{q}) = \int \tilde{B}(u,v)\Omega(u,v) \exp (iu.\hat{q} + iv.\hat{p}) \, du \, dv \; . \tag{2.12}$$

We see that $B(u,v)\Omega(u,v)$ can be considered as the "operator Fourier transform" of $\hat{B}^{\Omega}(\hat{p},\hat{q})$. Inserting (2.11) into (2.12) we obtain the first expression we are looking for

$$\hat{B}^{\Omega}(\hat{p},\hat{q}) = (2\pi)^{-2d} \int B(p,q)\Omega(u,v) \exp [-iu(q-\hat{q}) -iv.(p-\hat{p})] \, dp \, dq \, du \, dv \; . \tag{2.13}$$

Next, let $\hat{B}(\hat{p},\hat{q})$ be an operator representable by

$$\hat{B}(\hat{p},\hat{q}) = \int G(u,v) \exp (iu.\hat{q} + iv.\hat{p}) \, du \, dv. \tag{2.14}$$

Then

$$G(u,v) = (2\pi)^{-d} \operatorname{Tr} [B(\hat{p},\hat{q}) \exp (-iu.\hat{q} - iv.\hat{p})] \; . \tag{2.15}$$

(Tr stands for trace, $\operatorname{Tr} F(\hat{p},\hat{q}) = \int dq \langle q|F|q\rangle = \int dp \langle p|F|p\rangle$ )
We shall note the inverse of the correspondence rule $\Omega$ by $\Theta_{\Omega}$ (and the index $\Omega$ will be dropped when no confusion is possible)
$\hat{B}$ is mapped by $\Theta$ into

$$B^{\Theta}(p,q) = \int G(u,v)[\Omega(u,v)]^{-1} \exp (iu.q + iv.p) \, du \, dv \; . \tag{2.16}$$

Generalising the terminology introduced by Berezin [B4 ,B6] we call this the $\Theta$ -symbol of the operator $\hat{B}(\hat{p},\hat{q})$.
Inserting (2.15) into (2.16) we obtain

$$B^{\Theta}(p,q) = (2\pi)^{-d} \int \operatorname{Tr} \hat{B}(\hat{p},\hat{q}) \exp [iu(q-\hat{q})+ iv(p-\hat{p})]][\Omega(u,v)]^{-1} \, du \, dv \tag{2.17}$$

In terms of the mapping $\Delta$-operator

$$\Delta^{(\Omega)}(\bar{q}-\hat{q},\bar{p}-\hat{p}) = (2\pi)^{-d} \int \Omega(u,v) \exp[-iu.(\bar{q}-\hat{q})-iv.(\bar{p}-\hat{p})]\, du\, dv \qquad (2.18)$$

which is the operator associated to the 2d-dimensional $\delta$-function $\delta(\bar{q}-q)\delta(\bar{p}-p)$ formula's (2.13) and (2.17) can also be written as

$$\hat{B}^{\Omega}(\hat{p},\hat{q}) = \int B(p,q)\Delta^{(\Omega)}(q-\hat{q},p-\hat{p})\, dp\, dq \qquad (2.19)$$

and

$$B^{\Theta}(p,q) = 2\pi \,\mathrm{Tr}\, [\hat{B}(\hat{p},\hat{q})\Delta^{(\tilde{\Omega})}(q-\hat{q},p-\hat{p})] \qquad (2.20)$$

where

$$\tilde{\Omega}(u,v) = [\Omega(-u,-v)]^{-1}. \qquad (2.21)$$

Remark

Much of the present formalism was developed by Mehta [M11] and Cohen [C7] in relation to a statistical reformulation of quantum mechanics. We can explain the basic idea here: from (2.19) it follows that the expectation value of the operator $\hat{B}^{\Omega}(\hat{p},\hat{q})$ in a state $|\psi\rangle$ is given by

$$\langle\psi|\, \hat{B}^{\Omega}(\hat{p},\hat{q})\,|\psi\rangle = \int B(p,q)\, F^{\Omega}_{\psi}(p,q)\, dp\, dq \qquad (2.22)$$

where

$$F^{\Omega}_{\psi}(p,q) = \langle\psi|\Delta^{(\Omega)}(q-\hat{q},p-\hat{p})|\psi\rangle. \qquad (2.23)$$

The "distribution function" $F^{\Omega}_{\psi}(p,q)$ is a generalization of the Wigner distribution associated to Weyl's correspondence ($\Omega(u,v) = 1$). It was shown by Cohen that $F^{\Omega}_{\psi}$ for $\Omega(u,v)$ satisfying $\Omega(u,0) = \Omega(0,v) = 1$ is the most general distribution leading to the correct marginal distribution functions for q and p. In general however the distribution is neither positive definite nor does the expectation value of observables calculated using phase space integration give consistently the same value as the usual quantummechanical procedure.

### 2.1.3. Matrix elements of an operator

It will be useful to know the relation between two phase space functions $B^{\Theta_1}$ and $B^{\Theta_2}$ associated by two different mappings to the same operator $\hat{B}$. From (2.17) it is readily obtained that

$$B^{\Theta_2}(p,q) = \Omega_1(-i\frac{\partial}{\partial q}, -i\frac{\partial}{\partial p})\Omega_2^{-1}(-i\frac{\partial}{\partial q}, -i\frac{\partial}{\partial p})B^{\Theta_1}(p,q) . \tag{2.24}$$

In the derivation of the path integral representation we need an expression for matrix elements of the type $\langle q'|\hat{B}|q\rangle$. From the relation

$$\hat{B}(\hat{p},\hat{q}) = (2\pi)^{-2d} \int B^{\Theta_W}(p,q) \exp[-iu.(q-\hat{q})-iv(p-\hat{p})]dp\, dq\, du\, dv \tag{2.25}$$

where $B^{\Theta_W}(p,q)$ is the Weyl symbol of the operator $\hat{B}$ it is easily obtained that

$$\langle q'|\hat{B}(\hat{p},\hat{q}|q\rangle = (2\pi)^{-d} \int dp \exp[ip.\Delta]\, B^{\Theta_W}(p,\bar{q}) \tag{2.26}$$

where

$$\Delta \equiv q' - q \tag{2.27}$$

$$\bar{q} \equiv \frac{1}{2}(q' + q) . \tag{2.28}$$

From (2.26) and (2.24) for $\Theta_2 = \Theta_W$ it then follows that the matrix element of the operator $\hat{B}(\hat{p},\hat{q})$ can be related to its general $\Theta_\Omega$-symbol as follows:

$$\langle q'|\hat{B}(\hat{p},\hat{q})|q\rangle = (2\pi)^{-d} \int dp \exp[ip.\Delta]\Omega(-i\frac{\partial}{\partial\bar{q}}, -i\frac{\partial}{\partial p})B^{\Theta_\Omega}(p,\bar{q}). \tag{2.29}$$

By partial integration one obtains

$$\langle q'|\hat{B}(\hat{p},\hat{q})|q\rangle = (2\pi)^{-d} \int dp \exp[ip.\Delta]\Omega(-i\frac{\partial}{\partial\bar{q}}, -\Delta)\, B^{\Theta_\Omega}(p,\bar{q}) . \tag{2.30}$$

The function $b^{\Theta}(p,q',q)$ defined by

$$b^{\Theta}(p,q',q) \equiv \Omega(-i\frac{\partial}{\partial\bar{q}}, -\Delta)\, B^{\Theta_\Omega}(p,\bar{q}) \tag{2.31}$$

and appearing in (2.30) will play a fundamental role in the definition of the path integral representation associated to the correspondence rule $\Omega$ which will be derived in the next section. Remark that when the condition $\Omega(u,0) = 1$ is satisfied we have that

$$b^{\Theta}(p,q,q) = B^{\Theta_\Omega}(p,q). \tag{2.32}$$

In the next section the formalism is applied to the case of a time dependent operator and time dependent phase space functions. This can be done without problems. But we will always use time indeDendent correspondence rules($\Omega(u,v)$ time-independent).

A problem which is also of relevance is the determination of a correspondence rule that maps a given set of functions into a given set of operators. Since

$\Omega(u,v)$ is an analytic function we can always write in (2.13) its power series and determine the unknown coefficients by imposing the given correspondence. We will come back to this problem in section 2.5 but remark already that the conditions $f(q) \to f(\hat{q})$ and $f(p) \to f(\hat{p})$ lead to the restrictions $\Omega(u,0) = 1$ and $\Omega(0,v) = 1$ respectively. The correspondence rules listed in table 2.1 satisfy these conditions.

## 2.2. Correspondence rules and discretizations

### 2.2.1. In general

In section 1.4 we have derived a set of functional integral representations of the propagator $K(q,t;q_o,t_o)$ of equation (1.1) with $\mathcal{L}(q,\partial,t)$ given by (1.3). In this section we use the formalism developed in section 2.1 in order to obtain in a systematic way general classes of such functional integral representations. As in 1.4 we write the propagator as

$$K(q,t;q_o,t_o) = \int \prod_{i=1}^{N} dq_i \prod_{j=1}^{N+1} \langle q_j | U(t_j,t_{j-1}) | q_{j-1}\rangle \tag{2.33}$$

and we take as an approximation up to $\mathcal{O}(\varepsilon)$ of the short time propagator $\langle q_j|U(t_j,t_{j-1})|q_{j-1}\rangle$ the expression

$$\langle q_j|U(t_{j-1}+\varepsilon,t_{j-1})|q_{j-1}\rangle = \int\frac{dp}{(2\pi)^d}\exp[ip\Delta_j - i\varepsilon h^\gamma(p,q_j,q_{j-1},\bar{t}_j)] \tag{2.34}$$

where $h^\gamma(p,q',q,t)$ is a function (there are infinitely many as we have seen) such that

$$\langle q_j|\hat{H}(\hat{p},\hat{q},\bar{t}_j)|q_{j-1}\rangle = \int\frac{dp}{(2\pi)^d}\exp[ip\Delta_j]\, h^\gamma(p,q_j,q_{j-1},\bar{t}_j) \tag{2.35}$$

One obtains (2.34) writing $U(t_{j-1}+\varepsilon,t_{j-1}) = 1-i\varepsilon\hat{H}(\hat{p},\hat{q},\bar{t}_j)$ up to $\mathcal{O}(\varepsilon)$, then using (2.35) and writing $1-i\varepsilon h^\gamma$ as $\exp[-i\varepsilon h^\gamma]$ , all this being allowed since in the limit $N\to\infty$ only terms of $\mathcal{O}(\varepsilon) = \mathcal{O}(\frac{1}{N})$ contribute. We shall come back to this point in chapter 5. (See also [B4,K2,W7]). According to (2.30-31) a choice for $h^\gamma$ is given by the functions $h^\theta$ defined there i.e.

$$h^\theta(p,q',q;\tau) = \Omega(-i\frac{\partial}{\partial\bar{q}},-\Delta)\, H^{\theta_\Omega}(p,\bar{q},\tau), \tag{2.36}$$

$H^{\theta_\Omega}(p,\bar{q},\tau)$ being the $\theta_\Omega$-symbol of the operator $\hat{H}(\hat{p},\hat{q},\tau)$, and from (2.30) one has

$$\langle q_j|\hat{H}(\hat{p},\hat{q},\bar{t}_j)|q_{j-1}\rangle = \int\frac{dp}{(2\pi)^d}\exp[ip\Delta_j]\, h^\theta(p,q_j,q_{j-1},\bar{t}_j) \tag{2.37}$$

Consequently we obtain the following representation for the short time propagator

$$<q_j|U(t_j,t_{j-1})|q_{j-1}> = (2\pi)^{-d} \int dp_j \exp[\, ip_j.\Delta_j - i\varepsilon h^{\Theta}(p_j,q_j,q_{j-1},\bar{t}_j)\,] \tag{2.38}$$

and the following definition of the functional integral representations associated to a correspondence rule $\Omega$ with inverse $\Theta$ : $(q_{N+1} = q)$

$$K(q,t;q_\circ,t_\circ) = \lim_{N\to\infty} \int \Big( \prod_{i=1}^{N} dq_i \prod_{j=1}^{N+1} (2\pi)^{-d} dp_j \Big) \exp\{ i\varepsilon \sum_{j=1}^{N+1} [\, p_j.\frac{\Delta_j}{\varepsilon} - h^{\Theta}(p_j,q_j,q_{j-1},\bar{t}_j) ]\} \tag{2.39}$$

As explained in chapter I the short hand notation for (2.39) is

$$K(q,t;q_\circ,t_\circ) = \int_{\gamma(\Theta)} \mathcal{D}q\mathcal{D}p \;\exp\{ i\int_{t_\circ}^{t} d\tau\; p.\dot{q} - H^{\gamma(\Theta)}(p,q,\tau)\,\}\; \delta(q(t)-q)\,\delta(q(t_\circ)-q_\circ) \tag{2.40}$$

where

$$H^{\gamma(\Theta)}(p,q,\tau) = h^{\Theta}(p,q,q,\tau). \tag{2.41}$$

In general the phase space function $H^{\gamma(\Theta)}(p,q,\tau)$ appearing in the continuum expression of the functional integral doesnot need to coincide with the $\Theta$-symbol $H^{\Theta}(p,q,\tau)$ of the operator $\hat{H}(\hat{p},\hat{q},\tau)$ . It does however (see (2.32)) when $\Omega(u,0)=1$. We can therefore say that this condition determines correspondence rules that are associated in a natural way to functional integral representations in (p,q)-space. But other ones can be studied and we will meet examples in chapter IV on perturbation expansions. (section 4.3).

How do $H^{\Theta}$, $h^{\Theta}$ and $H^{\gamma(\Theta)}$ look like in the case we are interested in? Our operator $\hat{H}(\hat{p},\hat{q},\tau)$ is given by (1.68) which is explicitly

$$\hat{H}(\hat{p},\hat{q},\tau) = -\frac{1}{2}\, ic\, \hat{p}_\mu\hat{p}_\nu G^{\mu\nu}(\hat{q},\tau) - \hat{p}_\mu A^\mu(\hat{q},\tau) + \frac{i}{c} V(\hat{q},\tau) . \tag{2.42}$$

The anti-standard symbol of this operator is simply

$$H^{AS}(p,q,\tau) = -\frac{1}{2}\, ic\, p_\mu p_\nu G^{\mu\nu}(q,\tau) - p_\mu A^\mu(q,\tau) + \frac{i}{c} V(q,\tau). \tag{2.43}$$

According to (2.24) and the $\Omega$-function of anti-standard correspondence (see Table 2.1) a general $\Theta$-symbol of the operator $\hat{H}$ is given by

$$H^{\Theta}(p,q,\tau) = \bar{\Omega}\,(\, i\frac{\partial}{\partial q}\, ,\, i\frac{\partial}{\partial p}\,)\; H^{AS}(p,q,\tau) \tag{2.44}$$

where

$$\bar{\Omega}(u,v) = \exp\left[\frac{i}{2}\, u.v\right] \Omega(-u,-v)^{-1} . \tag{2.45}$$

It is clear that $H^{\Theta}(p,q,\tau)$ is of the form

$$H^{\Theta}(p,q,\tau) = -\frac{1}{2}\, ic\, p_\mu p_\nu G^{\mu\nu}_{\Theta}(q,\tau) - p_\mu A^{\mu}_{\Theta}(q,\tau) + \frac{i}{c}\, V_{\Theta}(q,\tau) . \tag{2.46}$$

From (2.31) and (2.41) it follows that $h^{\Theta}(p,q',q,\tau)$ and $H^{\gamma(\Theta)}(p,q,\tau)$ are of the form

$$h^{\Theta}(p,q',q,\tau) = -\frac{1}{2}\, ic\, p_\mu p_\nu g^{\mu\nu}_{\Theta}(q',q,\tau) - p_\mu a^{\mu}_{\Theta}(q',q,\tau) + \frac{i}{c}\, v_{\Theta}(q',q,\tau) \tag{2.47}$$

and

$$H^{\gamma(\Theta)}(p,q,\tau) = -\frac{1}{2}\, ic\, p_\mu p_\nu G^{\mu\nu}_{\gamma(\Theta)}(q',q,\tau) - p_\mu A^{\mu}_{\gamma(\Theta)}(q',q,\tau) + \frac{i}{c}\, V_{\gamma(\Theta)}(q',q,\tau). \tag{2.48}$$

When $\Omega(u,o) = 1$ one has in addition to $H^{\gamma(\Theta)} \equiv H^{\Theta}$ also $G^{\mu\nu}_{\Theta}(q,\tau) = G^{\mu\nu}(q,\tau)$.

### 2.2.2 Examples

- The mapping function $\Omega(u,v)$ related to the $\gamma_1(\alpha)$- discretization defined in (1.27-28) is given by

$$\Omega_1(u,v,\alpha) = \exp\{i(\tfrac{1}{2} - \alpha)u.v\}. \tag{2.49}$$

For $\alpha = 0, \frac{1}{2}$ and 1 this coincides with anti-standard, Weyl and standard correspondence respectively.
From (2.31) it follows that

$$\begin{aligned} h^{\Theta_1(\alpha)}(p,q',q,\tau) &= \exp\{-(\tfrac{1}{2} - \alpha)\Delta \,.\, \tfrac{\partial}{\partial \bar{q}}\}\, H^{\Theta_1(\alpha)}(p,\bar{q}) \\ &= H^{\Theta_1(\alpha)}(p,\bar{q} - (\tfrac{1}{2} - \alpha)\Delta) \\ &= H^{\Theta_1(\alpha)}(p,q^{(\alpha)}) \end{aligned} \tag{2.50}$$

and since $\Omega_1(u,o,\alpha) = 1$ we have $H^{\gamma_1(\alpha)} = H^{\Theta_1(\alpha)}$.

- The $\gamma_2(\alpha)$- discretization defined in (1.29) is obtained from the choice

$$\Omega_2(u,v,\alpha) = (1-\alpha)\exp(\tfrac{1}{2}\, i\, u.v) + \alpha \exp(-\tfrac{1}{2}\, iu.v). \tag{2.51}$$

For $\alpha = 0, \frac{1}{2}$ and 1 this coincides with anti-standard, symmetric and standard correspondence respectively. The proof follows immediately from (2.50) and the relation

$$\Omega_2(u,v,\alpha) = (1-\alpha)\Omega_1(u,v,0) + \alpha\, \Omega_1(u,v,1). \tag{2.52}$$

In table 2.2. we have listed the correspondence rules that appeared in table 2.1, together with the associated discretization. The polygonal path discretization related to the Born-Jordan rule of correspondence will be studied in section 2.4.

Table 2.2. Correspondence Rules and Discretizations

| Correspondence Rule | Discretization | |
|---|---|---|
| Anti-Standard | $\gamma_1(0) \equiv \gamma_2(0)$ | pre-point |
| Weyl | $\gamma_1(\frac{1}{2})$ | mid-point |
| Standard | $\gamma_1(1) \equiv \gamma_2(1)$ | post-point |
| Symmetric | $\gamma_2(\frac{1}{2})$ | |
| Born-Jordan | $\gamma(P)$ | polygonal path |

### 2.2.3 Remarks on generalizations

- In the definition of functional integrals given in chapter I there is a priori no relation between the discretization of different terms in the Hamiltonian. Accordingly, different correspondence rules can be used for each term in a sum of operators and a discretization $\gamma$ can depend on several $\Theta$'s.
- New mapping functions can be defined by taking linear combinations of others. Formula (2.52) is an example of this.
- A more original new mapping can be defined by using different mappings for different factors of an operator.

Suppose for example that

$$\hat{H}(\hat{p},\hat{q}) = \hat{H}_1(\hat{p},\hat{q})\, \hat{H}_2(\hat{p},\hat{q}). \tag{2.53}$$

We want to define $\Theta$ such that

$$H^{\Theta}(p,q) = H_1^{\Theta_1}(p,q)\, H_2^{\Theta_2}(p,q) \tag{2.54}$$

where $\Theta_1$ and $\Theta_2$ are of the type studied before.

In the framework of functional integration it should be possible to keep the relation

$$\langle q'|\hat{H}(\hat{p},\hat{q})|q\rangle = (2\pi)^{-d} \int dp\, \exp[ip.\Delta]h^{\Theta}(p,q',q) \tag{2.55}$$

with

$$h^{\Theta}(p,q',q) = h^{\Theta_1}(p,q',q)\, h^{\Theta_2}(p,q',q)$$
$$= [\Omega_1(-i \tfrac{\partial}{\partial\bar{q}}, -\Delta) H_1^{\Theta_1}(p,\bar{q})][\Omega_2(-i \tfrac{\partial}{\partial\bar{q}}, -\Delta) H_2^{\Theta_2}(p,q)] \,. \qquad (2.56)$$

This imposes severe restrictions on the possibilities there are, indeed, one always has

$$\langle q'|\hat{H}(\hat{p},\hat{q})|q\rangle = \int dq'' \langle q'|\hat{H}_1(\hat{p},\hat{q})|q''\rangle \langle q''|\hat{H}_2(\hat{p},\hat{q})|q\rangle$$
$$= (2\pi)^{-2d} \int dq'' dp dp' \exp[ip.(q'-q'')] h_1^{\Theta_1}(p,q',q'') .$$
$$. \exp[ip'(q''-q)] h_2^{\Theta_2}(p',q'',q) \,. \qquad (2.57)$$

To arrive at (2.55-56) one of the two p-integrations and the integration over q" has to be performed. This is possible when either $h_1^{\Theta_1}(p,q',q)$ or $h_2^{\Theta_2}(p,q',q)$ is independent of p. In the former case one has

$$\langle q'|\hat{H}(\hat{p},\hat{q})|q\rangle = (2\pi)^{-d} \int dp \exp[ip.\Delta] h_1^{\Theta_1}(q',q) h_2^{\Theta_2}(p,q',q) \qquad (2.58)$$

in the latter case

$$\langle q'|\hat{H}(\hat{p},\hat{q})|q\rangle = (2\pi)^{-d} \int dp \exp[ip.\Delta] h_1^{\Theta_1}(p,q',q)\, h_2^{\Theta_2}(q',q) \,. \qquad (2.59)$$

When $\hat{H}(\hat{p},\hat{q}) = \hat{H}_1(\hat{q})\hat{H}_2(\hat{p},\hat{q})$ or $\hat{H}(\hat{p},\hat{q}) = \hat{H}_1(\hat{p},\hat{q})\hat{H}_2(\hat{q})$ one obtains factorization immediately by writing

$$\langle q'|\hat{H}(\hat{p},\hat{q})|q\rangle = H_1(q') \langle q'|H_2(p,q)|q\rangle \qquad (2.60)$$

and

$$\langle q'|\hat{H}(\hat{p},\hat{q})|q\rangle = \langle q'|\hat{H}_1(\hat{p},\hat{q})|q\rangle H_2(q) \,. \qquad (2.61)$$

This method extends the schemes of Cohen and of Agarwal and Wolf in a useful way since it allows the treatment of "true divisors" [M19]. An important application is the correspondence between the Laplace-Beltrami operator and the free particle classical Hamiltonian in a curved space which will be constructed in section 5.8. The method has been used by us in [L6,L11-12,L14,L17,L19].

### 2.2.4. Configuration space functional integrals

The treatment based on correspondence rules is specially adapted to functional integral representations in phase space. Nevertheless, when considering systems with Hamiltonians quadratic in the momenta we can perform the Gaussian integration over the p-variables to arrive at a configuration space path integral representation associated to a correspondence rule $\Omega$ with inverse $\Theta$. In this way we obtain from (2.39) with $h^\Theta$ given by (2.47): $(q_{N+1}=q)$

$$K(q,t;q_\circ,t_\circ) = \lim_{N\to\infty} \int \Big( \prod_{i=1}^{N} dq_i \prod_{j=1}^{N+1} [(2\pi c\varepsilon)^d g_\Theta]^{-1/2} \Big) .$$

$$.\exp\{-\frac{\varepsilon}{c} \sum_{j=1}^{N+1} [\tfrac{1}{2} g^\Theta_{\mu\nu}(\frac{\Delta^\mu_j}{\varepsilon} + a^\mu_\Theta)(\frac{\Delta^\nu_j}{\varepsilon} + a^\nu_\Theta) - v_\Theta]\} \tag{2.62}$$

where $g_\Theta$, $g^\Theta_{\mu\nu}$, $a^\mu_\Theta$ and $v_\Theta$ are functions of $(q_j, q_{j-1}, \bar{t}_j)$, $g^\Theta_{\mu\nu}$ is the inverse matrix of $g^{\mu\nu}_\Theta$ and $g_\Theta = \det g^{\mu\nu}_\Theta$.
The short hand notation for (2.62) is

$$K(q,t;q_\circ,t_\circ) = \int_{\gamma(\Theta)} \mathcal{D}q\,(\frac{1}{\sqrt{cG_{\gamma(\Theta)}}})\ \exp\{-\frac{1}{c}\int_{t_\circ}^{t} d\tau\, L^{\gamma(\Theta)}(\dot{q}(\tau),q(\tau),\tau)\}\delta(q(t)-q)\,\delta(q(t_\circ)-q_\circ) \tag{2.63}$$

where

$$L^{\gamma(\Theta)}(\dot{q},q,\tau) = \frac{1}{2} G^{\gamma(\Theta)}_{\mu\nu}(q,\tau)(\dot{q}^\mu + A^\mu_{\gamma(\Theta)}(q,\tau)(\dot{q}^\nu + A^\nu_{\gamma(\Theta)}(q,\tau)) - V_{\gamma(\Theta)}(q,\tau) \tag{2.64}$$

$$G_{\gamma(\Theta)}(q,\tau) = g_\Theta(q,q,\tau) \tag{2.65a}$$

$$G^{\gamma(\Theta)}_{\mu\nu}(q,\tau) = g^\Theta_{\mu\nu}(q,q,\tau) \tag{2.65b}$$

$$A^\mu_{\gamma(\Theta)}(q,\tau) = a^\mu_\Theta(q,q,\tau) \tag{2.65c}$$

$$V_{\gamma(\Theta)}(q,\tau) = v_\Theta(q,q,\tau) . \tag{2.65d}$$

It should be remarked that in (1.30-1.37) we have been writing $[\det G_{\mu\nu}]^{+1/2}$ while here we have $[\det G^{\mu\nu}]^{-1/2}$, which is of course the same. Also there is a direct relation between $g_{\mu\nu}$ and $g$ because both follow from $g^{\mu\nu}$. In the general definition (1.30-1.37) this need not be the case.

## 2.3. Equivalence classes of discretizations

In section 1.2. we have given the definitions of functional integrals in phase space and in configuration space. We have also defined equivalence of discretizations there. Here we shall develop more fully this notion and find conditions for equivalence. We shall need these conditions in the next section.

Given a function $H(p,q,\tau)$ (or $L(\dot{q},q,\tau)$) the discretization $\gamma$ relates H (or L) in the continuum expression of a functional integral to the operator $\hat{H}^{\gamma}(\hat{p},\hat{q},\tau)$, such that the functional integral represents its propagator. Therefore we can say that a discretization $\gamma$ determines a correspondence between H and $\hat{H}^{\gamma}$. We have seen however that "small" changes in the discretization don't contribute in the limit $N \to \infty$. All discretizations related in this way determine the same correspondence and are called equivalent. In short, an equivalence class of discretizations characterizes a correspondence rule.

From our study of correspondence rules in section 2.1 we know that when restrictions are imposed on the form of $H(p,q,\tau)$ [e.g. only up to second degree in p and q] some correspondence rules that are different in general coincide. In the same way discretizations that are not equivalent in general can become equivalent when only special forms of H (or L) are considered. Examples will be given below.

### 2.3.1. Configuration space

It turns out that the conditions for equivalence can most easily be seen in the configuration space path integral.
The definition (1.37) is of the form

$$I_N = \int \prod_{i=1}^{N} dq_i \prod_{j=1}^{N+1} I(q_j,t_j;q_{j-1},t_{j-1}) \tag{2.66}$$

where

$$I(q_j,t_j;q_{j-1},t_{j-1}) = (2\pi c\varepsilon)^{-\frac{d}{2}} \exp\{ \frac{-\varepsilon}{2c} g^{[\gamma]}_{\mu\nu}(q_j,q_{j-1},\bar{t}_j) \frac{\Delta q^{\mu}_j}{\varepsilon} \frac{\Delta q^{\nu}_j}{\varepsilon} \} \cdot$$

$$\cdot F^{[\gamma]}(\Delta q_j,q_j,q_{j-1},\bar{t}_j), \tag{2.67}$$

$$F^{[\gamma]}(\Delta q_j,q_j;q_{j-1},\bar{t}_j) = [g^{[\gamma]}(q_j,q_{j-1},\bar{t}_j)]^{1/2} \exp\{-\frac{\varepsilon}{c} [a^{[\gamma]}_{\mu}(q_j,q_{j-1},\bar{t}_j) \frac{\Delta q^{\mu}_j}{\varepsilon}$$

$$+ v^{[\gamma]}(q_j,q_{j-1},\bar{t}_j)]\}. \tag{2.68}$$

Owing to the dominating exponential written explicitly in (2.67) and since (2.67) appears in the multidimensional integral (2.66) we can consider $\Delta q_j = q_j - q_{j-1}$ in $F^{[\gamma]}(\Delta q_j,q_j,q_{j-1},\bar{t}_j)$ as effectively being of order $\sqrt{\varepsilon}$ . Indeed, when c=1 the paths that contribute most are the ones for which $g^{[\gamma]}_{\mu\nu} \Delta q^{\mu}_j \Delta q^{\nu}_j / \varepsilon = O(1)$

($g_{\mu\nu}^{[\gamma]}$ is assumed to be positive definite), when c=i we have to use the stationary phase argument already given in section 1.1.: if $\Delta q_j$ is appreciably different from zero the exponential oscillates very rapidly and the integral over $q_j$ gives a very small value. Only if $\Delta q_j$ is of order $\sqrt{\varepsilon}$ we get important contributions. (Ref. [F ] p.77 and 276). In chapter 5 we shall make quantitative statements about this.

In the short time approximation (2.67) only terms up to order $\varepsilon$ contribute in the limit $N \to \infty$ . Therefore the value of the functional integral is unchanged when we replace $F^{[\gamma]}$ by $F^{[\tilde{\gamma}]}$ such that $(\Delta \equiv \Delta q_j)$

$$F^{[\gamma]} - F^{[\tilde{\gamma}]} = O(\varepsilon^{3/2}) = O(\Delta^3) . \tag{2.69}$$

It will also be unchanged when we replace in the dominating exponential $g_{\mu\nu}^{[\gamma]}$ by $g_{\mu\nu}^{[\gamma]}$ such that

$$g_{\mu\nu}^{[\gamma]} - g_{\mu\nu}^{[\tilde{\gamma}]} = O(\varepsilon^{3/2}) = O(\Delta^3). \tag{2.70}$$

Let $F^{[\tilde{\gamma}]}$ be given by (2.68) in terms of $g^{[\tilde{\gamma}]}$, $a_\mu^{[\tilde{\gamma}]}$ and $v^{[\tilde{\gamma}]}$, the condition (2.69) implies that $\gamma$ and $\tilde{\gamma}$ are equivalent when apart from (2.70) the following relations are satisfied:

$$g^{[\gamma]} - g^{[\tilde{\gamma}]} = O(\varepsilon^{3/2}) = O(\Delta^3) \tag{2.71}$$

$$a_\mu^{[\gamma]} - a^{[\tilde{\gamma}]} = O(\varepsilon) = O(\Delta^2) \tag{2.72}$$

$$v^{[\gamma]} - v^{[\tilde{\gamma}]} = O(\varepsilon^{1/2}) = O(\Delta) . \tag{2.73}$$

Equations (2.70-2.73) express sufficient conditions for equivalence of two discretizations $\gamma$ and $\tilde{\gamma}$ of an arbitrary configuration space path integral (1.30- 37).

Examples

- As a consequence of (2.73) all discretizations of the potential V(q,t) in (1.31) are equivalent to the prepoint discretization (cf. 1.35)

$$v^{[\gamma]} (q_j,q_{j-1},\bar{t}_j) - V(q_{j-1},\bar{t}_j) = O(\Delta) \tag{2.74}$$

Therefore as far as V is concerned all discretizations are equivalent.

- When are $\gamma_1(\alpha)$ and $\gamma_2(\alpha)$ equivalent? Let R(q) be any of the functions that have to be discretized as $r^{[\gamma]}(q',q)$. One has that

$$r^{[\gamma_1(\alpha)]}(q',q) = R(q^{(\alpha)}) = R(q) + \alpha\Delta^\mu\partial_\mu R(q) + \frac{1}{2}\alpha^2\Delta^\mu\Delta^\nu\partial_\mu\partial_\nu R(q) + O(\Delta^3) \tag{2.75}$$

$$r^{[\gamma_2(\alpha)]}(q',q) = (1-\alpha)R(q) + \alpha R(q') = R(q) + \alpha\Delta^\mu\partial_\mu R(q) + \frac{1}{2}\alpha\Delta^\mu\Delta^\nu\partial_\mu\partial_\nu R(q) + O(\Delta^3). \tag{2.76}$$

For $\alpha \neq 0,1$ the difference between (2.75) and (2.76) is of order $\Delta^2$. Consequently $\gamma_1(\alpha)$ and $\gamma_2(\alpha)$ are always equivalent for the terms linear in $\Delta$ i.e. $a_\mu^{[\gamma]}$ in (2.72). But they are not equivalent in general because (2.70) and (2.71) are not satisfied. They are equivalent when G and $G_{\mu\nu}$ are linear in q or independent of q.

- That discretization of the explicit time dependence in an arbitrary time $\bar{t}_j \in [t_{j-1},t_j]$ is allowed without changing the value of the functional integral partially follows from (2.70)-(2.73). Differences in the discretization of the explicit time dependence are of order $\varepsilon$. Therefore one has immediately from (2.72) and (2.73) that $A_\mu(q,\tau)$ and $V(q,\tau)$ can be discretized in any $\bar{t}_j \in [t_{j-1},t_j]$. But the same statement for $G(q,\tau)$ and $G_{\mu\nu}(q,\tau)$ does not follow from (2.70- 71) and deserves a separate proof, that we give now: (following [L17]) the condition that has to be satisfied is

$$g^{[\gamma]}(q_j,q_{j-1},\bar{t}_j) = \det g_{\mu\nu}^{[\gamma]}(q_j,q_{j-1},\bar{t}_j) + O(\varepsilon^{3/2}). \tag{2.77}$$

Set $A_\mu$ and V equal to zero and let $\gamma$ and $\tilde{\gamma}$ be such that (we drop the matrix indices $g_{\mu\nu}^{[\gamma]} = \underline{g}^{[\gamma]}$ )

$$\underline{g}^{[\gamma]} = \underline{g}^{[\tilde{\gamma}]} + \varepsilon\, \underline{g}_1^{[\gamma,\tilde{\gamma}]} \tag{2.78}$$

and that (2.77) is satisfied for both $\gamma$ and $\tilde{\gamma}$. We have then up to terms of $O(\varepsilon^{3/2})$

$$g^{[\gamma]} = \det \underline{g}^{[\gamma]} = \det \underline{g}^{[\tilde{\gamma}]} \det [1 + \varepsilon(\underline{g}^{[\tilde{\gamma}]})^{-1} \underline{g}_1^{[\gamma,\tilde{\gamma}]}]$$

$$= g^{[\tilde{\gamma}]}\exp\{\mathrm{Tr}\, \ln [1 + \varepsilon(\underline{g}^{[\tilde{\gamma}]})^{-1} \underline{g}_1^{[\gamma,\tilde{\gamma}]}]\} \tag{2.79}$$

and the short time approximation $I(q_j,t_j;q_{j-1},t_{j-1})$ can be written as (cf.(2.67)-(2.68))

$$I(q_j,t_j;q_{j-1},t_{j-1}) = [(2\pi c\varepsilon)^{-d}\, g^{[\gamma]}]^{1/2} \,.\, \exp\{-\frac{\varepsilon}{2c}\underline{g}^{[\gamma]}\frac{\Delta}{\varepsilon}\frac{\Delta}{\varepsilon} - \frac{1}{2c}\underline{g}_1^{[\gamma,\tilde{\gamma}]}\Delta\Delta +$$

$$+ \frac{\varepsilon}{2} \mathrm{Tr}\ [(\underline{g}^{\gamma})^{-1}\ \underline{g}_1^{[\gamma,\tilde{\gamma}]}] + O(\varepsilon^{3/2}) . \tag{2.80}$$

In chapter 5 it will be shown that in the second term of the argument of the exponential $\Delta\Delta$ can be replaced by $c\varepsilon(g^{[\tilde{\gamma}]})^{-1}$ without changing the value of the functional integral. It then follows that this term cancels against the last one and that the discretizations $\{g^{[\gamma]}, \underline{g}^{[\gamma]}\}$ and $\{g^{[\tilde{\gamma}]}, \underline{g}^{[\tilde{\gamma}]}\}$ related by (2.78) and both satisfying (2.77) are equivalent.

This calculation shows that the conditions (2.70-71) are sufficient but not necessary for the discretizations $\gamma$ and $\tilde{\gamma}$ to be equivalent.

It proofs that the definition (1.30-37) of functional integral in configuration space is independent of $\bar{t}_j$, $t_{j-1} \leq \bar{t}_j \leq t_j$ when (2.77) is satisfied.

### 2.3.2. Phase space

Conditions for equivalence of discretizations of phase space functional integrals of the form (1.18-25) for $H(p,q,\tau)$ quadratic in p follow from the conditions for equivalence of discretizations of configuration space functional integrals. Indeed, $h^{\gamma}(p,q',q,\tau)$ is of the form

$$h^{\gamma}(p,q',q,\tau) = \frac{1}{2}\, p_{\mu}p_{\nu}g^{\mu\nu}_{[\gamma]}\ (q',q,\tau) \quad - p_{\mu}a^{\mu}_{[\gamma]}\ (q',q,\tau) + v_{[\gamma]}(q',q,\tau) \tag{2.81}$$

and by performing the Gaussian integration over p-variables one obtains a configuration space functional integral. From the conditions (2.70-2.73) follow then immediately the conditions

$$g^{\mu\nu}_{[\gamma]} - g^{\mu\nu}_{[\tilde{\gamma}]} = O(\varepsilon^{3/2}) = O(\Delta^3) \tag{2.82}$$

$$a^{\mu}_{[\gamma]} - a^{\mu}_{[\tilde{\gamma}]} = O(\varepsilon) = O(\Delta^2) \tag{2.83}$$

$$v_{[\gamma]} - v_{[\tilde{\gamma}]} = O(\varepsilon^{1/2}) = O(\Delta) \tag{2.84}$$

sufficient for equivalence of $\gamma$ and $\tilde{\gamma}$. This shows that counting the order in $\varepsilon$ of (2.81) one must count $p_{\mu}$ as being of $O(\varepsilon^{-1/2})$. This can also be seen by remarking that in the p integration the dominating term is of the form $\exp[-\varepsilon p^2]$ and by the same reasoning as the one following (2.68) one has $p \sim O(\varepsilon^{-1/2})$. Furthermore, since (2.77) is satisfied by construction (i.e. as a consequence of the Gaussian integration) two discretizations $\gamma$ and $\tilde{\gamma}$ such that $g^{\mu\nu}_{[\gamma]} - g^{\mu\nu}_{[\tilde{\gamma}]} = \varepsilon g^{\mu\nu}_{1[\gamma,\tilde{\gamma}]}$ are also equivalent. (We deliberately don't write $g^{\mu\nu}_{[\gamma]} - g^{\mu\nu}_{[\tilde{\gamma}]} = O(\varepsilon)$, since in fact the result is not true for discretizations such that $g^{\mu\nu}_{[\gamma]} - g^{\mu\nu}_{[\tilde{\gamma}]} = \Delta^{\kappa}\Delta^{\lambda}a_{\kappa\lambda}$ ).

This proofs that the definition (1.18)-(1.25) is consistent in the sense that the value of the functional integral is independent of $\bar{t}_j, t_{j-1} \leq \bar{t}_j \leq t_j$. This arbitrariness in the discretization of the explicit time dependence is related to the fact that in the derivation of the path integral from the operator formalism an approximation of the short time evolution operator up to $\mathcal{O}(\varepsilon)$ can be used:

$$U(t_j,t_{j-1}) = 1 - i\,\varepsilon\hat{H}(\hat{p},\hat{q},\bar{t}_j) + \mathcal{O}(\varepsilon^2),\; t_{j-1} \leq \bar{t}_j \leq t_j \;. \tag{2.85}$$

Example

From (2.75)-(2.76) and (2.82)-(2.84) it follows that $\gamma_1(\alpha)$ and $\gamma_2(\alpha)$ are equivalent for $\alpha = 0$ and $\alpha = 1$ (then they coincide) and for the set of phase space functions which are at most quadratic in p and have a coefficient of the quadratic term which is constant or linear in q.

This implies that Weyl correspondence $(\gamma_1(\frac{1}{2}))$ and symmetric correspondence $(\gamma_2(\frac{1}{2}))$ give the same operator for all phase space functions of this form.

## 2.4. The inverse problem

Section 2.2 was devoted to the question how to construct in a systematic way functional integral representations for the propagator determined by a given operator $\hat{H}$. The inverse problem is the following: given a functional integral, what propagator does it represent? In other words given that

$$I(q,t;q_0,t_0) = \int_\gamma \mathcal{D}q\mathcal{D}p \;\exp\{i\int_{t_0}^{t} d\tau\,[p.\dot{q} - H(p,q,\tau)]\}\delta(q(t)-q)\,\delta(q(t_0)-q_0) \tag{2.86}$$

where $H(p,q,\tau)$ and $\gamma$ are fixed and given, what is the operator $\hat{H}^\gamma(\hat{p},\hat{q},\tau)$ such that $I(q,t;q_0,t_0)$ is the propagator associated to $\hat{H}^\gamma(\hat{p},\hat{q},\tau)$?

A first method to solve this problem is to see whether $\gamma$ is equivalent to some of the discretizations we have studied before. By definition this other discretization determines the same propagator, i.e. the same $\hat{H}^\gamma$. Very useful are of course the discretizations $\gamma(\Theta)$, $\Theta$ being the inverse of a correspondence rule, since then $\hat{H}^\gamma$ can be determined from (2.13).

A second method, which can be applied systematically for any $\gamma$ is to make a change in the discretization of (2.86) in such a way that we get a representation we know and can identify $\hat{H}^\gamma$ without further calculations. Here we will make this change of discretization in the phase space functional integral. In Ref. [L8] we did it in the configuration space functional integral. Then the calculation is longer, but can also be applied to study Feynman's definition of the configuration space functional integral (See Ref. [L13] and chapter 6 ).

## 2.4.1. First method

As an application of the first method we shall now study the polygonal path definition of the functional integral. This definition has been used and systematically developed in Truman's mathematical work on functional integrals [T6-T9] A review is given in[T10] .

Polygonal paths are defined by projection. Consider in (p,q)-space the continuous paths $(p(\tau),q(\tau))$ ,$t_\circ \leq \tau \leq t$. Associate to each path $(p(\tau),q(\tau))$ the projected paths defined by $(t_j = t_\circ + j\varepsilon\ ,\ t_{N+1} = t)$

$$(P_{N+1}\, p)(\tau) = p(\bar{t}_j) \equiv p_j\ ,\ \bar{t}_j \in [t_{j-1},t_j]\ , \tag{2.87a}$$

$$\begin{aligned}(P_{N+1}\, q)(\tau) &= q(t_{j-1})+(\tau-t_{j-1})(q(t_j)-q(t_{j-1}))/\varepsilon \\ &= q_{j-1} + (\tau - t_{j-1})\,\Delta q_j/\varepsilon \qquad t_{j-1}\leq \tau \leq t_j .\end{aligned} \tag{2.87b}$$

The paths $(P_{N+1}\, p(\tau),P_{N+1}\, q(\tau)\ )$ are discontinuous in p-space and continuous piecemeal linear ("polygonal") in q-space. The functional integral in phase space is defined as the limit (if it exists) $\lim_{N\to\infty} I_N$ with

$$I_N = \int\Big(\prod_{i=1}^{N} dq_i \prod_{j=1}^{N+1}\frac{dp_j}{(2\pi)^d}\Big)\exp\{i\int_{t_\circ}^{t} d\tau[P_{N+1}p.(P_{N+1}\dot{q})-H(P_{N+1}p,P_{N+1}q,\tau)]$$

$$=\int\Big(\prod_{i=1}^{N} dq_i \prod_{j=1}^{N+1}\frac{dp_j}{(2\pi)^d}\Big)\exp\{i\sum_{j=1}^{N+1}[\varepsilon\, p_j.\frac{\Delta q_j}{\varepsilon} - \int_{t_{j-1}}^{t_j} d\tau\, H(p_j,P_{N+1}q,\tau)]\ \}. \tag{2.88}$$

Let $H(p,q,\tau)$ be given by

$$H(p,q,\tau) = -\frac{1}{2}\, ic\, p_\mu p_\nu G^{\mu\nu}(q,\tau) - p_\mu A^\mu(q,\tau) + \frac{i}{c}\, V(q,\tau)\ , \tag{2.89}$$

then (2.88) becomes

$$\begin{aligned}I_N =\int\Big(\prod_{i=1}^{N} dq_i \prod_{j=1}^{N+1}\frac{dp_j}{(2\pi)^d}\Big)\exp\{i\sum_{j=1}^{N+1}[\varepsilon\, p_j.\frac{\Delta q_j}{\varepsilon} + \frac{ic}{2}\, p_{j\mu}p_{j\nu}\int_{t_{j-1}}^{t_j} d\tau G^{\mu\nu}(P_{N+1}q(\tau),\tau) \,+ \\ + p_{j\mu}\int_{t_{j-1}}^{t_j} d\tau\, A^\mu(P_{N+1}\, q(\tau),\tau)) - \frac{i}{c}\int_{t_{j-1}}^{t_j} d\tau V(P_{N+1}\, q(\tau),\tau)]\ \}\,.\end{aligned} \tag{2.90}$$

This can be considered as a generalization of our definition through discretization. The polygonal path discretization $\gamma(P)$ is defined by

$$\varepsilon\, h^{\gamma(P)}(p_j,q_j,q_{j-1},\bar{t}_j) = \int_{t_{j-1}}^{t_j} d\tau\, H(P_{N+1}p(\tau),P_{N+1}q(\tau),\tau) \qquad (2.91)$$

and in particular

$$\varepsilon\, g^{\mu\nu}_{\gamma(P)}(q_j,q_{j-1},\bar{t}_j) = \int_{t_{j-1}}^{t_j} d\tau\, G^{\mu\nu}(P_{N+1}\, q(\tau),\tau) \qquad (2.92a)$$

$$\varepsilon a^{\mu}{}_{\gamma(P)}(q_j,q_{j-1},\bar{t}_j) = \int_{t_{j-1}}^{t_j} d\tau\, A^{\mu}(P_{N+1}\, q(\tau),\tau) \qquad (2.92b)$$

$$\varepsilon v_{\gamma(P)}(q_j,q_{j-1},\bar{t}_j) = \int_{t_{j-1}}^{t_j} d\tau\, V(P_{N+1}\, q(\tau),\tau)\ . \qquad (2.92c)$$

To answer the question which propagator is represented by (2.90) we investigate whether $\gamma(P)$ is equivalent to one of the discretization $\gamma(\Theta)$. A discretization equivalent to $\gamma(P)$ is obtained by computing the integrals in (2.92a-c) up to the required order in $\varepsilon$ (See (2.82-84)).
One has that

$$\begin{aligned} G^{\mu\nu}(P_{N+1}q(\tau),\tau) &= G^{\mu\nu}(\bar{q}_j,\tfrac{1}{2}(t_{j-1}+t_j)) + (P_{N+1}q^{\lambda}(\tau)-\bar{q}^{\lambda}_j)(\partial_\lambda G^{\mu\nu})(\bar{q}_j,\bar{t}_j) \\ &+ \tfrac{1}{2}(P_{N+1}q^{\lambda}(\tau)-\bar{q}^{\lambda}_j)(P_{N+1}q^{\kappa}(\tau)-\bar{q}^{\kappa}_j)(\partial_\kappa\partial_\lambda G^{\mu\nu})(\bar{q}_j,\bar{t}_j) \\ &+[\ \tau - \tfrac{1}{2}\,(t_{j-1}+t_j)]\ (\frac{\partial}{\partial\bar{t}_j}\, G^{\mu\nu}(\bar{q}_j,\bar{t}_j)) + \mathcal{O}(\varepsilon^{3/2}) \end{aligned} \qquad (2.93a)$$

$$A^{\mu}(P_{N+1}\, q(\tau),\tau) = A^{\mu}(\bar{q}_j,\bar{t}_j) + (P_{N+1}q^{\nu}(\tau)-\bar{q}^{\nu}_j)(\partial_\nu A^{\mu})(\bar{q}_j,\bar{t}_j) + \mathcal{O}(\varepsilon) \qquad (2.93b)$$

$$V(P_{N+1}q(\tau),\tau) = V(\bar{q}_j,\bar{t}_j) + \mathcal{O}(\varepsilon^{1/2}) \qquad (2.93c)$$

In (2.93) $\bar{q}_j = \frac{1}{2}(q_{j-1}+q_j)$ and $\bar{t}_j$ indicates as before an arbitrary time between $t_{j-1}$ and $t_j$.
Using now

$$P_{N+1}\, q(\tau) - \bar{q}_j = (\tau - \tfrac{1}{2}(t_{j-1} + t_j))\, \frac{\Delta q_j}{\varepsilon} \qquad (2.94)$$

and

$$\int_{t_{j-1}}^{t_j} d\tau \, (\tau - \frac{1}{2}(t_{j-1} + t_j)) = 0 \tag{2.95a}$$

$$\int_{t_{j-1}}^{t_j} d\tau \, (\tau - \frac{1}{2}(t_{j-1} + t_j))^2 = \frac{1}{12}\varepsilon^3 \tag{2.95b}$$

we obtain that

$$g^{\mu\nu}_{\gamma(P)}(q_j,q_{j-1},\bar{t}_j) = G^{\mu\nu}(\bar{q}_j, \frac{1}{2}(t_{j-1}+t_j)) + \frac{1}{24}\Delta q_j^\lambda \Delta q_j^\kappa \; (\partial_\lambda \partial_\kappa G^{\mu\nu})(\bar{q}_j,\bar{t}_j) + \mathcal{O}(\Delta^3) \tag{2.96a}$$

$$a^\mu_{\gamma(P)}(q_j,q_{j-1},\bar{t}_j) = A^\mu(\bar{q}_j,\bar{t}_j) + \mathcal{O}(\Delta^2) \tag{2.96b}$$

$$v_{\gamma(P)}(q_j,q_{j-1},\bar{t}_j) = V(\bar{q}_j,\bar{t}_j) + \mathcal{O}(\Delta). \tag{2.96c}$$

As shown in the previous section the time $\frac{1}{2}(t_{j-1}+t_j)$ in (2.96a) can be changed to an arbitrary $\bar{t}_j$ without changing the value of the functional integral.

When there exists a correspondence rule $\Omega(u,v)$ satisfying $\Omega(u,o) = 1$ and such that the discretization $\gamma(\Theta)$ defined by

$$h^{\gamma(\Theta)}(p,q',q,\tau) = \Omega(-i\frac{\partial}{\partial \bar{q}}, -\Delta)H(p,\bar{q},\tau) \tag{2.97}$$

is equivalent to $\gamma(P)$ then it associates to H the same operator $\hat{H}$ as does the polygonal path definition of the functional integral. We show now that such a correspondence rule can be found in the class of $\Omega$-functions depending on u and v through the product u.v only. Then $\Omega(u,v)$ has the power series expansion ($a_0=1$)

$$\Omega(u,v) = \sum_{n=0}^{\infty} \frac{1}{n!} a_n (u.v)^n \tag{2.98}$$

where the coefficients $a_n$ are the n-th order derivatives with respect to u.v taken at u.v=0.

From (2.89), (2.97) and (2.98) one has that

$$g^{\mu\nu}_{\gamma(\Theta)}(q',q,\tau) = G^{\mu\nu}(\bar{q},\tau) + ia_1 \Delta q^\lambda (\partial_\lambda G^{\mu\nu})(\bar{q},\tau) - \frac{1}{2}a_2 \, \Delta q^\lambda \, \Delta q^\kappa \, (\partial_\kappa \partial_\lambda G^{\mu\nu})(\bar{q},\tau) + \mathcal{O}(\Delta^3). \tag{2.99a}$$

$$a^\mu_{\gamma(\Theta)}(q',q,\tau) = A^\mu(\bar{q},\tau) + i\, a_1 \Delta q^\lambda (\partial_\lambda A^\mu)(\bar{q},\tau) + \mathcal{O}(\Delta^2) \tag{2.99b}$$

$$v_{\gamma(\Theta)}\ (q',q,\tau) = V(\bar{q},\tau) + \mathcal{O}(\Delta). \tag{2.99c}$$

Comparing (2.96) and (2.99) one sees that $\gamma(P)$ and $\gamma(\Theta)$ will be equivalent when

$$a_1 = 0 \tag{2.100a}$$

$$a_2 = -\frac{1}{12} \tag{2.100b}$$

because then (2.82-84) will be satisfied. The coefficients $a_n, n \geq 3$ remain free because we are considering a Hamiltonian quadratic in p.

As anticipated in table 2.2 the $\Omega$ -function of the Born-Jordan rule of correspondence satisfies (2.100). Kerner and Sutcliffe showed in [K3] that the Born-Jordan rule of correspondence is uniquely determined by the polygonal path definition of functional integrals for arbitrary Hamiltonians. (They erronuously concluded from this that path integral quantization is in a unique way linked to this correspondence rule).

An analogous treatment can be given for functional integrals in configuration space.

### 2.4.2. Second method

Let us apply the second method to solve the inverse problem in general. The idea is to change (2.86) into a functional integral in prepoint or $\gamma_1(0)$-discretization i.e. to determine the phase space function $\bar{H}(p,q,\tau)$ such that

$$I(q,t;q_\circ,t_\circ) = \int_{\gamma_1(0)} \mathcal{D}q\mathcal{D}p\ \exp\{i \int_{t_\circ}^{t} d\tau[p.\dot{q} - \bar{H}(p(\tau),q(\tau),\tau)]\}\delta(q(t)-q)\ \delta(q(t_\circ)-q_\circ). \tag{2.101}$$

Then we know that $\hat{H}^\gamma$ is related to $\bar{H}$ by the antistandard rule of correspondence, which means that we obtain $\hat{H}$ by writing $\bar{H}$ with all functions p at the LHS of all functions q and then replacing p by $\hat{p}$ and q by $\hat{q}$.

Let $H(p,q,\tau)$ be given by (2.89). Through the discretization $\gamma$ we know a function $h^\gamma(p,q',q,\tau)$ such that $h^\gamma(p,q,q,\tau) = H(p,q,\tau)$ . It is of the form

$$h^\gamma(p,q',q,\tau) = -\frac{1}{2}\, ic\ p_\mu p_\nu g^{\mu\nu}_{[\gamma]}(q',q,\tau) - p_\mu a^\mu_{[\gamma]}(q',q,\tau) + \frac{i}{c}\, v_{[\gamma]}(q',q,\tau) \tag{2.102}$$

with $g^{\mu\nu}_{[\gamma]}, a^{\mu}_{[\gamma]}$ and $v_{[\gamma]}$ such that $g^{\mu\nu}_{[\gamma]}(q,q,\tau) = G^{\mu\nu}(q,\tau)$ , $a^{\mu}_{[\gamma]}(q,q,\tau) = A^{\mu}(q,\tau)$, $v_{[\gamma]}(q,q,\tau) = V(q,\tau)$.

A discretization $\tilde{\gamma}$ equivalent to $\gamma$ is defined by ($\partial' \equiv \partial/\partial q'$)

$$g^{\mu\nu}_{[\tilde{\gamma}]}(q',q,\tau) = G^{\mu\nu}(q,\tau) + \Delta^{\kappa}[\partial'_{\kappa} g^{\mu\nu}_{[\gamma]}(q',q,\tau)]\,|_{q'=q} + \frac{1}{2}\Delta^{\kappa}\Delta^{\lambda}[\partial'_{\kappa}\partial'_{\lambda}\, g^{\mu\nu}_{[\gamma]}(q',q,\tau)]\,|_{q'=q} \tag{2.103a}$$

$$a^{\mu}_{[\gamma]}(q',q,\tau) = A^{\mu}(q,\tau) + \Delta^{\kappa}[\partial'_{\kappa}\, a^{\mu}_{[\gamma]}(q',q,\tau)]\,|_{q'=q} \tag{2.103b}$$

$$v_{[\gamma]}(q',q,\tau) = V(q,\tau). \tag{2.103c}$$

Indeed, (2.103) is just the Taylor development of $g^{\mu\nu}_{[\gamma]}$ , $a^{\mu}_{[\gamma]}$ and $v_{[\gamma]}$ up to the required order to satisfy the conditions for equivalence (2.82-84). One has that $I(q,t;q_o,t_o)$ is given by the RHS of (2.86) with $\gamma$ replaced by $\tilde{\gamma}$ . The unusual factors $\Delta^{\kappa}$ and $\Delta^{\kappa}\Delta^{\lambda}$ make that $\tilde{\gamma}$ is not a prepoint discretization. We can get rid of these factors by using the following replacements, valid under the multidimensional integral in the sense that its value will be unchanged when $N \to \infty$ :

$$\varepsilon\, p_{\mu}\Delta^{\kappa} \doteq i\varepsilon\delta_{\mu}{}^{\kappa} \tag{2.104a}$$

$$\varepsilon\, p_{\mu}p_{\nu}\Delta^{\kappa} \doteq \varepsilon i p_{\mu}\delta_{\nu}{}^{\kappa} + \varepsilon i p_{\nu}\delta_{\mu}{}^{\kappa} \tag{2.104b}$$

$$\varepsilon\, p_{\mu}p_{\nu}\Delta^{\kappa}\Delta^{\lambda} \doteq -\varepsilon(\delta_{\mu}{}^{\kappa}\delta_{\nu}{}^{\lambda} + \delta_{\mu}{}^{\lambda}\delta_{\nu}{}^{\kappa}) . \tag{2.104c}$$

The symbol $\doteq$ denotes equality as far as use in the functional integral is concerned. The validity of the replacements (2.104) is related to the commutation relations in the operator formalism and is proved in chapter 5. There we will also study analogous replacements for powers of $\Delta$ in the configuration space functional integral.

By using (2.104) we make a change to an inequivalent discretization, not because the value of the functional integral changes but because $H(p,q,\tau)$ changes. The new discretization is the prepoint discretization, since the resulting h depends on p,q and $\tau$ only. Therefore we obtain in fact immediately $\bar{H}(p,q,\tau)$. It is given by

$$\bar{H}(p,q,\tau) = -\frac{1}{2}\, ic\; p_{\mu}p_{\nu}G^{\mu\nu}(q,\tau) - p_{\mu}\bar{A}^{\mu}(q,\tau) + \frac{i}{c}\,\bar{V}(q,\tau) \tag{2.105}$$

where

$$\bar{A}^{\mu}(q,\tau) = A^{\mu}(q,\tau) - c[\partial'_{\nu} \; g^{\mu\nu}_{[\gamma]}(q',q,\tau)] \,|_{q'=q} \tag{2.106a}$$

$$\bar{V}(q,\tau) = V(q,\tau) - c[\partial'_{\mu} \; a^{\mu}_{[\gamma]}(q',q,\tau)]\,|_{q'=q} + \frac{1}{2} c^{2}[\partial'_{\mu}\partial'_{\nu} \; g^{\mu\nu}_{[\gamma]}(q',q,\tau)] \,|_{q'=q} \tag{2.106b}$$

and the operator $\hat{H}^{\gamma}(\hat{p},\hat{q},\tau)$ we are looking for is

$$\hat{H}^{\gamma}(\hat{p},\hat{q},\tau) = -\frac{1}{2}\, ic \; \hat{p}_{\mu}\hat{p}_{\nu}G^{\mu\nu}(\hat{q},\tau) - \hat{p}_{\mu}\bar{A}^{\mu}(\hat{q},\tau) + \frac{i}{c}\,\bar{V}\,(\hat{q},\tau). \tag{2.107}$$

From the first step (change form $\gamma$ to $\tilde{\gamma}$) it is clear that $\hat{H}^{\gamma}$ is a function of the equivalence classes of discretizations.

## 2.5. Determination of a correspondence rule

At the end of section 2.1. we have already mentioned the problem of the determination of a correspondence rule that maps a given set of functions into a given set of operators.Within the class of correspondence rules defined by the $\Omega$ - function $\Omega(u,v)$ (cf. section 2.1) this problem can have many solutions or no solution at all.
A practical way to determine $\Omega(u,v)$ follows from formula (2.24) which can be written as

$$\Omega_{2}\left(-\, i \frac{\partial}{\partial q}, -i \frac{\partial}{\partial p}\right) B^{\Theta_2}(p,q) = \Omega_{1}\left(-\, i \frac{\partial}{\partial q}, -\, i \frac{\partial}{\partial p}\right) B^{\Theta_1}(p,q) \tag{2.108}$$

Let us assume that the $\Theta_1$-symbol of the operator $\hat{B}(\hat{p},\hat{q})$ can be determined and let us take for $B^{\Theta_2}(p,q)$ the given phase space function. Then $\Omega_{2}(u,v)$ is the mapping function that we have to determine. It is analytic in u and v, we can therefore write it as a power series with unknown coefficients. Formula (2.108) allows to determine conditions on these coefficients.
Examples will clarify this procedure.
A methode to determine $\Omega(u,v)$ based directly on the closed form expressions (2.13) and (2.17) has been given by Mizrahi in [M16]. There is of course a direct relation between his method and ours.

### Examples

Consider a two dimensional phase space. Then the general form of $\Omega(u,v)$ is

$$\Omega(u,v) = \sum_{k,\ell=o}^{\infty} a_{k\ell} \; u^{k}v^{\ell}. \tag{2.109}$$

1. What conditions on the coefficients $a_{k\ell}$ follow from the imposed relation $q^2p^2 \to \hat{q}\,\hat{p}^2\hat{q}$?

The anti-standard symbol of $\hat{q}\hat{p}^2\hat{q}$ is $p^2q^2 + 2ipq$. In (2.108) we use now $\Omega_1(u,v)$ $\Omega_1(u,v) = \exp \frac{i}{2}\, u\, v$ (cf. table 2.1) and $\Omega_2(u,v) = \Omega(u,v)$ as given by (2.109). The relation that follows leads to the conditions

$$
\begin{aligned}
&a_{00} = 1\\
&a_{10} = a_{01} = a_{11} = a_{02} = a_{20} = a_{12} = a_{21} = 0 \qquad (2.110)\\
&a_{22} = \frac{1}{8}
\end{aligned}
$$

All other coefficients remain free.

2. What restriction follows from the condition $q^n \to \hat{q}^n$ , $n \in \mathbb{N}$ ?

Now it is easiest to take for $\Omega_1$ the Weyl correspondence $\Omega_1(u,v) = 1$, and (2.108) reduces to

$$\sum_{k=0}^{\infty} a_{ko}\, (-1)^k \frac{n!}{(n-k)!}\, q^{n-k} = q^n \qquad (2.111)$$

which implies

$$
\begin{aligned}
&a_{00} = 1\\
&a_{k0} = 0,\ k > 0. \qquad (2.112)
\end{aligned}
$$

This is equivalent to

$$\Omega(u,0) = 1 \qquad (2.113)$$

In the same way the condition $p^m \to \hat{p}^m$ ,$m \in \mathbb{N}$ , leads to

$$\Omega(0,v) = 1 \qquad (2.114)$$

3. An important restriction follows from the condition that any monomial $p^mq^n$ is mapped into an operator which is a linear combination of permutations of products of m operators $\hat{p}$ and n operators $\hat{q}$. (All rules listed in table 2.1. are of this type). The antistandard symbol of this operator will always be of the form

$$\sum_{k=o}^{\min(m,n)} a_k\, p^{m-k}\, q^{m-k} \qquad (2.115)$$

since by writing the operator in anti-standard order we make commutations that make disappear operators $\hat{q}$ and $\hat{p}$ two by two. (See e.g. example 1); If we use now as in the first example $\Omega_1(u,v) = \exp \frac{i}{2} u\, v$ the RHS of (2.108) will also be of the form (2.115) (The coefficients $a_k$ change of course). Then one checks immediately that (with $B^{\Theta_2}(p,q) = p^m q^n$ , $\Omega_2(u,v)$ given by 2.109) there will follow always the condition

$$a_{k\ell} = 0 \text{ if } k \neq \ell \quad \text{and } k \leq n \text{ , } \ell \leq m \tag{2.116}$$

Therefore a correspondence rule that maps any monomial $p^m q^n$ into an operator which is a linear combination of permutations of products of m $\hat{p}$'s and n $\hat{q}$'s will only depend on the product uv, it will be of the form

$$\Omega(u,v) = \sum_{k=o}^{\infty} b_k (uv)^k . \tag{2.117}$$

The generalization to higher dimensions of this result is that $\Omega(u,v)$ can only depend on the products $u_\lambda v^\lambda$ (no summation over $\lambda$ ) $\lambda = 1,\ldots, d$.

In table 2.1. all $\Omega$ -functions depend in fact on $u.v = \sum_{\lambda=1}^{d} u_\lambda v^\lambda$ which is a more stringent condition.

4. In chapter 6 we will need the correspondence rule that relates the phase space function $p_\mu p_\nu B(q)$ ( $\mu$ and $\nu$ given and fixed) to the operator $\hat{p}_\mu B(\hat{q})\hat{p}_\nu$. From example 3 we know that this rule will be determined by an $\Omega$-function only depending on the products $u_\lambda v^\lambda, \lambda = 1, \ldots ,d$. Furthermore it will not be possible to get conditions for the coefficients of the terms of order $v^3$ since the function we are considering is only quadratic in p. (cf. example 1) Therefore we consider $\Omega(u,v)$ of the form

$$\Omega(u,v) = \Omega(0,0) + \sum_{\lambda=1}^{d} a_\lambda\, u_\lambda v^\lambda + \sum_{\lambda,\kappa=1}^{d} a_{\lambda\kappa}\, u_\lambda v^\lambda u_\kappa v^\kappa + O(v^3). \tag{2.118}$$

In the RHS of (2.108) we will again use the anti-standard rule of correspondence. The anti-standard symbol of $\hat{p}_\mu B(\hat{q})\hat{p}_\nu$ is $p_\mu p_\nu B(q) + i\, p_\mu \partial_\nu B(q)$, and

$$\Omega_{AS}\left( - i \frac{\partial}{\partial q} , - i \frac{\partial}{\partial p} \right)(p_\mu p_\nu B(q) + i\, p_\mu \partial_\nu B(q)) =$$

$$= \exp \left( - \frac{i}{2} \sum_{\lambda=1}^{d} \frac{\partial^2}{\partial q^\lambda \partial p_\lambda} \right)( p_\mu p_\nu B(q) + i\, p_\mu (\partial_\nu B(q))$$

$$= p_\mu p_\nu B(q) + \frac{i}{2} p_\mu (\partial_\nu B)(q) - \frac{i}{2} p_\nu (\partial_\mu B)(q) + \frac{1}{4} (\partial_\mu \partial_\nu\, B)(q) \tag{2.119}$$

On the other hand the LHS of (2.108) is given by

$$\Omega(-i\frac{\partial}{\partial q}, -i\frac{\partial}{\partial p})\, p_\mu p_\nu B(q) =$$

$$= \Omega(0,0) p_\mu p_\nu B(q) - p_\mu a_\nu (\partial_\nu B)(q) - p_\nu a_\mu (\partial_\mu B)(q) + (a_{\mu\nu} + a_{\nu\mu})(\partial_\mu \partial_\nu B)(q). \tag{2.120}$$

Identification of (2.119) and (2.120) leads to

$$\Omega(0,0) = 1 \tag{2.121a}$$

$$a_\mu = \frac{i}{2} \text{ if } \mu \neq \nu \tag{2.121b}$$

$$a_\nu = -\frac{i}{2} \text{ if } \mu \neq \nu \tag{2.121c}$$

$$a_\mu = a_\nu \quad \text{if } \mu = \nu \tag{2.121d}$$

$$a_{\mu\nu} = a_{\nu\mu} = \frac{1}{8} . \tag{2.121e}$$

The coefficients $a_\lambda$ , $\lambda \neq \mu,\nu$ and $a_{\lambda\kappa}$, $\lambda \neq \mu,\nu$ or $\kappa \neq \mu,\nu$ remain arbitrary. The simplest $\Omega$- function (*) defining the correspondence rule that maps $p_\mu p_\nu B(q)$ into $\hat{p}_\mu B(\hat{q})\hat{p}_\nu$ ( $\mu$ and $\nu$ given and fixed) is then given by

$$\Omega_{(\mu\nu)}(u,v) = 1 - \frac{i}{2} u_\nu v^\nu + \frac{i}{2} u_\mu v^\mu + \frac{1}{8} u_\nu v^\nu u_\mu v^\mu + O(v^3) . \tag{2.122}$$

One can check from table I that this is not one of the usual correspondence rules. An important application of this rule is given in section 5.8.

5. An interesting simplification occurs when B(q) is of the form $B^{\mu\nu}(q)$, a symmetric tensor in $\mu$ and $\nu$ and summation over $\mu$ and $\nu$ is performed. Then (2.119) reduces to $p_\mu p_\nu B^{\mu\nu}(q) + \frac{1}{4}\partial_\mu \partial_\nu B^{\mu\nu}(q)$, and the $\Omega$ -function defining the correspondence rule that maps $p_\mu p_\nu B^{\mu\nu}(q)$ into $\hat{p}_\mu B^{\mu\nu}(\hat{q})\hat{p}_\nu$ (summation over $\mu$ and $\nu$ ) is then found to be

$$\Omega(u,v) = 1 + \frac{1}{8}(u.v)^2 + O(v^3) . \tag{2.123}$$

(*) In Refs. [L17] and [L20] a slightly different method was followed to obtain $\Omega_{(\mu\nu)}$. The solution of (2.121) obtained there is less simple then the one obtained here.

## CHAPTER III: FUNCTIONAL INTEGRAL REPRESENTATIONS OF EXPECTATION VALUES. TIME ORDERED PRODUCTS.

In the first section of this chapter we define functional integrals of the form

$$<F[p(\tau),q(\tau),\tau]>^{\gamma}=\int_{\gamma} \mathcal{D}q\mathcal{D}p\ F[p(\tau),q(\tau),\tau]\ \exp\{i \int_{t_\circ}^{t} d\tau[p.\dot{q} - H^{\gamma}(p,q,\tau)]\}\{\delta\} \tag{3.1}$$

where $F[p(\tau),q(\tau),\tau]$ is a functional of the paths in phase space ($t_\circ\leqslant\tau\leqslant t$). The function $H^{\gamma}$ is assumed to depend on $\gamma$ in such a way that for $F \equiv 1$ and $\{\delta\} = \delta(q(t) - q)\ \delta(q(t_\circ)-q_\circ)$ (3.1) represents the propagator for the operator $\hat{H}(\hat{p},\hat{q},t)$. Functional integrals as in (3.1) arise in perturbation theory when $H^{\gamma}$ is written as $H_\circ^{\gamma} + H_1^{\gamma}$ and $\exp[-i \int_{t_\circ}^{t} H_1^{\gamma}]$ is developed in power series. (See chapter IV) Of basic importance is the case when $F[p(\tau),q(\tau),\tau]$ is just a product of p's and q's at a finite number of times $F[p(\tau),q(\tau),\tau] = G^{(m,n)}(t_1',\ldots,t_m';t_1,\ldots,t_n)$ with

$$G^{(m,n)}(t_1',\ldots,t_m';t_1,\ldots,t_n) = p_{\nu_1}(t_1')\ldots p_{\nu_m}(t_m')q^{\mu_1}(t_1)\ldots q^{\mu_n}(t_n) \tag{3.2}$$

$$t_\circ \leqslant t_i' \leqslant t\ ,\ i = 1,\ldots,m$$

$$t_\circ \leqslant t_j \leqslant t\ ,\ j = 1,\ldots,n\quad .$$

In chapter VII we shall see that correlation - and response functions of stochastic variables have functional integral representations of the form (3.1) with $F = G^{(m,n)}$ and $\{\delta\}= \delta(q(t_\circ)-q_\circ)$.

We do not define here the analogue of (3.1) for configuration space functional integrals since (3.1) is all we need to study the perturbation expansion from the phase space functional integral.(next chapter). In chapter 10 we shall consider the perturbation expansion from the configuration space functional integral.

In section 3.2. we relate the quantities (3.1) to the operator formalism. This allows us to infer in section 3.3. from the definition of the path integral through discretization a definition of time ordered product which is also defined for products of operators at equal times. In the case of $\gamma_1(\alpha)$ - discretization these relations between time ordering and functional formalism have been studied in detail by Leschke and Schmutz (Ref. [L32]). The approach presented here is based on our works [L4,L8,L10]. Other calculations showing how to handle functional integrals in a specific discretization can already be found in the books by Katz[K1] and Rosen [R2].

## 3.1. Definitions

The functional integral (3.1) is defined as the limit for $N \to \infty$ of the following expression

$$<F>_N^\gamma = \int \prod_{i=1}^{N} dq_i \prod_{j=1}^{N+1} (2\pi)^{-d} dp_j \; f^\gamma(\{q_i, i=0,\dots N+1\}\{p_j, j=1,\dots,N+1\}\{\bar{t}_j, j=1,\dots,N+1\}) .$$

$$. \exp \{i\varepsilon \sum_{j=1}^{N+1} [p_j \frac{\Delta_j}{\varepsilon} - h^\gamma (p_j, q_j, q_{j-1}, \bar{t}_j)] \} \tag{3.3}$$

where the function $f^\gamma$ is an approximation of the functional F defined by the discretization $\gamma$ , $h^\gamma$ is as before and we have taken $\{\delta\} = \delta(q(t)-q)\ \delta(q(t_\circ)-q_\circ)$ i.e. $q_{N+1}$ and $q_\circ$ fixed.

### 3.1.1. Examples

1. A simple example is

$$F = \int_{t_\circ}^{t} d\tau\ q(\tau)\delta(t'-\tau) = q(t') \ , \ t_\circ < t' < \tau$$

For any N there is a k such that $t_{k-1} < t' \leqslant t_k$ and (3.1) for $F = q(t')$ is defined by (3.3) for $f^\gamma = q_k$ , $t_{k-1} < t' \leqslant t_k$ .
In the same way one can define (3.1) with $F = p(t')$ and one can generalize this definition to $F = G^{(m,n)}$ at least if all times $t_j, j=1,\dots,n$, are different from all times $t_i'$ ,$i=1,\dots,m$.
In the case of coinciding times the discretization $\gamma$ has to give all necessary information. E.g. for $F = p_\mu(t')q^\mu(t')$ one has that $<F>_N^{\gamma_1(0)}$ is defined by (3.3) for $f^{\gamma_1(0)} = p_{k\mu}q^\mu_{k-1}$ ,$t_{k-1} < t' \leqslant t_k$ while $<F>_N^{\gamma_1(1)}$ is defined by $f^{\gamma_1(1)} = p_{k\mu}q^\mu_k$.
From our experiences of chapter II we do not expect $<F>^\gamma$ to be $\gamma$ -independent in general.

2. Consider the following three functionals:

$$F_1[p(\tau),q(\tau),\tau] = \int_{t_0}^{t} d\tau\ p_\mu(\tau)p_\nu(\tau)\ B^{\mu\nu}(q(\tau),\tau)\ , \tag{3.4}$$

$$F_2[p(\tau),q(\tau),\tau] = \int_{t_0}^{t} d\tau\ p_\mu(\tau)B^{\mu}(q(\tau),\tau)\ , \tag{3.5}$$

$$F_3[p(\tau),q(\tau),\tau] = \int_{t_0}^{t} d\tau\ W(q(\tau),\tau). \tag{3.6}$$

If we are given a discretization $\gamma$ , this now means that we are given a discrete approximation for these functionals since they are of the same form as the terms appearing in $\int d\tau H^\gamma$ . (Again $H^\gamma$ is assumed to be quadratic in p). The functional integral (3.1) for a functional $F[\ p(\tau),q(\tau),\tau\ ]$ which only depends on a finite number of functionals $F_1,F_2$ or $F_3$ is then defined by $(F\,[p(\tau),q(\tau),\tau\,] \equiv g\ (\{F_1\},\{F_2\},\{F_3\}\ ))$

$$\langle F\rangle^\gamma = \lim_{N\to\infty} \int (\prod_{i=1}^{N} dq_i \prod_{j=1}^{N+1} (2\pi)^{-d}\ dp_j)g\ (\{f_1^\gamma\},\{f_2^\gamma\}\ ,\{f_3^\gamma\}\ )\ .$$

$$.\ \exp\ \{i\,\varepsilon \sum_{j=1}^{N+1} [p_j\,.\,\frac{\Delta_j}{\varepsilon} - h^\gamma(p_j,q_j,q_{j-1},\bar{t}_j)]\ \} \tag{3.7}$$

where

$$f_1^\gamma = \varepsilon \sum_{j=1}^{N+1} p_{j\mu}p_{j\nu}b^{\mu\nu}_{[\gamma]}(q_j,q_{j-1},\bar{t}_j) \tag{3.8}$$

$$f_2^\gamma = \varepsilon \sum_{j=1}^{N+1} p_{j\mu}b^{\mu}_{[\gamma]}(q_j,q_{j-1},\bar{t}_j) \tag{3.9}$$

$$f_3^\gamma = \varepsilon \sum_{j=1}^{N+1} w_{[\gamma]}(q_j,q_{j-1},\bar{t}_j) \tag{3.10}$$

$$\bar{t}_j \in [t_{j-1},t_j] \tag{3.11}$$

$$b^{\mu\nu}_{[\gamma]}(q,q,\tau) = B^{\mu\nu}(q,\tau)\ , \tag{3.12}$$

$$b^{\mu}_{[\gamma]}(q,q,\tau) = B^{\mu}(q,\tau) \tag{3.13}$$

$$w_{[\gamma]}(q,q,\tau) = W(q,\tau) \tag{3.14}$$

The proof that the limit (3.7) (if it exists) is independent of the value of $\bar{t}_j$ satisfying (3.11) follows from the corresponding result for the phase space functional integral representation of the propagator (cf. end of section 2.3) for all cases where g $(\{F_1\},\{F_2\},\{F_3\})$ arises from a perturbation expansion.

### 3.1.2. Generating functional

Usually [A2,M3] functional integrals of the type (3.1) are defined formally by functional derivation of a generating functional. The common procedure is the following: define a generating functional $Z[J,J^\star]$ by

$$Z[J,J^\star]=\int_\gamma \mathcal{D}q\mathcal{D}p \;\; \exp\{i \int_{t_0}^{t} d\tau[p.\dot{q}- H^\gamma(p,q,\tau) + J^\star(\tau)p(\tau) + J(\tau).q(\tau ]\}\{\delta\} \quad . \tag{3.15}$$

The sources $J^\star$ and J have components $J^{\star 1},\ldots,J^{\star d}$ and $J_1,\ldots,J_d$ respectively and since for any $\gamma$ the discretization of $\int d\tau[J^\star.p + J.q]$ is equivalent to e.g. $\varepsilon\;\Sigma[J^\star(\bar{t}_j).p_j + J(\bar{t}_j).q_j]$ , $Z[J,J^\star]$is in fact the propagator for the Hamiltonian operator

$$\hat{H}_{JJ^\star}(\hat{p},\hat{q},\tau) = \hat{H}(\hat{p},\hat{q},\tau) - J^\star(\tau)\hat{p} - J(\tau)\hat{q} \tag{3.16}$$

and is therefore discretization independent.

The functional integral (3.1) is then defined as

$$\langle F\rangle= F[\frac{1}{i} \frac{\delta}{\delta J^\star(\tau)}, \frac{1}{i} \frac{\delta}{\delta J(\tau)},\tau]\; Z[J,J^\star]\Big|_{J=J^\star= 0} \quad . \tag{3.17}$$

Written as it stands (3.17) implies that $\langle F\rangle$ is discretization independent. However, in order to be well-defined and equivalent to (3.3) $\langle F\rangle$ in (3.17) has to be interpreted as a limit.

We have for example in the case that F is a function of $F_1$, $F_2$ and $F_3$ only (cf. example 2)

$$\langle g\;(\{F_i(p(\tau),q(\tau),\tau),i=1,2,3.\}\;)\rangle^\gamma$$

$$= \lim_{\delta\to 0} \langle g\;(\{f_i^\gamma\;(p(\tau),q(\tau),q(\tau-\delta))\;,\; i=1,2,3\;\}\;\rangle \tag{3.18}$$

where on the right hand side the brackets < > have the meaning of (3.17).

Although the formal procedure (3.17) is a very useful one allowing for a very elegant treatment of perturbation theory (cf. next chapter) the reminiscence of the discretization in the form (3.18) has to be taken as part of the procedure in order to avoid ambiguities.

## 3.2. Representation of matrix elements by functional integrals

### 3.2.1. Operator formalism

In chapter I we have introduced in (1.69-70) the evolution operator $U(t,t')$ associated to the differential equation (1.1). Its inverse is given by

$$[U(t,t')]^{-1} \equiv U^{-1}(t,t') = U(t',t). \tag{3.19}$$

Let us use the notation

$$U(t) \equiv U(t,0) \tag{3.20}$$

and define the time-dependent operators

$$\hat{p}(t) = U^{-1}(t)\hat{p}\, U(t), \tag{3.21}$$

$$\hat{q}(t) = U^{-1}(t)\hat{q}\, U(t), \tag{3.22}$$

the bra-vector

$${}^{L}\langle q,t| = \langle q|U(t) \tag{3.23}$$

and the ket-vector

$$|q,t\rangle^{R} = U^{-1}(t)|q\rangle . \tag{3.24}$$

If $\hat{H}(\hat{p},\hat{q},\tau)$ is hermitean then $U(t,t')$ is unitary and $U^{-1}(t) = U^{+}(t)$. Then ${}^{L}\langle q,t|$ is the vector in the dual space corresponding to $|q,t\rangle^{R}$ and we can simply write $\langle q,t|$ and $|q,t\rangle$ .

From (3.21-22) in their infinitesimal form follow the equations of motion for $\hat{p}(t)$ and $\hat{q}(t)$ :

$$\dot{\hat{p}}(t) = i\,[H(\hat{p}(t),\hat{q}(t),t)\ ,\ \hat{p}(t)] \tag{3.25}$$

$$\dot{\hat{q}}(t) = i\,[H(\hat{p}(t),\hat{q}(t),t)\ ,\ \hat{q}(t)]\ . \tag{3.26}$$

Using (1.69) and $U(t,t_0) = U(t)U^{-1}(t_0)$ the propagator $K$ of (1.1) can be written as

$$K(q,t;q_0,t_0) = {}^{L}\langle q,t|q_0,t_0\rangle^{R}. \tag{3.27}$$

The quantities in which we will be interested are the matrix elements (or expectation values) between $|q_0,t_0\rangle^{R}$ and ${}^{L}\langle q,t|$ of time-ordered products of operators.

Let $\hat{z}(\tau)$ stand for either $\hat{p}(\tau)$ or $\hat{q}(\tau)$. The time ordered product of $\ell$ operators $\hat{z}_i(t_i)$, $i = 1,\dots,\ell$ is defined by

$$P\,\hat{z}_1(t_1)\hat{z}_2(t_2)\dots\hat{z}_\ell(t_\ell) = \hat{z}_{\pi(1)}(t_{\pi(1)})\,\hat{z}_{\pi(2)}(t_{\pi(2)})\dots\hat{z}_{\pi(\ell)}(t_{\pi(\ell)}) \tag{3.28}$$

where $\pi$ is the permutation of $1,2,\dots,\ell$ such that $t_{\pi(1)} > t_{\pi(2)} > \dots > t_{\pi(\ell)}$. By this definition the P-product is not defined when two or more of the times $t_k$ coincide. The generalization of time-ordered product to the case of equal times will follow from the path integral representation of the matrix elements

$${}^{L}\langle q,t|P\,\hat{z}(t_1)\hat{z}(t_2)\,\dots\,\hat{z}(t_\ell)|q_\circ,t_\circ\rangle^{R}\,. \tag{3.29}$$

### 3.2.2. Functional integral representations

We start by computing ${}^{L}\langle q,t|\hat{q}^\mu(t')|q_\circ,t_\circ\rangle^{R}$ where $t_\circ < t' < t$. The derivation that follows is almost the same as the one leading to the path integral representation of the propagator. We divide again the time interval $[t_\circ,t]$ into N+1 pieces of length $\varepsilon$ : $t_j = t_\circ + j\varepsilon$ , $j = 0,1,\dots,N+1$, $t_{N+1}=t$ and we suppose $t'=t_k$ for some k, $0 \leqslant k \leqslant N+1$. (If this is not the case we take k as in example 1 in page 44, the difference vanishes in the limit $N \to \infty$, as the calculation shows). Then using (3.22-24) and (3.19-20):

$$\begin{aligned}{}^{L}\langle q,t|\hat{q}^\mu(t')|q_\circ,t_\circ\rangle^{R} &= \langle q|U(t)U^{-1}(t_k)\hat{q}^\mu U(t_k)U^{-1}(t_\circ)|q_\circ\rangle \\ &= \langle q|U(t,t_k)\hat{q}^\mu U(t_k,t_\circ)|q_\circ\rangle\,.\end{aligned} \tag{3.30}$$

Inserting N times $\int dq|q\rangle\langle q| = 1$ and using $\hat{q}^\mu|q_k\rangle = q_k^\mu\,|q_k\rangle$ we obtain

$${}^{L}\langle q,t|\hat{q}^\mu(t')|q_\circ,t_\circ\rangle^{R} = \int \prod_{i=1}^{N} dq_i\; q_k{}^{\mu} \prod_{j=1}^{N+1} \langle q_j|U(t_j,t_{j-1})|q_{j-1}\rangle\,. \tag{3.31}$$

For the short time propagator we can again use any of the approximations up to $\mathcal{O}(\varepsilon)$

$$\langle q_j|U(t_j,t_{j-1})|q_{j-1}\rangle = \int (2\pi)^{-d}\, dp_j\, \exp\{i\varepsilon[p_j\frac{\Delta_j}{\varepsilon} - h^\gamma(p_j,q_j,q_{j-1},\bar{t}_j)]\}\,. \tag{3.32}$$

In the limit $N \to \infty$ one obtains then for ${}^{L}\langle q,t|\hat{q}^\mu(t')|q_\circ,t_\circ\rangle^{R}$ precisely the

the functional integral $\langle q^\mu(t')\rangle^\gamma$ defined in example 1 :

$$ {}^L\langle q,t|\hat{q}^\mu(t')|q_\circ,t_\circ\rangle^R = \int_\gamma \mathcal{D}q\mathcal{D}p\; q^\mu(t')\; \exp\{i\int_{t_\circ}^{t} d\tau[p.\dot{q} - H^\gamma(p(\tau),q(\tau),\tau)]\} $$
$$ \delta(q(t)-q)\delta(q(t_\circ)-q_\circ) \quad (3.33) $$

The procedure that lead us to (3.33) is straightforwardly generalised to ${}^L\langle q,t|\hat{q}^{\mu_1}(t_1)\hat{q}^{\mu_2}(t_2)\ldots\hat{q}^{\mu_n}(t_n)|q_\circ,t_\circ\rangle^R$ when $t > t_1 > t_2 \ldots > t_n$ and therefore, by definition of time ordered product to

$$ {}^L\langle q,t|P\,\hat{q}^{\mu_1}(t_1)\ldots\hat{q}^{\mu_n}(t_n)|q_\circ,t_\circ\rangle^R $$
$$ = \int_\gamma \mathcal{D}q\mathcal{D}p\; q^{\mu_1}(t_1)\ldots q^{\mu_n}(t_n)\exp\{i\int_{t_\circ}^{t} d\tau[p.\dot{q} - H^\gamma(p,q,\tau)]\} $$
$$ \delta(q(t)-q)\delta(q(t_\circ)-q_\circ) \quad (3.34) $$

for $t_i \neq t_j$ , $i \neq j$.

Let us now also see how to represent ${}^L\langle q,t|\,\hat{p}_\mu(t')|q_\circ,t_\circ\rangle^R$ , $t_\circ < t' < t$. In the same way as for $\hat{q}^\mu(t')$ we can write this as

$$ \langle q|U(t,t_k)\hat{p}_\mu U(t_k,t_\circ)|q_\circ\rangle = \int \prod_{i=1}^{N} dq_i \,(\prod_{j=k+1}^{N+1} \langle q_j|U(t_j,t_{j-1})|q_{j-1}\rangle\,). $$
$$ .\;\langle q_k|\hat{p}_\mu U(t_k,t_{k-1})|q_{k-1}\rangle(\prod_{j=1}^{k-1}\langle q_j|U(t_j,t_{j-1})|q_{j-1}\rangle\,). \quad (3.35) $$

One has that

$$ \langle q_k|\hat{p}_\mu\; U(t_k,t_{k-1})|q_{k-1}\rangle = -\,i\frac{\partial}{\partial q_k^\mu}\langle q_k|U(t_k,t_{k-1})|q_{k-1}\rangle $$
$$ = -\,i\frac{\partial}{\partial q_k^\mu}\int(2\pi)^{-d}\,dp_k\,\exp\,[ip_k.(q_k-q_{k-1}) - i\varepsilon\, h^\gamma(p_k,q_k,q_{k-1},\bar{t}_k)] $$
$$ = \int(2\pi)^{-d}\,dp_k\,[p_{k_\mu} - \varepsilon\frac{\partial}{\partial q_k^\mu}\, h^\gamma(p_k,q_k,q_{k-1},\bar{t}_k)]\;. $$
$$ .\;\exp\,[ip_k.(q_k-q_{k-1}) - i\varepsilon h^\gamma(p_k,q_k,q_{k-1},\bar{t}_k)]. $$

In the limit $N \to \infty$, $\varepsilon \to 0$ only $p_{k_\mu} = p_\mu(t')$ remains and one obtains finally

$$ {}^L\langle q,t|\hat{p}_\mu(t')|q_\circ,t_\circ\rangle^R = \int_\gamma \mathcal{D}q\mathcal{D}p\; p_\mu(t')\; \exp\{i\int_{t_\circ}^{t} d\tau[p.\dot{q} - H^\gamma(p,q,\tau)]\}\,\delta(q(t)-q) $$
$$ \delta(q(t_\circ)-q_\circ) $$

$$= \langle p_\mu(t')\rangle^\gamma \ . \qquad (3.36)$$

From the combination of (3.34) and (the generalization of) (3.36) there follows that

$$^L\langle q,t|\,P\,\hat{z}_1(t_1)\ \dots\ \hat{z}_\ell(t_\ell)|\ q_\circ,t_\circ\rangle^R = \langle z_1(t_1)\dots\ z_\ell(t_\ell)\rangle^\gamma \qquad (3.37)$$

for all times $t_i$, $i = 1,\dots,n$ different from one another.

The right hand side of (3.37) represents a functional integral representation in the $\gamma$ -discretization. From the equality (3.37) it follows in fact that this functional integral is independent of the discretization. In the next section we show that this need no longer be true when the left hand side is not defined i.e. when some of the times $t_i$ coincide.

The quantities in (3.37) are obtained by functional derivation of a generating functional $Z[J,J^\star]$ which is in the operator formalism given by

$$Z[J,J^\star] = {}^L\langle q,t|\,P\,\exp\{+\,i\int_{t_\circ}^{t} d\tau[J(\tau)\ .\hat{q}(\tau) + J^\star(\tau).\hat{p}(\tau)]\}\ |q_0,t_\circ\rangle^R . \qquad (3.38)$$

## 3.3. A generalized time-ordered product

The time-ordered product associated to the discretization $\gamma$ is denoted by $P_\gamma$ and is defined by

$$^L\langle q,t|\ P_\gamma\hat{z}_1(t_1)\dots\hat{z}_\ell(t_\ell)|q_\circ,t_\circ\rangle^R = \langle z_1(t_1)\dots z_\ell(t_\ell)\rangle^\gamma . \qquad (3.39)$$

First of all we remark that when all times $t_i, i=1,\dots,\ell$ are different from another the $P_\gamma$-product coincides with the P-product defined in section 3.3. When $t_1 = t_2 = \dots = t_\ell = t'$ the product $z_1(t_1)\dots\ z_\ell(t_\ell)$ is a monomial of the form $(m+n = \ell)$

$$M(p(t'),q(t')) = q^{\mu_1}(t')\dots\ q^{\mu_n}(t')\ p_{\nu_1}(t')\ \dots\ p_{\nu_m}(t'). \qquad (3.40)$$

The functional integral $\langle M(p(t'),q(t')\rangle^\gamma$ is by definition

$$\langle M(p(t'),q(t')\rangle^\gamma = \lim_{N\to\infty}\int\prod_{i=1}^{N} dq_i\ \prod_{j=1}^{N+1}\ (2\pi)^{-d}dp_j\ m^\gamma(p_k,q_k,q_{k-1})\ \exp\{i\varepsilon\sum_{j=1}^{N+1}[p_j.\ \frac{\Delta_j}{\varepsilon}\ -$$

$$- h^{\gamma}(p_j,q_j,q_{j-1},\bar{t}_j)] \} \qquad (3.41)$$

where $q_k$ is determined by $t' \in (t_{k-1},t_k]$ and $m^{\gamma}(p,q',q)$ is such that $m^{\gamma}(p,q,q) = M(p,q)$.

The definition of $P_{\gamma}$-product (3.39) is then, using (3.41), given by

$$P_{\gamma}M(\hat{p}(t'),\hat{q}(t')) = \lim_{\delta\to 0} \; P\, m^{\gamma}(\hat{p}(t'),\hat{q}(t'+\delta),\hat{q}(t'-\delta)) \qquad (3.42)$$

where at the right hand side P is the usual time ordering operator. If some, but not all the times in (3.39) coincide, the generalization is obvious.

Example

The $P_{\gamma_2(\alpha)}$-product associated to the $\gamma_2(\alpha)$-discretization orders time-dependent operators as the usual P-product (greater times at the left hand side) and the ordered product of operators at the same time $t'$ is given by

$$\begin{aligned} P_{\gamma_2(\alpha)}\, M(\hat{p}(t'),\hat{q}(t')) &= \lim_{\delta\to 0} P[(1-\alpha)M(\hat{p}(t'),\hat{q}(t'-\delta)) + \alpha\, M(\hat{p}(t'),\hat{q}(t'+\delta)] \\ &= (1-\alpha)\, \hat{p}_{\nu_1}(t')\ldots \hat{p}_{\nu_m}(t')\hat{q}^{\mu_1}(t')\ldots \hat{q}^{\mu_n}(t') \\ &+ \alpha\hat{q}^{\mu_1}(t')\ \ldots\ \hat{q}^{\mu_n}(t')\hat{p}_{\nu_1}(t')\ldots \hat{p}_{\nu_m}(t'). \qquad (3.43) \end{aligned}$$

## CHAPTER IV: PERTURBATION EXPANSIONS

In this chapter we investigate the functional integral formalism for perturbation theory. All calculations are done for a two-dimensional phase space. The generalization to any finite number of dimensions is straightforward. In the last section we comment on the extension to field theory.

First we describe in section 4.1. how to set up a perturbation expansion of the generating functional $Z[J,J^\star]$ by splitting from the Hamiltonian operator $\hat{H}(\hat{p},\hat{q})$ a quadratic part $\hat{H}_o(\hat{p},\hat{q})$, determining the free generating functional $Z_o[J,J^\star]$ and treating $\hat{H}_I = \hat{H} - \hat{H}_o$ as a perturbation. We calculate explicitly the free generating functional.

In section 4.2. we show that the multiplicity of functional integral representations of $Z[J,J^\star]$ according to the discretizations $\gamma$ is reflected in the perturbation expansion in a multiplicity of Feynman rules. We also show that the mechanism that makes the value of the series discretization independent order by order in the coupling constants in $\hat{H}_I$ is a cancellation between discretization dependent vertices and contributions from tadpoles. In other words, the definition of the functional integral through a formal series expansion has the same problems as the definition by discrete approximation and limiting procedure. The specification of a definite ordering scheme or equivalence class of discretizations is necessary and sufficient to make all terms uniquely defined.

In section 4.3. we generalize the treatment of section 4.2. to functional integral representations determined by $\Omega$-functions not satisfying $\Omega(u,o) = 1$. This includes in particular the normal and anti-normal correspondence rules. Section 4.4. contains two remarks, the first one on the extension of our results to field theory the second on the definition of functional integrals without limiting procedure through the formal series expansion.

Most of the results presented in this chapter are contained in our works [L4?,L8, L10].

### 4.1. Setting up a perturbation expansion.

Let be given as in (1.74) a time-independent operator $\hat{H}(\hat{p},\hat{q})$ of the form

$$\hat{H}(\hat{p},\hat{q}) = -\frac{ic}{2}\hat{p}^2 G(\hat{p}) - \hat{p}\,A(\hat{q}) + \frac{i}{c}V(\hat{q}) \tag{4.1}$$

and let us split from $\hat{H}(\hat{p},\hat{q})$ a quadratic part writing

$$\hat{H}(\hat{p},\hat{q}) = \hat{H}_o(\hat{p},\hat{q}) + \hat{H}_1(\hat{p},\hat{q}) \tag{4.2}$$

where

$$\hat{H}_o(\hat{p},\hat{q}) = -\frac{ic}{2}\hat{p}^2 G_o - \hat{p}A_o(\hat{q}) + \frac{i}{c}V_o(\hat{q}) \tag{4.3}$$

with $G_o$ a constant and $A_o(\hat{q})$ and $V_o(\hat{q})$ respectively linear and quadratic in $\hat{q}$.

The free Hamiltonian $\hat{H}_o(\hat{p},\hat{q})$ defines a solvable model and by considering $\hat{H}_1(\hat{p},\hat{q})$ as a perturbation we can compute quantities in the full theory defined by $\hat{H}(\hat{p},\hat{q})$ order by order in the coupling constants in $\hat{H}_1(\hat{p},\hat{q})$.

In chapter III we have defined in (3.15) the generating functional $Z[J,J^\star]$ by its phase space functional integral representations:

$$Z[J,J^\star] = \int_\gamma \mathcal{D}q\mathcal{D}p \; \exp\{i\int_{t_o}^{t} d\tau[p\dot{q} - H^\gamma(p,q) + Jq + J^\star p]\} \; \delta(q(t)-q)\delta(q(t_o)-q_o). \tag{4.4}$$

At zero sources this quantity is the propagator associated with $\hat{H}$ and by functional derivation one obtains expectation values of functions or functionals of $q(\tau)$ and $p(\tau)$, $t_o < \tau < t$. A perturbation expansion of this quantity leads to a perturbation expansion of the propagator and of expectation values.
For any discretization $\gamma$ the splitting (4.2) of $\hat{H}$ into $\hat{H}_o$ and $\hat{H}_1$ induces a corresponding splitting of $H^\gamma$ into $H_o^\gamma$ and $H_1^\gamma$. The functional integral (4.4) with $H^\gamma$ replaced by $H_o^\gamma$ is nothing but the generating functional $Z_o[J,J^\star]$ associated with $\hat{H}_o(\hat{p},\hat{q})$ and is a discretization independent quantity. Since $\hat{H}_o$ is quadratic in $\hat{p}$ and $\hat{q}$ also $H_o^\gamma(p,q)$ is quadratic in $p$ and $q$ and $Z_o[J,J^\star]$ is a Gaussian functional integral which can be computed exactly. Then the complete generating functional $Z[J,J^\star]$ can be obtained from the free generating functional as follows: (4.4) is equal to

$$Z[J,J^\star] = \int_\gamma \mathcal{D}q\mathcal{D}p \; [\exp -i\int_{t_o}^{t} H_1^\gamma(p,q)d\tau]\exp\{i\int_{t_o}^{t} d\tau[p\dot{q}-H_o^\gamma(p,q) + Jq + J^\star p]\} \; . \; \delta(q(t)-q)\delta(q(t_o)-q_o) \tag{4.5}$$

which formally can be written as

$$Z[J,J^\star] = \exp\{-i\int_{t_o}^{t} H_1^\gamma\left(\frac{1}{i}\frac{\delta}{\delta J^\star(\tau)}, \frac{1}{i}\frac{\delta}{\delta J(\tau)}\right)d\tau\} \, Z_o[J,J^\star] \; . \tag{4.6}$$

Developing $\exp\{-i\int_{t_o}^{t} d\tau \, H_1^\gamma\}$ one obtains an explicit expression as an infinite series for the functional integral (4.4). No ambiguities will arise if $H_1^\gamma(p,q)$ is treated as the limit

$$H_1^\gamma(p(\tau),q(\tau) = \lim_{\delta\to o} h_1^\gamma(p(\tau),q(\tau+\delta),q(\tau-\delta)) \tag{4.7}$$

which contains all necessary information on the discretization $\gamma$ .

Functional integrals of the form (3.1) can then also be computed in perturbation theory by applying (3.17-18) to (4.6).

In Appendix 4.1. we give some more details on the relation to the operator formalism and in Appendix 4.2. we rewrite (4.6) so as to obtain the functional version of Wick's theorem.

Let us compute now the free generating functional $Z_\circ[J,J^\star]$ in the example where

$$G_\circ = 1 \tag{4.8}$$

$$A_\circ(\hat{q}) = \mu\,\hat{q} \tag{4.9}$$

$$V_\circ(\hat{q}) = -\frac{1}{2}\,\omega^2\hat{q}^2. \tag{4.10}$$

The Weyl-symbol of the free Hamiltonian $\hat{H}_\circ$ is

$$H_\circ^{\Theta_W}(p,q) = -\frac{ic}{2}\,p^2 - \mu pq - \frac{i}{2c}\,\omega^2 q^2 + \frac{i}{2}\,\mu \tag{4.11}$$

and the mid-point functional integral representation of $Z_\circ[J,J^\star]$ is

$$Z_\circ[J,J^\star] = \int_{\gamma_1(\frac{1}{2})} \mathcal{D}q\mathcal{D}p \;\; \exp i\int_{t_\circ}^{t} d\tau[p\dot{q} - H_\circ^W(p,q) + Jq + J^\star p]$$

$$\delta(q(t)-q)\delta(q(t_\circ)-q_\circ). \tag{4.12}$$

There are several methods available to compute $Z_\circ[J,J^\star]$ . The one we will use here starts from (4.12) and uses the formula (1.14) for the variables $(x_1,\ldots,x_{2N+1}) \equiv (p_1,\ldots p_{N+1},q_1,\ldots,q_N)$, which means that we integrate p's and q's at the same time.

Using equivalence of discretizations we change to post-point discretization in the p-independent terms in the argument of the exponential in (4.12). The resulting discretized expression can be written as

$$Z_\circ[J,J^\star] = \lim_{N\to\infty}\int\Big(\prod_{i=1}^{N} dq_i \prod_{j=1}^{N+1} \frac{dp_j}{2\pi}\Big) \exp\Big\{-\frac{1}{2}\sum_{k,\ell=1}^{N+1} \varepsilon^2\,[p_j p_k\,\frac{c}{\varepsilon}\,\delta_{jk} + q_j q_k\,\frac{\omega^2}{c\varepsilon}\,\delta_{jk}] -$$

$$-\frac{1}{2}\sum_{j=1}^{N+1}\varepsilon[-2i\,p_j(\frac{q_j-q_{j-1}}{\varepsilon} + \mu\,\frac{q_j+q_{j-1}}{2})] + \varepsilon\sum_{J=1}^{N+1}[i\,J_j^\star\,p_j + i\,J_j q_j + \frac{1}{2}\,\varepsilon\mu]\Big\}, \tag{4.13}$$

Rewriting the term $\sum_{j=1}^{N+1} p_j(q_j-q_{j-1})$ as

$$p_{N+1}q - p_1 q_\circ + \sum_{j=1}^{N+1} \sum_{i=1}^{N} p_j q_i \delta_{ji} - \sum_{j=1}^{N+1} \sum_{i=1}^{N} p_j q_i \delta_{i,j-1}$$

we obtain the argument of the exponential in the form $-\frac{1}{2}\varepsilon^2 x_i x_j A_{ij} + \varepsilon b_i x_i$ where

$$\underline{A} = \begin{vmatrix} A^{11} & A^{12} \\ A^{21} & A^{22} \end{vmatrix} \tag{4.14}$$

$$A^{11}_{ij} = \frac{c}{\varepsilon}\delta_{ij} \quad , i,j = 1,\ldots, N+1 \tag{4.15a}$$

$$A^{22}_{ij} = \frac{\omega^2}{c\varepsilon}\delta_{ij} \quad , i,j=1, \ldots, N \tag{4.15b}$$

$$A^{12}_{ji} = -i\left[\left(\frac{1}{\varepsilon^2} + \frac{\mu}{2\varepsilon}\right)\delta_{ji} + \left(-\frac{1}{\varepsilon^2} + \frac{\mu}{2\varepsilon}\right)\delta_{j-1,i}\right] \quad ,i=1,\ldots,N \quad j=1,\ldots,N+1 \tag{4.15c}$$

$$A^{21}_{ij} = A^{12}_{ji} \tag{4.15d}$$

$$b = \left(i J^\star_j - iq_\circ\left(1-\tfrac{1}{2}\varepsilon\mu\right)\frac{1}{\varepsilon}\delta_{1j} + iq\left(1-\tfrac{1}{2}\varepsilon\mu\right)\frac{1}{\varepsilon}\delta_{N+1,j}, iJ_k\right)$$

$$j=1,\ldots,N+1, \; k = 1,\ldots,N \,. \tag{4.16}$$

In the limit $\varepsilon \to 0$, $N \to \infty$ the matrix $\underline{A}$ and the vector b formally tend to

$$A^\infty(\tau,\tau') = \begin{vmatrix} c\,\delta(\tau-\tau') & -i\left(\frac{\partial}{\partial\tau} + \mu\right)\delta(\tau-\tau') \\ -i\left(-\frac{\partial}{\partial\tau} + \mu\right)\delta(\tau-\tau') & \frac{\omega^2}{c}\delta(\tau-\tau') \end{vmatrix} \tag{4.17}$$

$$b^\infty(\tau) = \left(i J^\star(\tau) - iq_\circ\delta(\tau-\tau_\circ) + iq\delta(\tau-t), \; i\, J(\tau)\right) \tag{4.18}$$

To find $Z_\circ[J,J^\star]$ we need $\lim_{N\to\infty}(b,A^{-1}b)$ . This is equal to $(b^\infty,(A^\infty)^{-1}b^\infty)$ except for the fact that the finite matrix $\underline{A}$ has a unique inverse $\underline{A}^{-1} = \underline{D}$ while boundary conditions are required to compute the inverse of $\underline{A}^\infty$ . These conditions follow from (4.15). Indeed $\underline{D}$ is also of the form

$$\underline{D} = \begin{vmatrix} D^{11} & D^{12} \\ D^{21} & D^{22} \end{vmatrix}$$

and one of the four sets of equations for it is

$$\sum_{j=1}^{N+1} A_{ij}^{11} D_{jk}^{12} + \sum_{j=1}^{N} A_{ij}^{12} D_{jk}^{22} = 0 \qquad i=1,\dots,N+1, k=1,\dots,N \tag{4.19}$$

which is using (4.15)

$$c\, D_{ik}^{12} - i\,[(\frac{1}{\varepsilon} + \frac{\mu}{2})D_{ik}^{22} + (-\frac{1}{\varepsilon} + \frac{\mu}{2})\, D_{i-1,k}^{22}\,] = 0 , \qquad i=2,\dots,N \quad k=1,\dots,N \tag{4.20a}$$

$$c\, D_{1k}^{12} - i\ (\frac{1}{\varepsilon} + \frac{\mu}{2})D_{1k}^{22} = 0 , \tag{4.20b}$$

$$c\, D_{N+1,k}^{12} - i\ (-\frac{1}{\varepsilon} + \frac{\mu}{2})D_{N,k}^{22} = 0 . \tag{4.20c}$$

The first equation of (4.20) represents the differential equation for $D^{\infty 12}(\tau,\tau')$ namely

$$c\, \underline{D}^{\infty 12}(\tau,\tau') - i[\,\frac{\partial}{\partial\tau} + \mu\,]\, \underline{D}^{\infty 22}(\tau,\tau') = 0 \tag{4.21}$$

(It has to be remarked that $\underline{A} \to \underline{A}^{\infty}$ and $\sum_j A_{ij}D_{jk} = \delta_{ik}$ ,$\int d\tau'\ \underline{A}^{\infty}(\tau,\tau')\underline{D}^{\infty}(\tau',\tau'')= \delta(\tau-\tau'')$ imply $\underline{D}/\varepsilon^2 \to \underline{D}^{\infty}$)

The boundary condition compatible with (4.20b-c) is

$$\underline{D}^{\infty 22}(t_{\circ},\tau') = \underline{D}^{\infty 22}(T,\tau') = 0 \tag{4.22}$$

since $D_{ok}^{22}$ and $D_{N+1,k}^{22}$ are precisely the quantities that do not appear ("are zero") in (4.20 b-c).

These conditions are enough to determine the unique inverse of $\underline{A}^{\infty}$ corresponding to the inverse of $\underline{A}$. Let us drop then from now on the superscript "∞". The set of differential equations for $\underline{D}$ is (4.21) and

$$c\ D^{11}(\tau,\tau') - i[\,\frac{\partial}{\partial\tau} + \mu]D^{21}(\tau,\tau') = \delta(\tau-\tau') \tag{4.23}$$

$$-\ i\ [-\,\frac{\partial}{\partial\tau} + \mu]D^{11}(\tau,\tau') + \frac{\omega^2}{c}\ D^{21}(\tau,\tau') = 0 \tag{4.24}$$

$$-\ i\ [-\,\frac{\partial}{\partial\tau} + \mu]D^{12}(\tau,\tau') + \frac{\omega^2}{c}\ D^{22}(\tau,\tau') = \delta(\tau-\tau'). \tag{4.25}$$

Eliminating from (4.21) and (4.25) $D^{12}(\tau,\tau')$ we obtain an equation for $D^{22}(\tau,\tau')$ :

$$\left[\frac{\partial^2}{\partial\tau^2} - \rho^2\right] D^{22}(\tau,\tau') = - c\ \delta(\tau-\tau') \tag{4.26}$$

where

$$\rho^2 = \mu^2 + \omega^2 . \tag{4.27}$$

We see that $D^{22}(\tau,\tau')$ is a Green's function for the operator $\left(\frac{\partial^2}{\partial\tau^2} - \rho^2\right)$ . Once $D^{22}$ is known, $D^{12}$ and $D^{11}$ can be obtained from (4.21) and (4.23) using also $D^{12}(\tau,\tau') = D^{21}(\tau',\tau)$ . ($\underline{A}$ and $\underline{D}$ are symmetric). $D^{11}(\tau,\tau')$ and $D^{22}(\tau,\tau')$ are continuous in $\tau = \tau'$, $D^{12}(\tau,\tau')$ has a jump at the diagonal $\tau = \tau'$. Explicitly $\underline{D}$ is given by:

$$D^{22}(\tau,\tau') = \frac{c}{\rho\ \sinh\rho(t-t_o)}\ [\theta(\tau-\tau')\sinh\rho(\tau'-t_o)\sinh\rho(t-\tau) + \theta(\tau'-\tau)\sinh\rho(\tau-t_o)\sinh\rho(t-\tau')] \tag{4.28}$$

$$D^{12}(\tau,\tau') = \frac{i}{\rho\ \sinh\ \rho(t-t_o)}\ [\ \theta(\tau-\tau')\sinh\rho(\tau'-t_o)(-\rho\cosh\rho(t-\tau)+\mu\sinh\rho(t-\tau)) + \theta(\tau'-\tau)\ (\rho\cosh\rho(\tau-t_o)+\ \mu\sinh\ \rho(\tau-t_o))\sinh\rho(t-\tau')] \tag{4.29}$$

$$D^{11}(\tau,\tau') = \frac{1}{c\,\rho\ \sinh\ \rho(t-t_o)} \cdot [\ \theta(\tau-\tau')(\rho\cosh\rho(\tau'-t_o)+\mu\sinh\rho(\tau'-t_o))(\rho\cosh\rho(t-\tau)-\mu\sinh\rho(t-\tau)) + \theta(\tau-\tau')(\rho\cosh\rho(\tau-t_o)+\mu\sinh\rho(\tau-t_o))(\rho\cosh\rho(t-\tau)-\mu\sinh\ \rho(t-\tau'))] . \tag{4.30}$$

One can check easily that $D^{12}(\tau,\tau')$ satisfies

$$D^{12}(\tau,t_o) = D^{12}(\tau,t) = 0 \tag{4.31}$$

and $$\lim_{\delta\to o}\ D^{12}(\tau,\tau+\delta) - D^{12}(\tau,\tau-\delta) = i . \tag{4.32}$$

In terms of the solutions (4.28-4.30) one obtains for $Z_o[J,J^\star]$:

$$Z_\circ[J,J^\star] = K \exp\{-\frac{1}{2}\int_{t_\circ}^{t} d\tau \int_{t_\circ}^{t} d\tau' \, [(J^\star(\tau) - q_\circ\delta(\tau-t_\circ) + q\delta(\tau-t))D^{11}(\tau,\tau') .$$

$$. (J^\star(\tau') - q_\circ\delta(\tau'-t_\circ) + q\delta(\tau'-t)) \,]$$

$$- \int_{t_\circ}^{t} d\tau \int_{t_\circ}^{t} d\tau' \, [(J^\star(\tau) - q_\circ\delta(\tau-t_\circ) + q\delta(\tau-t))D^{12}(\tau,\tau')J(\tau')]$$

$$- \frac{1}{2}\int_{t_\circ}^{t} d\tau \int_{t_\circ}^{t} d\tau' \, J(\tau)D^{22}(\tau,\tau') \, J(\tau') \,\} \tag{4.33}$$

where the normalization factor is independent of $J$ and $J^\star$ and given by (see Appendix 4.3)

$$K = Z_\circ[0,0]\Big|_{q_\circ=q=0}$$

$$= \left[\frac{\rho}{2c\pi \sinh\rho(t-t_\circ)}\right]^{1/2} \exp\left[\tfrac{1}{2}\mu(t-t_\circ)\right] . \tag{4.34}$$

It is convenient to eliminate the $\delta$-functions in (4.33) by writing $Z_\circ[J,J^\star]$ as

$$Z_\circ[J,J^\star] = K' \exp\{-\int_{t_\circ}^{t} d\tau d\tau' [\,\frac{1}{2} J^\star(\tau)D^{11}(\tau,\tau') \, J^\star(\tau')$$

$$+ J^\star(\tau) \, D^{12}(\tau,\tau')J(\tau')$$

$$+ \frac{1}{2} J(\tau)D^{22}(\tau,\tau') \, J(\tau')]$$

$$+ \int_{t_\circ}^{t} d\tau' \, S_1(\tau') \, J^\star(\tau') + \int_{t_\circ}^{t} d\tau' \, S_2(\tau')J(\tau') \,\} \tag{4.35}$$

where

$$S_1(\tau') = q_\circ \, D^{11}(t_\circ,\tau') - q \, D^{11}(t,\tau') \tag{4.36}$$

$$S_2(\tau') = q_\circ \, D^{12}(t_\circ,\tau') - q \, D^{12}(t,\tau') \tag{4.37}$$

and

$$K' = K \exp[-\frac{1}{2} q_\circ^2 \, D^{11}(t_\circ,t_\circ) + q_\circ q \, D^{11}(t_\circ,t) - \frac{1}{2} q^2 D^{11}(t,t)] . \tag{4.38}$$

Now we are ready to investigate the perturbation expansion generated by (4.6). We introduce the graphs in figure 4.1. to denote the functions $D^{11}(\tau,\tau')$,

$D^{12}(\tau,\tau')$, $D^{22}(\tau,\tau')$ and $S_1(\tau')$ , $S_2(\tau')$ . From the figure and (4.35) we see that an undulated line is associated with $p(\tau)$ while a straight line is associated with $q(\tau)$.

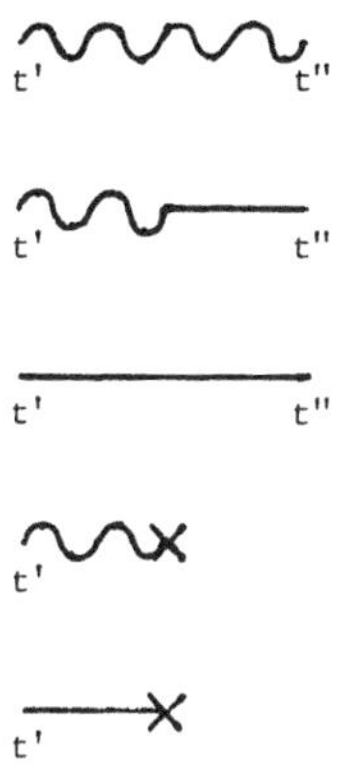


$= D^{11}(t',t'')$

$= D^{12}(t',t'')$

$= D^{22}(t',t'')$

$= S_1(t')$

$= S_2(t')$

Figure 4.1.: Graphs representing the functions $D^{ij}(t',t'')$ and $S_i(t')$, $i=1,2$ appearing in $Z_\circ[J,J^\star]$ as given by (4.35).

The graphs in the expansion obtained from (4.6) are constructed from the graphs of figure 4.1 joined in interaction vertices determined by the function $H_1^\gamma(p,q)$.

## 4.2. Perturbation expansions and discretizations

We have in (4.4) infinitely many functional integral representations for $Z[J,J^\star]$. For each equivalence class of discretizations a different phase space function $H^\gamma$ appears. We have then also in general a different $H_1^\gamma$ and consequently different interaction vertices in the graphical representation of the series expansion (4.6). Nevertheless the value of the expansion should be $\gamma$- independent in each order of the expansion parameters in $H_1^\gamma$ . This will turn out to be possible owing to the $\gamma$-dependent values taken by the simplest closed loops or tadpoles in the expansion.

Indeed, in general one has in $H_1^\gamma$ $(p(\tau),q(\tau))$ products $p(\tau)q(\tau)$ at the same time $\tau$ . These give when $p(\tau) \to \frac{1}{i}\,\frac{\delta}{\delta J^\star(\tau)}$ and $q(\tau) \to \frac{1}{i}\,\frac{\delta}{\delta J(\tau)}$ a term

$$- \frac{\delta^2}{\delta J^\star(\tau)\delta J(\tau)}\; Z_\circ[J,J^\star]$$

which contains a contribution $D^{12}(\tau,\tau)$ which is not defined ($D^{12}(\tau,\tau')$ is discontinuous at the diagonal). We shall see first in an example how the prescription (4.7) resolves this ambiguity and then study systematically the perturbation expansion from functional integrals in $\gamma(\Omega)$ - discretization.

### 4.2.1. Example

In the case of $\gamma_2(\alpha)$ -discretization the phase space function associated with the interaction Hamiltonian

$$\hat{H}_1(\hat{p},\hat{q}) = -\frac{ic}{2}\hat{p}^2G_1(\hat{q}) - \hat{p}A_1(\hat{q}) + \frac{i}{c}V_1(\hat{q}) \tag{4.39}$$

is given by (cf. section 1.4)

$$\begin{aligned} H^{\gamma_2(\alpha)}(p,q) &= -\frac{ic}{2}p^2G_1(q) - p[A_1(q) + \alpha c G_1'(q)] \\ &\quad + \frac{i}{c}[V_1(q) - \alpha c\, A_1'(q) + \alpha(\alpha-\tfrac{1}{2})\, c^2 G_1''(q)] \\ &= -\frac{ic}{2}p^2G_1(q) - p\, A_1^\alpha(q) + \frac{i}{c}V_1^\alpha(q) . \end{aligned} \tag{4.40}$$

From the definition of the functional integral through the discretized Hamiltonian $h^{\gamma_2(\alpha)}(p,q',q) = (1-\alpha)H^{\gamma_2(\alpha)}(p,q) + \alpha H^{\gamma_2(\alpha)}(p,q')$, where q' is at a time slightly bigger than p and q at a time slightly smaller we infer that in the right hand side of (4.6) $H_1^{\gamma_2(\alpha)}(p(\tau),q(\tau))$ should be interpreted as

$$\begin{aligned} \lim_{\delta\to o} &-\frac{ic}{2}p^2(\tau)[(1-\alpha)G_1(q(\tau-\delta)) + \alpha G_1(q(\tau+\delta)) - \\ &- p(\tau)[(1-\alpha)A_1^\alpha(q(\tau-\delta)) + \alpha A_1^\alpha(q(\tau+\delta))] + \frac{i}{c}V_1^\alpha(q(\tau)). \end{aligned} \tag{4.41}$$

This gives an unambiguous value to all the terms $D^{12}(\tau,\tau)$ arising in the expansion (4.6).

Let

$$B(\tau) \equiv \lim_{\delta\to o} D^{12}(\tau,\tau-\delta). \tag{4.42}$$

According to (4.32) one has

$$\lim_{\delta\to o} D^{12}(\tau,\tau+\delta) = B(\tau) + i. \tag{4.43}$$

From (4.41) it follows that the product of two terms $D^{12}(\tau,\tau)$ coming from $p^2(\tau)D_1(q(\tau))$ in $H_1^{\gamma 2(\alpha)}$ has the value $[(1-\alpha)B(\tau)^2+\alpha(B(\tau)+i)^2]$ . When one term $D^{12}(\tau,\tau)$ comes either from $p^2(\tau)D_1(q(\tau))$ or $p(\tau)A_1(q(\tau))$ the value is $(1-\alpha)B(\tau)+\alpha(B(\tau)+i)$.

We see that these terms take $\alpha$ -dependent values. This $\alpha$-dependence should combine with the explicit $\alpha$ -dependence of $A_1^\alpha$ and $V_1^\alpha$ in such a way that the $\alpha$ -dependence cancels completely in order to obtain the final result $Z[J,J^*]$ independent of $\alpha$ . We shall study this cancellation mechanism below.

### 4.2.2. Feynman rules and cancellation mechanism

After this example we can now proceed to the study of the perturbation expansion (4.6) for arbitrary discretizations defined by a correspondence rule $\Omega$ , characterized by an $\Omega$-function $\Omega(u,v)$.

We have seen in section 2.2. that the correspondence rules associated in a natural way with phase space functional integrals are the ones satisfying $\Omega(u,0)=1$, since then $H^{\gamma(\Theta)}\equiv H^\Theta$ . In section 2.5. we have proved that correspondence rules which map any of the monomials $q^n p^m$ into a linear combination of permutations of products of n operators $\hat{q}$ and m operators $\hat{p}$ can only depend on the product uv. Let us restrict ourselves first to correspondence rules of this type and denote from now on the $\Omega$-function by $\Omega(uv)$. We also impose $\Omega(0)=1$.

#### A. Values of the $D^{12}$-tadpoles

In the formula (4.6) appears the phase space function (compare with 2.44)

$$H_1^{\Theta_\Omega}(p,q) = \bar{\Omega}\left(-\frac{\partial^2}{\partial q\partial p}\right) H_1^{AS}(p,q) \tag{4.44}$$

which has to be interpreted as (compare with 2.31, 2.41)

$$H_1^\Theta(p,q) = h_1^\Theta(p,q,q) = \Omega\left(i\Delta\frac{\partial}{\partial\bar{q}}\right)H_1^\Theta(p,\bar{q})\Big|_{q'=q} \tag{4.45}$$

with $\quad \bar{q} = \frac{1}{2}(q'+q), \quad \Delta = q'-q.$

Let $\hat{H}_1(\hat{p},\hat{q})$ again be given by (4.39), then $h_1^\Theta(p,q',q)$ is of the form

$$h_1^\Theta(p,q',q) = -\frac{ic}{2}p^2 g_1^\Theta(q',q) - p\, a_1^\Theta(q',q) + \frac{i}{c}v_1^\Theta(q',q) \tag{4.46}$$

with

$$g_1^{\Theta}(q',q) = \Omega(i\,\Delta\,\frac{\partial}{\partial\bar{q}})G_1(\bar{q}) \qquad (4.47)$$

$$a_1^{\Theta}(q',q) = \Omega(i\,\Delta\,\frac{\partial}{\partial\bar{q}})A_1^{\Theta}(\bar{q}) \qquad (4.48)$$

$$v_1^{\Theta}(q',q) = \Omega(i\,\Delta\,\frac{\partial}{\partial\bar{q}})V_1^{\Theta}(\bar{q})\,. \qquad (4.49)$$

In contrast with the situation in the example it is not easy to read from (4.47-49) the rules for the tadpoles in the expansion. Therefore we use an equivalent discretization $\gamma \approx \gamma(\Omega)$ . By definition of equivalence it has $H_1^{\gamma}(p,q) \equiv H_1^{\Theta}(p,q)$ i.e. the same interaction vertices in the perturbation expansion. From the rules for the tadpoles in the $\gamma$ -discretization and the conditions for equivalence will follow the rules for the tadpoles in the $\gamma(\Omega)$ - discretization. Let $h_1^{\gamma}(p,q',q)$ be given by

$$h_1^{\gamma}(p,q',q) = (1-c_1)\,H_1^{\Theta}(p,\bar{q}+c_2\Delta) + c_1\,H_1^{\Theta}(p,\bar{q}+c_3\Delta). \qquad (4.50)$$

$\gamma$ is equivalent to $\gamma(\Omega)$ when the conditions (2.82-84) are satisfied . To derive equations for the constants $c_1, c_2, c_3$ we develop in 4.47-49 $\Omega(uv)$ in powers of uv and expand in (4.50) around $\bar{q}$. The conditions (2.82-84) are satisfied when ($\Omega'(0) = \left.\frac{\partial\Omega(x)}{\partial x}\right|_{x=0}$ , etc... )

$$(1-c_1)c_2 + c_1c_3 = i\Omega'(0) \qquad (4.51)$$

$$(1-c_1)c_2^2 + c_1c_3^2 = -\ \Omega''(0). \qquad (4.52)$$

From these equations $c_2$ and $c_3$ can be solved in terms of $c_1, \Omega'(0)$ and $\Omega''(0)$, but we do not need this solution explicitly.

We have then that in the formula

$$Z[J,J^{\star}] = \exp\{-\,i\int_{t_0}^{t} d\tau\, H_1^{\Theta_\Omega}(\frac{1}{i}\,\frac{\delta}{\delta J^{\star}(\tau)}\,,\,\frac{1}{i}\,\frac{\delta}{\delta J(\tau)})\}\,Z_0[J,J^{\star}] \qquad (4.53)$$

$H_1^{\Theta_\Omega}(p(\tau),q(\tau)) \equiv H_1^{\gamma}(p(\tau),q(\tau))$ can be interpreted as $\lim_{\delta\to 0} h_1^{\gamma}(p(\tau),q(\tau+\delta),q(\tau-\delta))$.

Tadpoles come from $p^2\,G_1(q)$ and $p\,A_1^{\Theta}(q)$ , so, relevant is the interpretation of $G_1(q)$ and $A_1^{\Theta}(q)$ which is ($F_1$ is either $G_1$ or $A_1^{\Theta}$ )

$$F_1(q(\tau)) = \lim_{\delta\to 0}\ (1-c_1)\,[\,F_1((\tfrac{1}{2}-c_2)\,(q(\tau-\delta)+(\tfrac{1}{2}+c_2)q(\tau+\delta))]$$
$$+\ c_1\,[F_1((\tfrac{1}{2}-c_3)q(\tau-\delta) + (\tfrac{1}{2}+c_3)q(\tau+\delta))]\ . \qquad (4.54)$$

This formula defines unambiguously the expansion (4.53). The rules for the $D^{12}$ tadpoles that one obtains from (4.54) and using (4.51-52) are as follows:

- a double tadpole $D^{12}D^{12}$ coming from $p^2(\tau)G_1(q(\tau))$ takes the value

$$(1-c_1)[(\tfrac{1}{2} - c_2)B(\tau)+(\tfrac{1}{2} + c_2)(B(\tau)+i\ )]^2 + c_1[(\tfrac{1}{2} - c_3)B(\tau)+(\tfrac{1}{2} + c_3)(B(\tau) + i\ )]^2$$

$$=[B(\tau) + \tfrac{1}{2} i - \Omega'(0)]^2 + \Omega''(0) - \Omega'(0)^2 \qquad (4.55)$$

- a single tadpole $D^{12}$ coming from $p(\tau)^2G_1(q(\tau))$ or form $p(\tau)A_1^{\Theta}(q(\tau))$ takes the value

$$(1-c_1)[B(\tau)+ (\tfrac{1}{2} + c_2)i] + c_1[B(\tau) + (\tfrac{1}{2} + c_3)i] = B(\tau) + \tfrac{1}{2} i - \Omega'(0)\ . \qquad (4.56)$$

It follows from the fact that the discretization of both $G_1(q)$ and $A_1^{\Theta}(q)$ is determined by the same correspondence rule that the single tadpole takes the same value for both . We remark that the value of a double tadpole is equal to the square of a single tadpole only when

$$\Omega'(0)^2 = \Omega''(0). \qquad (4.57)$$

Example

In section 2.2. we have seen that the $\gamma(\Omega)$ - discretization equivalent to $\gamma_2(\alpha)$-discretization is determined by the $\Omega$-function

$$\Omega(u,v) =(1-\alpha)\exp\ (\tfrac{1}{2}\ iuv) + \alpha\ \exp\ (- \tfrac{1}{2}\ i\ uv). \qquad (4.58)$$

Since here $\Omega'(0) = (\frac{1}{2} -\alpha)\ i$ and $\Omega''(0) = - \frac{1}{4}$ the value of a double tadpole is

$$[B(\tau) + \alpha i]^2 - \tfrac{1}{4} + (\tfrac{1}{2} - \alpha)^2 = (1-\alpha)\ B(\tau)^2 + \alpha(B(\tau) + i\ )^2 \qquad (4.59)$$

while a single tadpole has the value

$$B(\tau) + \alpha\ i, \qquad (4.60)$$

which are the results we have obtained before.
Since all the correspondence rules listed in table 2.1. (chapter 2) are of the type we have studied here we can list them in table 4.1 together with the associated rules for the tadpoles:

Table 4.1 : Correspondence rules and associated values of the tadpoles $D^{12}(\tau,\tau)$ in the perturbation expansion

| correspondence rule | single tadpole | double tadpole |
|---|---|---|
| Weyl | $B(\tau) + \frac{1}{2}i$ | $(B(\tau) + \frac{1}{2}i)^2$ |
| Symmetric | $B(\tau) + \frac{1}{2}i$ | $(B(\tau) + \frac{1}{2}i)^2 - \frac{1}{4}$ |
| Anti-standard | $B(\tau)$ | $B(\tau)^2$ |
| Standard | $B(\tau) + i$ | $(B(\tau) + i)^2$ |
| Born-Jordan | $B(\tau) + \frac{1}{2}i$ | $(B(\tau) + \frac{1}{2}i)^2 - \frac{1}{12}$ |

B. Cancellation mechanism

We have shown in A. that the $D^{12}$-tadpoles in the expansion take $\Omega$-dependent values. In order to obtain order by order in the coupling constants in $\hat{H}_1$ a generating functional $Z[J,J^\star]$ independent of $\Omega$ there must exist a cancellation mechanism between the explicit $\Omega$-dependence of the interaction vertices in $H_1^{\Theta_\Omega}$ and the $\Omega$-dependence of the tadpoles. To illustrate this we consider now the interaction Hamiltonian

$$\hat{H}_I(\hat{p},\hat{q}) = -\frac{ic\lambda}{4}\hat{p}^2\hat{q}^2. \tag{4.55}$$

The antistandard symbol of this operator is simply

$$H_1^{AS}(p,q) = -\frac{ic\lambda}{4}p^2q^2. \tag{4.56}$$

Using (4.44) the $\Theta_\Omega$ -symbol for the case of $\Omega$-functions depending only on uv can be computed. It is given by

$$H_1^{\Theta_\Omega}(p,q) = -\frac{ic\lambda}{4}p^2q^2 + ic\lambda\left(\frac{i}{2} - \Omega'(0)\right)pq - ic\lambda\left(-\frac{1}{8} - \frac{1}{2}\Omega''(0) - \frac{i}{2}\Omega'(0) + \Omega'(0)^2\right). \tag{4.57}$$

We see that $A_1^{\Theta_\Omega}(q)$ and $V_1^{\Theta_\Omega}(q)$ indeed depend on $\Omega(uv)$ explicitly through $\Omega'(0)$ and $\Omega''(0)$.

The interaction vertices in $-i\,H_1^{\Theta_\Omega}(p(\tau),q(\tau))$ are shown in figure 4.2. One has that

$$g_4 = -\frac{c\lambda}{4}, \qquad g_{12} = c\lambda\left(\frac{i}{2} - \Omega'(0)\right), \text{ and}$$

$$g_\circ = -c\lambda\left(-\frac{1}{8} - \frac{1}{2}\Omega''(0) - \frac{i}{2}\Omega'(0) + \Omega'(0)^2\right)$$

Fig 4.2. Interaction vertices in $-iH_1^{\Theta_\Omega}(p(\tau),q(\tau))$ for $\hat{H}_1(\hat{p},\hat{q}) = -\frac{ic\lambda}{4}\hat{p}^2\hat{q}^2$.

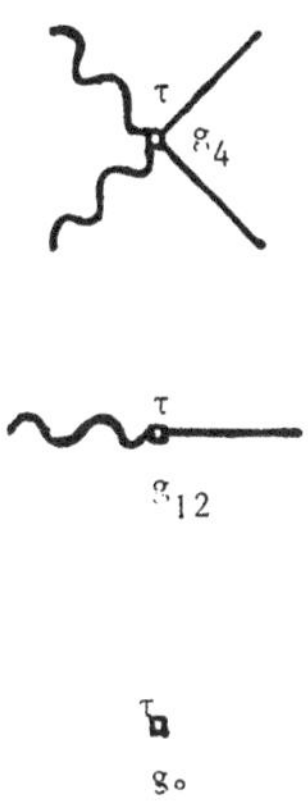


The contribution of first order in $\lambda$ in the expansion of (4.51) and taken at zero sources is represented graphically in figure 4.3. The graphs 2 and 5 have to be multiplied by combinational factor 2 and 4. Graphs 1-6 are proportional to $g_4$, graphs 7 and 8 proportional to $g_{12}$ and graph 9 proportional to $g_\circ$.

In first order in $\lambda$ all closed loops are tadpoles. The graphs containing the a priori undefined tadpole $D^{12}(\tau,\tau)$ are 2,5 and 7. The sum of these with other graphs depending explicitly on $\Omega(uv)$ (i.e. the graphs 8 and 9) must give an $\Omega$-independent result. In fact, inspection of figure 4.3. shows that the $\Omega$ - dependence should cancel in the graphs without crosses (2,7 and 9) and in the ones with crosses(5 and 8). By the rules derived in part A of this section the former sum up to

$$\int_{t_0}^{t} d\tau\{2g_4[(B(\tau) + \frac{1}{2} i - \Omega'(0))^2 + \Omega''(0) - \Omega'(0)^2] + g[B(\tau) + \frac{1}{2} i - \Omega'(0)] + g_0\}$$

$$= \int_{t_0}^{t} d\tau\{-\frac{c\lambda}{2}[(B(\tau) + \frac{1}{2} i - \Omega'(0))^2 + \Omega''(0) - \Omega'(0)^2] + c\lambda(\frac{i}{2} - \Omega'(0))[B(\tau) + \frac{1}{2} i - \Omega'(0)] - c\lambda(-\frac{1}{8} - \frac{1}{2}\Omega''(0) - \frac{i}{2}\Omega'(0) + \Omega'(0)^2)\}$$

$$= \int_{t_0}^{t} d\tau\{-\frac{c\lambda}{2}(B(\tau) + \frac{1}{2} i)^2 + \frac{i}{2} c\lambda(B(\tau) + \frac{i}{2}) + \frac{1}{8} c\lambda\} \qquad (4.58)$$

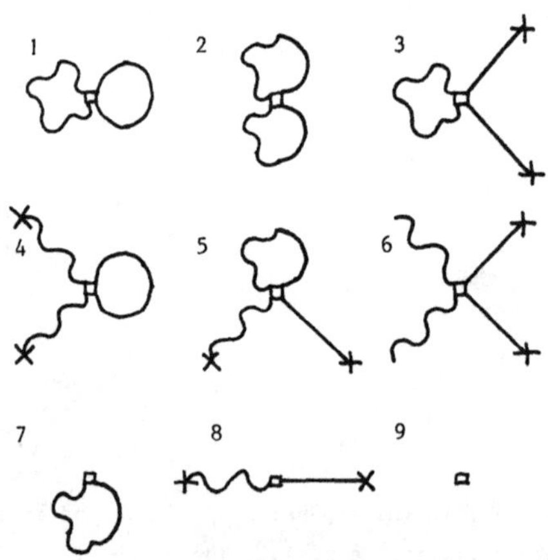


Figure 4.3.: Graphs representing the contribution of first order in $\lambda$ in the expansion of $Z[0,0]$ with $H_1^{\Theta_\Omega}$ given by (4.57).

and the latter to

$$\int d\tau \quad S_1(\tau)S_2(\tau)\ [4g_4(B(\tau) + \frac{1}{2}\, i - \Omega'(0)) + g_{12}]$$

$$= \int d\tau \ S_1(\tau)S_2(\tau)\ [- c\ \lambda(B(\tau) + \frac{1}{2}\, i - \Omega'(0)) + c\lambda\ (\frac{1}{2}\, i - \Omega'(0))\ ]$$

$$= \int d\tau \ S_1(\tau)S_2(\tau)\ [- c\lambda(B(\tau) + \frac{1}{2}\, i) + \frac{i}{2}\, c\lambda\ ] \quad . \tag{4.59}$$

In both cases the $\Omega$-dependence cancels and the result equals that obtained by taking $\Omega'(0) = \Omega''(0) = 0$ (e.g. Weyl correspondence) from the beginning.

C. Conclusion

For any correspondence rule $\Omega(uv)$ the generating functional $Z[J,J^\star]$ has a diagrammatic perturbation expansion in terms of the free propagators in $Z_\circ[J,J^\star]$ and the $\Omega$ -dependent vertices in $H_1^{\Theta_\Omega}(p,q)$. The a priori undefined tadpoles $D^{12}(\tau,\tau)$ have unambiguously defined values once the underlying discretization or correspondence rule is taken into account. The value of the perturbation expansion is independent of $\Omega(uv)$ order by order in the coupling constants in the operator $\hat{H}_1(\hat{p},\hat{q})$ owing to a cancellation between the discretization dependent vertices and the contributions from $\Omega$-dependent tadpoles. It's immediately clear that these conclusions remain valid for $1 \leq d < \infty$ (d the dimension of q-space). The case $d\to\infty$ is considered in section 4.4.
In applications the choice of correspondence rule is dictated by convenience only. E.g. in the example we have been studying ($\hat{H}_1(\hat{p},\hat{q}) = - \frac{ic\lambda}{4}\hat{p}^2\hat{q}^2$) the simplest thing to do is use the anti-standard rule of correspondence since then $H^{\Theta AS}(p,q) = - \frac{ic\lambda}{4}\ p^2q^2$ and only the graphs 1-6 in figure 4.2. appear. Furthermore, as one can read from table 4.1. all $D^{12}$ tadpoles take the value $B(\tau)$.

## 4.3. Correspondence rules not satisfying $\Omega(u,0) = 1$.

### 4.3.1. Normal and anti-normal ordering and its generalization

The property $H^{\Upsilon\Theta L} \equiv H^\Theta$, valid when $\Omega(u,0) = 1$ (See section 2.2) shows that this condition defines the restricted class of correspondence rules associated in a natural way with functional integrals in (p,q)-space.

Nevertheless other examples can be treated as we have done in chapter 2 . The most prominent among them are the so-called normal and anti-normal rules of association which we shall explain first. [A3,M11]

If we introduce two new independent variables

$$a = (q+ip)/\sqrt{2} \qquad (4.60)$$

$$a^{\star} = (q-ip)/\sqrt{2}$$

then, as is well-known, the corresponding operators $\hat{a}$ and $\hat{a}^{+}$ may be interpreted as creation and annihilation operators and satisfy the commutation relation

$$[\hat{a},\hat{a}^{+}] = \hat{1} . \qquad (4.61)$$

The operator associated with a function by the normal (anti-normal) rule of correspondence is obtained by writing the function in terms of the variables a and $a^{\star}$ and replacing them by $\hat{a}$ and $\hat{a}^{+}$ respectively but putting in products all operators $\hat{a}$ at the right (left) hand side of the operators $\hat{a}^{+}$.
The $\Omega$-functions defining these correspondence rules are given by

$$\Omega_N(u,v) = \exp[\tfrac{1}{4}(u^2 + v^2)] \qquad (4.62)$$

and

$$\Omega_{AN}(u,v) = \exp[-\tfrac{1}{4}(u^2+v^2)] . \qquad (4.63)$$

This can be derived from (2.2) as follows. By the normal rule of correspondence $\exp(iuq + ivp) = \exp\{[ia(u-iv)+ ia^{\star}(u+iv)]/\sqrt{2}\}$ is mapped into $\exp[i\,\hat{a}^{+}(u+iv)/\sqrt{2}]\,\exp[i\,\hat{a}(u - iv)/\sqrt{2}]$ .
Using the formula $\exp\hat{A}\exp\hat{B} = \exp\frac{1}{2}[\hat{A},\hat{B}]\exp(\hat{A} + \hat{B})$, valid when $\hat{A}$ and $\hat{B}$ both commute with the commutator $[\hat{A},\hat{B}]$ and $[\hat{a}^{+},\hat{a}] = -\hat{1}$ this can be written as

$$\exp[\tfrac{1}{4}(u^2+v^2)]\,\exp[iu\hat{q} + iv\hat{p}]$$

from which one can read $\Omega_N(u,v)$. For $\Omega_{AN}(u,v)$ it works in the same way.

These rules are of particular interest in connection with field-theoretical calculations, where the normal ordered products are usually employed. They have also been found very useful in investigations related to the connection between semiclassical and quantum description of optical coherence phenomena [S 8].
In this context general ordering techniques were developed in [C1] and [A3].

From the fact that $\hat{a}$ and $\hat{a}^{+}$ are involved one can expect already that normal and anti-normal ordering and its generalizations are associated in a rather natural way with the holomorphic form of the functional integral in $(a,a^{\star})$-space also called coherent state functional integral.[F1,S11]

But here we shall show how to use these rules in a phase space functional integral of the type defined in chapter 2.

We shall consider now the functions $\Omega(u,v)$ of the form

$$\Omega(u,v) = \exp\{ \tfrac{1}{4} \alpha u^2 + \tfrac{1}{4} \beta v^2\} \, \Omega_1(uv) \tag{4.64}$$

where $\alpha$ and $\beta$ are free parameters and $\Omega_1(uv)$ is an analytic function of the product uv. For $\alpha = \beta = s$ and $\Omega_1(uv) = 1$ this coincides with the s-ordering introduced by Cahill and Glauber[C1] which reduces for s = +1 and s = -1 to normal and anti-normal ordering respectively. For $\alpha = \beta = 0$ and $\Omega_1(uv)$ one of the functions listed in tabel 2.1. we obtain all other common rules of correspondence. The $\lambda$ -ordering used by Leschke et al.[L31] is given by (4.64) with $\Omega_1(uv) \equiv \exp [\tfrac{i}{2}\gamma \, uv]$ . In fact, Leschke et al. studied the perturbation theory using operator methods. Here we corroborate and extend their results using the functional integral formalism.

### 4.3.2. Functional integral in phase space

Let $\hat{H}(\hat{p},\hat{q})$ be a given operator. Its $\Theta_\Omega$ -symbol for $\Omega(u,v)$ given by (4.64) is (see (2.44-45))

$$H^{\Theta_\Omega}(p,q) = \exp[ \tfrac{1}{4} \alpha \frac{\partial^2}{\partial q^2} + \tfrac{1}{4} \beta \frac{\partial^2}{\partial p^2}] \; \bar{\Omega}_1 (- \frac{\partial^2}{\partial q \partial p} ) \, H^{AS}(p,q) \tag{4.65}$$

with

$$\bar{\Omega}_1(uv) = \exp[ \tfrac{i}{2} uv] \, .[\Omega_1(uv)]^{-1} \tag{4.66}$$

and $H^{AS}(p,q)$ its anti-standard symbol.

According to (2.36) the function $h^{\Theta_\Omega}(p,q',q)$ appearing in the discretized form of the functional integral representation of the propagator and the generating functional is

$$h^{\Theta_\Omega}(p,q',q) = \exp [- \tfrac{1}{4} \alpha \frac{\partial^2}{\partial \bar{q}^2} + \tfrac{1}{4}\beta\Delta^2]\Omega_1(i\Delta \frac{\partial}{\partial \bar{q}})H^{\Theta_\Omega}(p,\bar{q}). \tag{4.67}$$

Let us define $H^{\Theta_{\Omega_1}}(p,q)$ by

$$H^{\Theta_{\Omega_1}}(p,q) = \bar{\Omega}_1(- \frac{\partial^2}{\partial q \partial p} ) \, H^{AS}(p,q). \tag{4.68}$$

Combining (4.65-68) one obtains that

$$h^{\Theta_\Omega}(p,q',q) = \exp [ \tfrac{1}{4} \beta (\Delta^2 + \frac{\partial^2}{\partial p^2} ) ] \, \Omega_1(i\Delta \frac{\partial}{\partial \bar{q}} ) \, H^{\Theta_{\Omega_1}}(p,\bar{q}) \tag{4.69}$$

and consequently

$$H^{\gamma(\Theta_\Omega)}(p,q) = h^{\Theta_\Omega}(p,q,q) = \exp\left[\frac{1}{4}\beta\frac{\partial^2}{\partial p^2}\right] H^{\Theta_{\Omega_1}}(p,q) \tag{4.70}$$

which is independent of $\alpha$. For $\beta = 0$, $H^{\gamma(\Theta_\Omega)} \equiv H^{\Theta_{\Omega_1}}$ and we have obtained the result thatthe discretization associated with the $\Omega$-function (4.64) for $\beta = 0$ is equivalent to the discretization associated with $\Omega_1(uv)$.(Since also $H^{\Theta_{\Omega 1}} \equiv H^{\gamma(\Theta_{\Omega 1})}$ )

Remark: Formula (4.69) is of the form

$$h^{\Theta_\Omega}(p,q',q) = \exp\left[\frac{1}{4}\beta\left(\Delta^2 + \frac{\partial^2}{\partial p^2}\right)\right] h^{\Theta_{\Omega_1}}(p,q',q).$$

For Hamiltonians quadratic in p $h^{\Theta_\Omega}$ and $h^{\Theta_{\Omega_1}}$ are (up to equivalence) of the form

$$h^{\Theta_{\Omega_1}}(p,q',q) = -\frac{ic}{2}p^2 g^{\Theta_1}(q',q) - p\, a^{\Theta_1}(q',q) + \frac{i}{c} v^{\Theta_1}(q',q) \tag{4.71}$$

$$h^{\Theta_\Omega}(p,q',q) = -\frac{ic}{2}\left[p^2\left(1+\frac{1}{4}\beta\Delta^2\right) + \frac{1}{2}\beta\right] g^{\Theta_1}(q',q) - p a^{\Theta_1}(q',q) + \frac{i}{c} v^{\Theta_1}(q',q) \tag{4.72}$$

and one can change from $\gamma(\Omega)$ - discretization to $\gamma(\Omega_1)$-discretization using $p^2\Delta^2 \doteq -2$ (see section 2.4).

### 4.3.3. Perturbation expansions

We study now the perturbation expansions that can be obtained from the general formula (4.6) for $Z[J,J^\star]$ in the case that $\gamma$ is $\gamma(\Theta_\Omega)$ with $\Omega(u,v)$ given by (4.64).

First of all we remark that the methods of the previous section can easily be generalized to this case. The interaction vertices in the expansion can be read from $H^{\gamma(\Theta_\Omega)}(p,q)$ given by (4.70) and the values of the $D^{12}$-tadpoles are determined by $h^{\Theta_\Omega}(p,q',q)$ given by (4.69). In fact, the rules for these tadpoles are exactly the same for $\gamma(\Theta_\Omega)$ as for $\gamma(\Theta_{\Omega 1})$-discretization except for the value of the double tadpole $D^{12}(\tau,\tau)D^{12}(\tau,\tau)$ which is given by

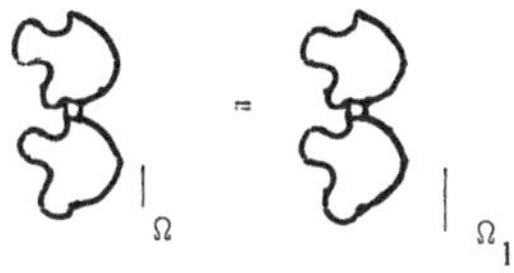


$$- \frac{1}{2}\beta \qquad (4.73)$$

Thenall $\beta$ -dependence cancels.

Note that in the construction the phase space function $H^{\Theta_\Omega}(p,q)$ only appears as an intermediate step between $\hat{H}(\hat{p},\hat{q})$ and $H^{\gamma(\Theta_\Omega)}(p,q)$. Now we shall introduce in a more relevant way the $\Theta_\Omega$-symbol $H_1^{\Theta_\Omega}$ of $\hat{H}_1$ in the perturbation expansion by showing that the generating functional $Z[J,J^\star]$ can also be written as

$$Z[J,J^\star] = \exp\left[-i\int_{t_0}^{t} d\tau\, H_1^{\Theta_\Omega}(p,q)\right]\Bigg|_{\substack{p=\frac{1}{i}\frac{\delta}{\delta J^\star}\\ q=\frac{1}{i}\frac{\delta}{\delta J}}} \tilde{Z}_0[J,J^\star] \qquad (4.74)$$

where $H_1^{\Theta_\Omega}(p,q) = \lim\limits_{q'\to q} \Omega_1\left(i\Delta \frac{\partial}{\partial \bar{q}}\right) H_1^{\Theta_\Omega}(p,\bar{q})$ and $\tilde{Z}_0[J,J^\star]$ is the modified free generating functional

$$\tilde{Z}_0[J,J^\star] = K'\exp\{-\int d\tau d\tau' [\tfrac{1}{2} J^\star(\tau)\tilde{D}^{11}(\tau,\tau')J^\star(\tau') + J^\star(\tau)\tilde{D}^{12}(\tau,\tau')J(\tau')$$

$$+ \tfrac{1}{2} J(\tau)\tilde{D}^{22}(\tau,\tau')J(\tau')] + \int d\tau [S_1(\tau)J^\star(\tau) + S_2(\tau)J(\tau)]\} \qquad (4.75)$$

where

$$\tilde{D}^{11}(\tau,\tau') = \begin{cases} D^{11}(\tau,\tau') & , \tau \neq \tau' \\ D^{11}(\tau,\tau') - \frac{1}{2}\beta, & \tau = \tau' \end{cases} \qquad (4.76a)$$

$$\tilde{D}^{12}(\tau,\tau') = D^{12}(\tau,\tau') \qquad (4.76b)$$

$$\tilde{D}^{22}(\tau,\tau') = \begin{cases} D^{22}(\tau,\tau') & , \tau \neq \tau' \\ D^{22}(\tau,\tau') - \frac{1}{2}\alpha, & \tau = \tau'. \end{cases} \qquad (4.76c)$$

Formulae (4.74-76) mean that given any $\Omega$-function of the form (4.64) one can generate , using $H_1^{\Theta_\Omega}(p,q)$ as the interaction Hamiltonian, the perturbation series expansion of the full generating functional $Z[J,J^\star]$ if one modifies the $D^{11}(\tau,\tau)$ and $D^{22}(\tau,\tau)$ equal time contractions (tadpoles) as follows:

 $= \tilde{D}^{11}(\tau,\tau) = D^{11}(\tau,\tau) - \frac{1}{2}\beta$

 $= \tilde{D}^{22}(\tau,\tau) = D^{22}(\tau,\tau) - \frac{1}{2}\alpha .$

The values of the $D^{12}$-tadpoles are the same as for $\Omega_1$-discretization and independent of $\alpha$ and $\beta$. As before (section 4.2) the mechanism that makes the perturbation expansion independent of the correspondence rule used is the cancellation between the explicitly discretization dependent vertices and contributions from discretization dependent tadpoles.
The proof of the statement (4.74-76) is given in Appendix 4.2. Here we shall just illustrate it by an example.

Example

Let $\hat{H}_1(\hat{p},\hat{q})$ again be given by (4.55). Using (4.65) (4.68) and (4.57) applied for $\Omega_1$ one obtains

$$H_1^{\Theta_\Omega}(p,q) = H_1^{\Theta_{\Omega_1}}(p,q) - \frac{ic\lambda}{8}\alpha p^2 - \frac{ic\lambda}{8}\beta q^2 - \frac{ic\lambda}{16}\alpha\beta . \tag{4.77}$$

The interaction vertices in $-iH_1^{\theta_{\Omega_1}}$ are the ones shown in figure 4.2. and the three new vertices are drawn in figure 4.4. ( $g_{11} = -\frac{c\lambda}{8}\alpha$, $g_{22} = -\frac{c\lambda}{8}\beta$ , $g_{oo} = -\frac{c\lambda}{16}\alpha\beta$)

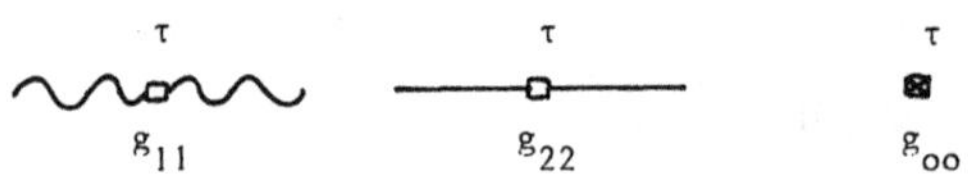


figure 4.4.: $\alpha$ - and $\beta$- dependent vertices in $-\,i\,H_1^{\Theta_\Omega}(p(\tau),q(\tau))$ (See (4.77) ).

These vertices contribute in first order in $\lambda$ in the expansion (4.74) of $Z[J,J^*]$ the graphs drawn in figure 4.5. The appearance of the new graphs does not have influence on the cancellation between the graphs 2,7 and 9 and between the graphs 5 and 8 since the latter five graphs only depend on $\Omega_1(uv)$ (No $D^{11}$ or $D^{22}$ tadpoles are involved and the coupling constants $g_4, g_{12}$ and $g_0$ depend on $\Omega_1'(0)$ and $\Omega_1''(0)$ only. The new cancellation that shows up is between the graphs 1 and 10, 11,12, between 3 and 14 and between 4 and 13. Indeed, the first group sums up to

$$\int d\tau [g_4\, \tilde{D}^{11}(\tau,\tau)\tilde{D}^{22}(\tau,\tau) + g_{11}\, \tilde{D}^{11}(\tau,\tau) + g_{22}\, \tilde{D}^{22}(\tau,\tau) + g_{oo}]$$

$$= \int d\tau [-\frac{c\lambda}{4}(D^{12}(\tau,\tau) - \frac{1}{2}\beta)(D^{22}(\tau,\tau) - \frac{1}{2}\alpha) - \frac{c\lambda}{8}\alpha\,(D^{12}(\tau,\tau) - \frac{1}{2}\beta)$$

$$- \frac{c\lambda}{8}\beta\,(D^{22}(\tau,\tau) - \frac{1}{2}\alpha) - \frac{c\lambda}{16}\alpha\beta\,]$$

- 73 -

$$= \int d\tau \; [-\frac{c\lambda}{4} \; D^{11}(\tau,\tau) \; D^{22}(\tau,\tau)] \tag{4.78}$$

the second to

$$\int d\tau [g_4 \; \tilde{D}^{11}(\tau,\tau) + g_{22}] \; S_2(\tau) S_2(\tau) =$$

$$= \int d\tau \; g_4 \; D^{11}(\tau,\tau) \; S_2(\tau) S_2(\tau) \tag{4.79}$$

the third to

$$\int d\tau [g_4 \; \tilde{D}^{22}(\tau,\tau) + g_{11}] \; S_1(\tau) \; S_1(\tau) = \int d\tau \; g_4 \; D^{22}(\tau,\tau) S_1(\tau) S_1(\tau) \tag{4.80}$$

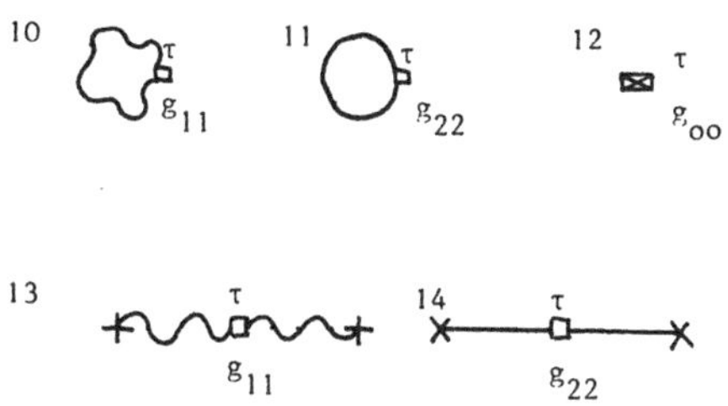


Figure 4.5.: Graphs representing the contribution to first order of the vertices of figure 4.4.

## 4.4. Extension to field theory and definition without limiting procedure

### 4.4.1. Extension to field theory

We have seen that for finite dimensional systems the generating functional $Z[J,J^*]$ has in general many different functional integral representations defined in terms of discretization $\gamma$ . In the case that $\gamma$ is associated with a correspondence rule $\Omega$ between operators $\hat{p}$ and $\hat{q}$ and functions on the phase space we have shown in detail that every representation leads to a diagrammatic perturbation expansion in terms of the free propagators in $Z_\circ[J,J^*]$ and the $\Omega$-dependent vertices in $H_1^{\gamma(\Theta_\Omega)}(p,q)$ in which the a priori undefined tadpoles $D^{12}(\tau,\tau)$ have values determined by the discretization or correspondence rule.

In the case of field theory ($d \to \infty$) the tadpoles become undefined because of divergent integrals and it becomes necessary to specify, in addition to a definite ordering scheme a regularization scheme.

Suzuki, Hirshfeld and Leschke [S12] have investigated (with operator methods) whether the extra freedom associated with the choice of a regularization procedure for the infinite quantities could be exploited in order to eliminate the ordering terms. They find that this is the case when the generalization to fields of $\gamma_1(\alpha)$ - ordering ("u-ordering" in their terminology) is combined with dimensional regularization. And in fact this conclusion remains true for all orderings which are the generalization to fields of the orderings depending on u and v through u.v only.

This explains why calculations in field theory in which ordering problems are neglected can lead to correct results provided dimensional regularization is used to treat all divergent quantities.

But the principal conclusion remains thatthe problems of ordering and discretization have to be taken into account also in field theoretic calculations, as has been noticed already in the context of soft-pion theorems in chiral dynamics ([D21], [S 9]) and more recently in the quantum formalism of soliton scattering ([G9]) . By field theoretic we mean all cases where an infinite number of degrees of freedom is involved. This can be e.g. the local study of the critical behaviour of a reaction-diffusion model system (See [L24, L25]).

An interesting review of canonical and functional methods in quantum mechanics and quantum field theory especially devoted to the ordering problems in the perturbation theory was given by Hirshfeld, Suzuki and Leschke in [H5]. A recent work on the ordering problems in Yang-Mills theory is [C6].

## 4.4.2. Phase space path integrals without limiting procedure

Phase space path integrals without limiting procedure for quantum mechanics were introduced by Mizrahi in [M 15]. The procedure is to construct the most general Gaussian measure in phase space through its Fourier transform. (This is just the free generating functional) and to define a general path integral by treating the non-Gaussian part as a perturbation i.e. by the formal expansion (4.6).

Also in field theory calculations where the functional integrals are widely used to derive properties of the perturbation expansion this "algebraic definition" has been adopted (See Zinn-Justin [Z1]).

However, as we have seen this procedure is in general not sufficient to give a well-defined meaning to the terms of the perturbation expansion.

In [M16] Mizrahi has considered the ordering problem and given an procedure to get a well-defined expansion. What we have shown systematically in our work [L8] and in this chapter is that the definition without limiting procedure by a formal series expansion has the same problems as the definition through discretization. No definition of path integrals reveals which correspondence rule is preferred and once a rule is chosen path integrals can be unambiguously defined with or without limiting procedure ([M18]).

## Appendix 4.1. Perturbation expansion in the operator formalism

In the operator formalism the perturbation expansion for the generating functional is constructed as follows.

Let $U(t,t')$ and $U_\circ(t,t')$ be the evolution operators associated with the complete and the free Hamiltonian respectively. Define (remember (3.20) $U(t) \equiv U(t,0)$)

$$U_1(t) = U_\circ^{-1}(t)U(t) \quad , \tag{A4.1}$$

$$U_1(t',t) = U_1(t')U_1^{-1}(t). \tag{A4.2}$$

Operators and states in the interaction picture ip are then defined by

$$\hat{q}^{ip}(t) = U_\circ^{-1}(t)\ \hat{q}\ U_\circ(t) \tag{A4.3a}$$

$$\hat{p}^{ip}(t) = U_\circ^{-1}(t)\ \hat{p}\ U_\circ(t) \tag{A4.3b}$$

$$|q,t>_{ip}^{R} = U_\circ^{-1}(t)|q> \tag{A4.4a}$$

$${}_{ip}^{L}<q,t| = <q|\ U_\circ(t) \tag{A4.4b}$$

and $U_1(t',t)$ satisfies

$$i\ \frac{\partial}{\partial t'}\ U_1(t',t) = \hat{H}_1^{ip}(t')\ U_1(t',t)\ ,\ U_1(t,t) = 1 \tag{A4.5}$$

with

$$\hat{H}_1^{ip}(t\ ) = U_\circ^{-1}(t)\ \hat{H}_1\ U_\circ(t)\ . \tag{A4.6}$$

The generating functional $Z[J,J^\star]$ is given in the operator formalism by (3.38) where the states are the Heisenberg picture states defined in (3.23-3.24). Using that $|q_\circ,t_\circ>^R = U_1^{-1}(t_\circ)\ |q_\circ,t_\circ>_{ip}^R$ and ${}^L<q,t| = {}_{ip}^{L}<q,t\ |U_1(t)$ one obtains

$$Z[J,J^\star] = {}_{ip}^{L}<q,t\ |U_1(t) \quad P\exp\{+\ i\int_{t_\circ}^{t} d\tau\ J(\tau)\hat{q}(\tau) + J^\star(\tau)\hat{p}(\tau)\ \}$$
$$U_1^{-1}(t_\circ)\ |q_\circ,t_\circ>_{ip}^{R}. \tag{A4.7}$$

We meet now the problem of time-ordering of products of noncommuting operators at equal time.

(A4.5) is solved by

$$U_1(t) = P \exp [- i \int_0^t d\tau \, \hat{H}_1(\hat{p}(\tau),\hat{q}(\tau))] \qquad (A4.8)$$

where the time ordering at equal times follows from the way we have written $\hat{H}_1$. Let this ordering be fixed and let $P_\gamma$ be the corresponding time ordered product (see chapter 3). Then $Z[J,J^\star]$ is given by

$$Z[J,J^\star] = {}^{L}_{ip}\langle q,t| \, P_\gamma \exp [-i \int_{t_o}^{t} d\tau \, \hat{H}_1(\hat{p}(\tau),\hat{q}(\tau)) + J(\tau)\hat{q}(\tau) + J^\star(\tau)\hat{p}(\tau)] \, |q_o,t_o\rangle^{R}_{ip} \qquad (A4.9)$$

which has (cf. the methods of chapter 3) the functional integral representation (4.4) with $H^\gamma = H^\gamma_o + H^\gamma_1$ .

## Appendix 4.2. Wick's theorem

Formula (4.6) for the generating functional is of the form

$$Z[J,J^\star] = E \, [\frac{1}{i}\frac{\delta}{\delta J^\star(\tau)}, \frac{1}{i} \frac{\delta}{\delta J(\tau)}] \, Z_o[J,J^\star] \qquad (A4.10)$$

with

$$E[p(\tau),q(\tau)] = \exp[-i \int_{t_o}^{t} d\tau \, H^\gamma_1(p(\tau),q(\tau))] \qquad (A4.11)$$

and can also be written as

$$Z[J,J^\star] = E \, [\frac{1}{i} \frac{\delta}{\delta J^\star(\tau)} , \frac{1}{i} \frac{\delta}{\delta J(\tau)}] \, Z_o[ \frac{1}{i}\frac{\delta}{\delta q(\tau)} , \frac{1}{i} \frac{\delta}{\delta p(\tau)}] \exp \{i \int_{t_o}^{t} d\tau \, [J(\tau)q(\tau) + J^\star(\tau)p(\tau)]\} \Big|_{q=p=0} \qquad (A4.12)$$

$$Z[J,J^\star] = Z_o \, [\frac{1}{i} \frac{\delta}{\delta q(\tau)} , \frac{1}{i} \frac{\delta}{\delta p(\tau)}] \, E \, [p(\tau),q(\tau)] \exp \{i \int_{t_o}^{t} d\tau[J(\tau)q(\tau) + J^\star(\tau)p(\tau)]\}\Big|_{q=p=0} . \qquad (A4.13)$$

Consider the quantities $\langle F\rangle$ defined by (3.17), applying the same manipulation as that from (A4.10) to (A4.13) and setting $J=J^\star= 0$ one obtains

$$\langle F\rangle = Z_o[ \frac{1}{i} \frac{\delta}{\delta q(\tau)} , \frac{1}{i} \frac{\delta}{\delta p(\tau)}] \, F \, [p(\tau),q(\tau)] \, E \, [p(\tau),q(\tau)]\Big|_{q=p=0} . \qquad (A4.14)$$

This is the functional form of Wick's theorem ([H6]). $Z_\circ$ is the exponential of a quadratic form which generates all possible contractions of p and q, in the perturbation expansion of $\langle F \rangle$, obtained by expanding the exponential $E[p(\tau),q(\tau)]$.

All ordering problems are contained in the factor $F[p(\tau),q(\tau)]E[p(\tau),q(\tau)]$ i.e. one is given the function $h_1^\gamma(p,q',q)$ such that $h_1^\gamma(p,q,q) = H_1^\gamma(p,q)$ and also when two or more times in F coincide the discretization $\gamma$ gives the good interpretation. Therefore each equal time contraction $D^{12}(\tau,\tau)$ in the expansion is well-defined.

It should be remarked that we can speak about a Dyson-Wick expansion in the original sense only when the double tadpole equals the square of the single tadpole. The most general correspondence rule having this feature is defined by

$$\Omega(u,v) = \exp\{\frac{1}{4}\alpha u^2 + \frac{1}{4}\beta v^2 + \frac{i}{2}\gamma uv\}, \tag{A4.15}$$

The Dyson-Wick expansions associated with this ordering function were studied by Leschke, Hirshfeld and Suzuki in the operator formalism ([L30,L31], by us in the functional integral formalism ([L 9] and this chapter).

## Appendix 4.3. Integration by parts in the action, normalization of the free generating functional.

Consider the following configuration space path integral in $\gamma_1(\alpha)$ discretization

$$\int_{\gamma_1(\alpha)} \mathcal{D}q \exp\{-\int_{t_\circ}^{t} d\tau\,[\frac{1}{2c}\dot{q}^2 + \dot{q}A(q)]\}\delta(q(t)-q)\,\delta(q(t_\circ)-q_\circ). \tag{A4.16}$$

We shall show how to integrate by parts the term $\dot{q}A(q)$. The integral $\int_{t_\circ}^{t} d\tau\,\dot{q}A(q)$ in (A4.1) has the meaning

$$\sum_{j=1}^{N+1} \varepsilon(q_j - q_{j-1})A(q_{j-1}^\alpha), \quad q_{j-1}^\alpha = (1-\alpha)q_{j-1} + \alpha q_j. \tag{A4.17}$$

Let B(q) be such that $dB/dq = A(q)$, one has that

$$B(q) - B(q_\circ) = \sum_{j=1}^{N+1} B(q_j) - B(q_{j-1})$$

$$= \sum_{j=1}^{N+1} (q_j - q_{j-1})A(q^\alpha_{j-1}) + (\frac{1}{2} - \alpha)\Delta_j^2 \, A'(q^\alpha_{j-1}). \quad (A4.18)$$

In the functional integral (A4.16) $\Delta_j^2$ can be replaced by $\varepsilon c$ (See section 5.3) and since $B(q) - B(q_\circ) = \int_{q_\circ}^{q} dq\, A(q)$ one has that

$$\int_{t_\circ}^{t} d\tau\, \dot{q}\, A(q(\tau)) = \int_{q_\circ}^{q} dq\, A(q) - (\frac{1}{2} - \alpha) \int_{t_\circ}^{t} d\tau\, A'(q(\tau)). \quad (A4.19)$$

We see that integration by parts is allowed in the usual way, that is without the second term, when $\alpha = \frac{1}{2}$. In fact, the integral we meet here is a stochastic integral, which is for $\alpha = 0$ the Ito integral and for $\alpha = \frac{1}{2}$ the Stratonovich integral (See [A 7]). The generalization of (A4.19) to higher dimensions and explicitly time dependent A is ($A_\mu = \partial_\mu D$)

$$\int_{t_o}^{T} dt\dot{q}^\mu A_\mu(q,t) = D(q(T),T) - D(q(t_o),t_o) - \int_{t_o}^{T} \frac{\partial}{\partial t} D(q,t)dt + (\alpha - \frac{1}{2})\int_{t_o}^{T} cg^{\mu\rho}(q)\partial_\rho A_\mu(q,t)dt \quad (A4.19')$$

when the dominating exponential is $\exp[-\frac{\varepsilon}{2c} g_{\mu\nu}(q_j,q_{j-1}) \frac{\Delta_j^\mu}{\varepsilon} \frac{\Delta_j^\nu}{\varepsilon}]$.

We go back now to formula (4.12) for the free generating functional $Z_\circ[J,J^*]$. Integrating there over the p-variables and setting $J=J^* = q=q_\circ=0$ we obtain

$$K = Z_\circ[0,0]\Big|_{q=q_\circ=0} = \int_{\gamma_1(\frac{1}{2})} \mathcal{D}q \quad \exp\{-\int_{t_\circ}^{t} d\tau[\frac{1}{2c} (\dot{q} + \mu q)^2 + \frac{1}{2c}\omega^2 q^2 - \frac{1}{2}\mu] \}$$
$$\delta(q(t)) \quad \delta(q(t_\circ)). \quad (A4.20)$$

The factor $\exp \int_{t_\circ}^{t} d\tau\, \mu$ does not depend on q and can be put apart, in the factor $\exp -\frac{1}{c}\int_{t_\circ}^{t} \mu\dot{q}q\, d\tau$ we integrate by parts, since $q = q_\circ=0$ it is just 1. Hence

$$K = \exp \mu(t-t_\circ) \int_{\gamma_1(\frac{1}{2})} \mathcal{D}q \quad \exp\{-\frac{1}{c}\int_{t_\circ}^{t} d\tau[\frac{1}{2}\dot{q}^2 + (\mu^2+\omega^2)q^2]\}\delta(q(t))\delta(q(t_\circ)) \quad (A4.21)$$

and for c=i we recognize the harmonic oscillator propagator given in section 1.1 but for $\hbar = 1$, $q = q_\circ=0$ and $m^2 = -(\mu^2 + \omega^2) = -\rho^2$. Comparing with (1.8) we find K as given by (4.34).

In fact the functional integral (A4.20) is almost the same as the one computed explicitly in section 1.3 and the same method as was used there could also have been used here.

Appendix 4.4.: Proof of (4.74-4.76)

Let the $\Theta_\Omega$ -symbol of the operator $\hat{H}(\hat{p},\hat{q})$ be given by

$$H^{\Theta_\Omega}(p,q) = -\frac{ic}{2}\, p^2 G^{\Theta_\Omega}(q) - p\, A^{\Theta_\Omega}(q) + \frac{i}{c}\, V^{\Theta_\Omega}(q). \tag{A4.22}$$

For $\gamma \equiv \gamma(\Theta_\Omega)$ (4.4) has the meaning

$$Z[J,J^\star] = \lim_{N\to\infty} \int \Big( \prod_{i=1}^{N} dq_i \prod_{j=1}^{N+1} \frac{dp_j}{2\pi} \Big) \exp \Big\{ i\varepsilon \sum_{j=1}^{N+1} \Big[ p_j \frac{\Delta_j}{\varepsilon} - h^{\Theta_\Omega}(p_j,q_j,q_{j-1}) + + J_j q_j + J_j^\star p_j \Big] \Big\} \tag{A4.23}$$

where

$$h^{\Theta_\Omega}(p,q',q) = -\frac{ic}{2}\, p^2 g^{\Theta_\Omega}(q',q) - p\, a^{\Theta_\Omega}(q',q) + \frac{i}{c}\, v^{\Theta_\Omega}(q',q) \tag{A4.24}$$

with (up to equivalence)

$$g^{\Theta_\Omega}(q',q) = \Big( 1 + \frac{1}{4}\beta\, \Delta^2 \Big) \exp \Big[ -\frac{1}{4}\alpha \frac{\partial}{\partial \bar{q}^2} \Big] \Omega_1 \Big( i\Delta \frac{\partial}{\partial \bar{q}} \Big) G^{\Theta_\Omega}(\bar{q}), \tag{A4.25a}$$

$$a^{\Theta_\Omega}(q',q) = \exp \Big[ -\frac{1}{4}\alpha \frac{\partial}{\partial \bar{q}^2} \Big] \Omega_1 \Big( i\Delta \frac{\partial}{\partial \bar{q}} \Big) A^{\Theta_\Omega}(\bar{q}), \tag{A4.25b}$$

$$v^{\Theta_\Omega}(q',q) = \exp \Big[ -\frac{1}{4}\alpha \frac{\partial}{\partial \bar{q}2} \Big] \Omega_1 \Big( i\Delta \frac{\partial}{\partial \bar{q}} \Big) V^{\Theta_\Omega}(\bar{q}). \tag{A4.25c}$$

We have developed the factor $\exp[\frac{1}{4}\beta\Delta^2]$ to be able to use the replacement $p^2\Delta^2 \doteq -2 = -\partial^2/\partial p^2\ p^2$. (See 2.104) . By this replacment we make a change to an inequivalent discretization in the sense that the phase space function appearing in the continuum expression of the functional integral will change. (This is precisely what we want, we want to arrive at $H^{\Theta_\Omega}$ instead of $H^{\gamma(\Theta_\Omega)}$ . )

(A4.23) can then also be written as

$$Z[J,J^\star] = \lim_{N\to\infty} \int \Big( \prod_{i=1}^{N} dq_i \prod_{j=1}^{N+1} \frac{dp_j}{2\pi} \Big) \exp \Big\{ i\varepsilon \sum_{j=1}^{N+1} \Big[ p_j \frac{\Delta_j}{\varepsilon} - - \exp \Big[ -\frac{\alpha}{4} \frac{\partial^2}{\partial \bar{q}_j^2} - \frac{\beta}{4} \frac{\partial^2}{\partial p_j^2} \Big] \Omega_1 \Big( i\Delta_j \frac{\partial}{\partial \bar{q}_j} \Big) H^{\Theta_\Omega}(p_j,\bar{q}_j) + J_j q_j + J_j^\star p_j \Big] \Big\}, \tag{A4.26}$$

When we split a quadratic part $\hat{H}_\circ$ from the Hamiltonian $\hat{H}$ (See (4.2)), then we shall obtain a representation of the form (4.6) with $H_1^\gamma(p,q)$ now given by $\exp[-\frac{\alpha}{4}\frac{\partial^2}{\partial \bar{q}^2} - \frac{\beta}{4}\frac{\partial^2}{\partial p^2}] H_1^{\Theta_\Omega}(p,q)$ and our task is to incorporate the exponential in a redefinition of the free generating functional.

The perturbation series is given by

$$Z[J,J^\star] = \{\sum_{k=0}^{\infty} \frac{(-i)^k}{k!} \int d\tau_1 \ldots d\tau_k \prod_{\ell=1}^{k} \exp[-\frac{\alpha}{4}\frac{\partial^2}{\partial q(\tau_\ell)^2} - \frac{\beta}{4}\frac{\partial^2}{\partial p(\tau_\ell)^2}]$$

$$H_1^{\Theta_\Omega}(p(\tau_\ell),q(\tau_\ell))\}\Big|_{\substack{p=\frac{1}{i}\frac{\delta}{\delta J^\star} \\ q=\frac{1}{i}\frac{\delta}{\delta J}}}$$

$$\cdot \int_{\gamma(\Theta_\Omega)} \mathcal{D}q\mathcal{D}p \; \exp\{i\int_{t_\circ}^{t} d\tau \; p\dot{q} - H_\circ^{\gamma\,(\Theta_\Omega)}(p,q) + Jq + J^\star p]\}$$

$$\delta(q(t)-q)\;\delta(q(t_\circ)-q_\circ). \qquad \text{(A4.27)}$$

We want to show that the generating functional $\tilde{Z}[J,J^\star]$ defined by

$$\tilde{Z}[J,J^\star] = \exp[-i\int_{t_\circ}^{t} d\tau \; H_1^{\Theta_\Omega}(p,q)]\Big|_{\substack{p=\frac{1}{i}\frac{\delta}{\delta J^\star} \\ q=\frac{1}{i}\frac{\delta}{\delta J}}} \tilde{Z}_\circ[J,J^\star] \qquad \text{(A4.28)}$$

($\tilde{Z}_\circ[J,J^\star]$ given by (4.75-4.76)) satisfies $\mathcal{L}(\tilde{Z}) = \mathcal{L}(Z)$ (A4.29)

with

$$\mathcal{L}(Z) = \frac{\delta^{n+m}}{\delta J(\tau_1)\ldots\delta J(\tau_n)\delta J^\star(\tau_1')\ldots\delta J^\star(\tau_m')} Z\Big|_{J=J^\star=0} \qquad \text{(A4.30)}$$

in particular also

$$\tilde{Z}[0,0] = Z[0,0], \qquad \text{(A4.31)}$$

since then $\tilde{Z}$ and Z are equal.

Formally the step from (A4.27) to (A4.28) follows from

$$\mathcal{L}\;\exp[-\frac{\alpha}{4}\frac{\partial^2}{\partial q(\tau')^2} - \frac{\beta}{4}\frac{\partial^2}{\partial p(\tau')^2}] H^{\Theta_\Omega}(p(\tau'),q(\tau'))\Big|_{\substack{p=\frac{1}{i}\frac{\delta}{\delta J^\star} \\ q=\frac{1}{i}\frac{\delta}{\delta J}}} \exp[i\int d\tau \; J(\tau)q(\tau) + J^\star(\tau)p(\tau)]$$

$$= \mathcal{L} H^{\Theta_\Omega}(p(\tau')q(\tau'))\Big|_{\substack{p=\frac{1}{i}\frac{\delta}{\delta J^\star} \\ q=\frac{1}{i}\frac{\delta}{\delta J}}} \exp\{\frac{\alpha}{4}\int d\tau d\tau'' J(\tau)\frac{\delta(\tau-\tau'')}{\delta(0)} J(\tau'') + \frac{\beta}{4}\int d\tau d\tau'' J^\star(\tau)\frac{\delta(\tau-\tau'')}{\delta(o)} J^\star(\tau'')\}$$

$$.\ \exp[i \int d\tau\ J(\tau)q(\tau) + J^\star(\tau)p(\tau)] \tag{A4.32}$$

which is a generalization of the simple formula

$$\frac{\partial}{\partial J}[\exp(\frac{\partial^2}{\partial x^2})V(x)]\exp(xJ)\Big|_{J=0} = \frac{\partial}{\partial J}V(\frac{\partial}{\partial J})\exp(\alpha J^2)\exp(xJ)\Big|_{J=0}. \tag{A4.33}$$

Let us now show how it works in a discrete calculation. The proof we give here generalizes the one given in[L10] which was restricted to $H^{\Theta_\Omega}$ of the form $-\frac{ic}{2}p^2 + \frac{i}{c}V^{\Theta_\Omega}(q)$. We only exhibit the relevant parts of the formulae. The discrete form of

$$\exp[-i\int_{t_\circ}^{t} d\tau\ H_1^{\gamma(\Theta)}(p,q)]\Big|_{\substack{p=\frac{1}{i}\frac{\delta}{\delta J^\star} \\ q=\frac{1}{i}\frac{\delta}{\delta J}}} \exp[i\int_{t_\circ}^{t} d\tau\ J(\tau)q(\tau)+J^\star(\tau)p(\tau)] \tag{A4.34}$$

is

$$\exp\{-i\varepsilon\sum_{j=1}^{N+1} h_1^{\Theta_\Omega}(p_j,q_j,q_{j-1})\}\Big|_{\substack{p=\frac{1}{i\varepsilon}\frac{\partial}{\partial J^\star} \\ q=\frac{1}{i\varepsilon}\frac{\partial}{\partial J}}} \exp\{i\varepsilon\sum_{j=1}^{N+1}[J_j q_j + J_j^\star p_j]\} \tag{A4.35}$$

which after the replacement of $p_j^2\Delta_j^2$ by $-\frac{\partial^2}{\partial p_j^2}p_j^2$ is

$$\exp\{-i\varepsilon\sum_{J=1}^{N+1}\exp[-\frac{1}{4}\frac{\partial^2}{\partial \bar{q}_j^2} - \frac{1}{4}\beta\frac{\partial^2}{\partial p_j^2}]\,\Omega_1(i\Delta\frac{\partial}{\partial \bar{q}_j})\,H_1^{\Theta_\Omega}(p_j,\bar{q}_j)\}\Big|_{\substack{p=\frac{1}{i\varepsilon}\frac{\partial}{\partial J^\star} \\ q=\frac{1}{i\varepsilon}\frac{\partial}{\partial J}}}$$

$$\exp\{i\varepsilon\sum_{J=1}^{N+1}[J_j q_j + J_j^\star p_j]\}. \tag{A4.36}$$

Developing the first exponential one obtains

$$\sum_{k=o}^{\infty} \frac{(-i\varepsilon)^k}{k!} \prod_{\ell=1}^{k} \sum_{j_\ell=1}^{N+1} \exp[-\frac{1}{4}\alpha \frac{\partial^2}{\partial \bar{q}^2_{j\ell}} - \frac{1}{4}\beta \frac{\partial^2}{\partial p^2_{j\ell}}]\Omega_1(i\Delta \frac{\partial}{\partial \bar{q}_{j\ell}})H_1^{\Theta_\Omega}(p_{j\ell},\bar{q}_{j\ell})\Big|_{\substack{p=\frac{1}{i\varepsilon}\frac{\partial}{\partial J^\star} \\ q=\frac{1}{i\varepsilon}\frac{\partial}{\partial J}}} *$$

$$* \exp\{i\,\varepsilon \sum_{j=1}^{N+1} [J_j q_j + J_j^\star p_j]\} \qquad (A4.37)$$

and using the generalization of (A4.33) one has that $\mathcal{L}$ (with $\delta/\delta J \to \frac{1}{i\varepsilon}\frac{\partial}{\partial J}$ etc.) applied to (A4.37) equals $\mathcal{L}$ applied to

$$\sum_{k=0}^{\infty} \frac{(-i\varepsilon)^k}{k!} \prod_{\ell=1}^{k} \sum_{j_\ell=1}^{N+1} [\Omega_1(i\Delta \frac{\partial}{\partial \bar{q}_{j\ell}})H_1^{\Theta}\Omega(p_{j\ell},\bar{q}_{j\ell})]\Big|_{\substack{p=\frac{1}{i\varepsilon}\frac{\partial}{\partial J^\star} \\ q=\frac{1}{i\varepsilon}\frac{\partial}{\partial J}}} \exp[\varepsilon^2 \frac{1}{4}\alpha \bar{J}^2_{j_\ell} + \varepsilon^2 \frac{1}{4}\beta J^{\star 2}_{j_\ell}]$$

$$\exp\{i\varepsilon \sum_{j=1}^{N+1} [J_j q_j + J_j^\star p_j]\}, \qquad (A4.38)$$

Without loss of generality, suppose $j_\ell \neq j_{\ell'}$ for $\ell \neq \ell'$ (remember we are treating discretised integrals and the modification amounts to a change of the integrand in a set of measure zero) , we can replace then in (A4.38) $\exp[\varepsilon^2 \frac{1}{4}\alpha \bar{J}^2_{j\ell} + \varepsilon^2 \frac{1}{4}\beta J^{\star 2}_{j\ell}]$ by $\exp\{\sum_{j=1}^{N+1} [\varepsilon^2 \frac{1}{4}\alpha \bar{J}^2_j + \varepsilon^2 \frac{1}{4}\beta J^{\star 2}_j]\}$ and resume the series which arose from the development of the exponential to obtain that $\mathcal{L}$ applied to (A4.38) equals $\mathcal{L}$ applied to

$$\exp\{-i\varepsilon \sum_{j=1}^{N+1} \Omega_1(i\Delta \frac{\partial}{\partial \bar{q}_j}) H_1^{\Theta_\Omega}(p_j,\bar{q}_j)\}\Big|_{\substack{p=\frac{1}{i\varepsilon}\frac{\partial}{\partial J^\star} \\ q=\frac{1}{i\varepsilon}\frac{\partial}{\partial J}}} \exp\{\varepsilon^2 \sum_{J=1}^{N+1} [\frac{1}{4}\alpha \bar{J}_j^{\,2} + \frac{1}{4}\beta J_j^{\star\,2}]\}\cdot$$

$$\cdot \exp\{i\varepsilon \sum_{j=1}^{N+1} [J_j q_j + J_j^\star p_j]\}. \qquad (A4.39)$$

We recall now that the last exponential in (A4.39) represents the free generating functional $Z_o[J,J^\star]$. When it is computed we arrive at

$$\exp\{-i\varepsilon\sum_{j=1}^{N+1}\Omega_1(i\Delta\frac{\partial}{\partial\bar{q}_j})H_1^{\Theta_\Omega}(p_j,\bar{q}_j)\}\Big|_{\substack{p=\frac{1}{i\varepsilon}\frac{\partial}{\partial J^\star}\\ q=\frac{1}{i\varepsilon}\frac{\partial}{\partial J}}}\exp\{\varepsilon^2\sum_{j=1}^{N+1}[\frac{1}{4}\alpha\bar{J}_j^2+\frac{1}{4}\beta J_j^{\star 2}]\}\cdot$$

$$\cdot\exp\{-\varepsilon^2\sum_{j=1}^{N+1}\sum_{k=1}^{N+1}[\frac{1}{2}J_j^\star\frac{D_{jk}^{11}}{\varepsilon^2}J_k^\star+J_j^\star\frac{D_{jk}^{12}}{\varepsilon^2}J_k+$$

$$+\frac{1}{2}J_j\frac{D_{jk}^{22}}{\varepsilon^2}J_k]\}\tag{A4.40}$$

$$=\exp\{-i\varepsilon\sum_{j=1}^{N+1}\Omega_1(i\Delta\frac{\partial}{\partial\bar{q}_j})H_1^{\Theta_\Omega}(p_j,\bar{q}_j)\}\Big|_{\substack{p=\frac{1}{i\varepsilon}\frac{\partial}{\partial J^\star}\\ q=\frac{1}{i\varepsilon}\frac{\partial}{\partial J}}}\cdot$$

$$\cdot\exp\{-\varepsilon^2\sum_{j=1}^{N+1}\sum_{k=1}^{N+1}[\frac{1}{2}J_j^\star(\frac{D_{jk}^{11}}{\varepsilon^2}-\frac{1}{2}\beta\delta_{jk})J_j^\star+J_j^\star\frac{D_{jk}^{12}}{\varepsilon^2}J_k+\frac{1}{2}J_j(\frac{D_{jk}^{22}}{\varepsilon^2}-\frac{1}{2}\alpha\delta_{jk})J_k]\}\tag{A4.41}$$

which is indeed the discrete version of (4.74-4.76) (See the remark after (4.21)) As stated in the text all $D^{11}$ and $D^{22}$ tadpoles are modified at equal times and the values of the $D^{12}$ tadpoles are completely determined by $\Omega_1$.

## CHAPTER V : SHORT TIME PROPAGATORS AND THE RELATIONS BETWEEN THEM

### 5.1. Introduction

Derivations of the path integral representation of the propagator by a limiting procedure have in common that a lattice in time ($t_j = t_o + j\varepsilon$ , $j=0,1,\dots, N+1$, $t_{N+1} = t$) is used and that repeated use is made of the semi-group property (1.4). In this way the propagator $K(q,t;q_o,t_o)$ of (1.1) is represented by ($q_{N+1} = q$)

$$K(q,t;q_o,t_o) = \lim_{N\to\infty} \int \prod_{i=1}^{N} dq_i \prod_{j=1}^{N+1} K(q_j,t_j;q_{j-1},t_{j-1}). \tag{5.1}$$

Differences occur because of the different derivations used for determining an appropriate approximation for the short time propagator $K(q_j,t_j;q_{j-1},t_{j-1})$. The straightforward approach that we have applied in chapters I-IV is based on the approximation

$$K(q_j,t_j;q_{j-1},t_{j-1}) = \langle q_j|U(t_j,t_{j-1})|q_{j-1}\rangle \simeq \langle q_j|(1 - i\varepsilon \hat{H}(\hat{p},\hat{q},\bar{t}_j)|q_{j-1}\rangle$$
$$, \quad t_{j-1} \leqslant \bar{t}_j \leqslant t_j \tag{5.2}$$

and we have studied in detail the consequences of this basic relation.

A priori however there need not be any freedom in choosing a short time propagator since the propagator $K(q+\Delta,t+\varepsilon;q,t)$ can be computed as a power series in $\Delta$ and the unique approximation up to order $\varepsilon$ (taking into account that $\Delta = \mathcal{O}(\sqrt{\varepsilon})$) can be determined exactly. In this chapter we shall clarify this paradox by doing the calculation in two different ways and comparing the result with the approximation (5.2).

The first method (section 5.2.) starts by scaling the operator $\hat{p}$ as $\hat{p} = \hat{P}/\sqrt{\varepsilon}$ and considering $\hat{P}$ as a variable of order $\varepsilon^{0}$ (A scaling used first by Graham in [G15]). The short time propagator obtained in this way turns out to be related to the short time propagator used in the previous chapters by some relations between powers of $\Delta$ and powers of $\varepsilon$ in the configuration space functional integral. In section 5.3. we derive these relations from their analogues in the phase space functional integral and show that they are just a manifestation of the commutations relations between $\hat{p}$ and $\hat{q}$. In general these relations link all the different short time propagators that can be used in the functional integral and provide a direct way to change from one discretization to another inequivalent one.

The second method to compute the propagator $K(q+\Delta,t+\varepsilon;q,t)$ as a power series in $\Delta$ up to $O(\varepsilon) = O(\Delta^2)$ (presented in section 5.5) relies on a WKB-type expansion. Covariance under changes of coordinates q plays an important role. Therefore we study first in section 5.4. the transformation properties of equation (1.1-1.3) and introduce so-called normal coordinates. It turns out that the matrix $G^{\mu\nu}$ in (1.3) can be interpreted as the metric of a Riemannian manifold. Because of the use of normal coordinates special attention can easily be paid to the role played by the Riemannian curvature of the manifold.

The final result for the propagator $K(q,\varepsilon;q_o,0)$ up to order $\varepsilon$ in the limit $\Delta = O(\sqrt{\varepsilon})$ is given in formula (5.115). The extension to time-dependent metric is studied in subsection 5.5.4.

Special cases are (5.25-5.26) (one dimension) and (5.110) (normal coordinates). To these formulae will be referred in the next chapter where we study covariant definitions of functional integrals.

## 5.2. On the validity of $\hat{U} = 1 - i\varepsilon\hat{H}$

Formally the propagator from $(q,0)$ to $(q',\varepsilon)$ of the equation

$$\dot{\phi}\,(q,t) = [\tfrac{1}{2}\,c\,\partial_\mu\partial_\nu G^{\mu\nu}(q,t) + \partial_\mu A^\mu(q,t) + \tfrac{1}{c}\,V(q,t)]\phi(q,t) \tag{5.3}$$

is

$$\langle q'|\,U(\varepsilon,0)\,|q\rangle = \langle q'|\;P\,\exp\,[-i\int_0^\varepsilon d\tau\;\hat{H}(\hat{p},\hat{q},\tau)d\tau]\,|q\rangle \tag{5.4}$$

where P is the time ordering operator (cf.appendix 4.1) and

$$\hat{H}(\hat{p},\hat{q},t) = -\frac{ic}{2}\;\hat{p}_\mu\hat{p}_\nu\;G^{\mu\nu}(q,t) - \hat{p}_\mu A^\mu(\hat{q},t) - \frac{i}{c}\,V(q,t). \tag{5.5}$$

In the functional integral we need an approximation of this propagator up to $O(\varepsilon)$, and the method of the previous chapters was to use as an intermediate step

$$U(\varepsilon,0) = 1 - i\varepsilon\;\hat{H}(\hat{p},\hat{q},\bar{t}) + O(\varepsilon^2) \tag{5.6}$$

with $\bar{t} \in [0,\varepsilon]$ and to write for example $(\Delta \equiv q' - q)$

$$\langle q'|\,U(\varepsilon,0)\,|q\rangle = \int \frac{dp}{(2\pi)^d}\;\exp\,[ip.(q'-q) - i\,\varepsilon\,H(p,q,\bar{t})] \tag{5.7}$$

$$= (2\pi\varepsilon c)^{-d/2}(\det G^{\mu\nu}(q,\bar{t}))^{-1/2}\exp[-\tfrac{\varepsilon}{c}\,G_{\mu\nu}(q,\bar{t})(\tfrac{\Delta^\mu}{\varepsilon} + A^\mu(q,\bar{t}))(\tfrac{\Delta^\nu}{\varepsilon} + A^\nu(q,\bar{t}))$$

$$- \frac{\varepsilon}{c} \; V(q,\bar{t}) \; ] \; . \qquad (5.8)$$

In the limit $\varepsilon \to 0$, $\langle q' | \; U(\varepsilon,0) | q\rangle \to \delta(q'-q)$.
This singular behavior is represented in (5.8) by the preexponential factor and the dominating exponential, and the analogue of $\delta(q'-q)f(q') = \delta(q'-q)f(q)$ is that in all quantities multiplying the dominating exponential $q'-q$ can be considered as a quantity of order $\sqrt{\varepsilon}$. This also means that the dominant contribution to the integration over $p$ in (5.7) comes from the paths with $p = O(\varepsilon^{-1/2})$ (cf. Section 2.3). But then it is not obvious that (5.6) gives a correct approximation up to $O(\varepsilon)$.

We shall now justify this procedure by computing (5.4) treating from the beginning $p$ as a quantity of order $\varepsilon^{-1/2}$ and $\Delta$ as a quantity of order $\varepsilon^{+1/2}$ and showing that the result is related to that obtained using (5.6) by some relations between powers of $\Delta$ and powers of $\varepsilon$. We follow our works [L6,L10], another reference where this problem is treated is [W3].

When we define, following Graham([G15]) a scaled momentum operator by

$$\hat{P} = \sqrt{\varepsilon}\; \hat{p} \qquad (5.9)$$

then the corresponding variable $P$ in the path integral will be of order unity. In terms of $\hat{P}$ and $\hat{q}$ the short time evolution operator $U(\varepsilon, 0)$ which we denote from now on by $\hat{U}(\hat{p},\hat{q};\varepsilon)$ is given by ( $\bar{t} \in [0,\varepsilon]$ )

$$\hat{U}(\hat{P}/\sqrt{\varepsilon},\hat{q},\varepsilon) = \exp\,[-\frac{c}{2}\,\hat{P}_\mu \hat{P}_\nu G^{\mu\nu}(\hat{q},\,\frac{\varepsilon}{2}) + i\sqrt{\varepsilon}\hat{P}_\mu A^\mu(\hat{q},\bar{t}) - \frac{\varepsilon}{c}\,V(\hat{q},\bar{t}) + O(\varepsilon^{3/2})] \qquad (5.10)$$

In the quadratic term in $P$, $\frac{\varepsilon}{2}$ appears because only then the integral in (5.4) is approximated up to $O(\varepsilon)$. (Compare with 2.92a - 2.96a)

Let $U^{AS}(P/\sqrt{\varepsilon},q;\varepsilon)$ be the anti-standard symbol of $\hat{U}(\hat{P}/\sqrt{\varepsilon},\hat{q};\varepsilon)$. According to the general theory of section 2.1 we have that

$$\langle q' | \hat{U}(\hat{P}/\sqrt{\varepsilon},\hat{q};\varepsilon) | q\rangle = \int \frac{dP}{(2\pi\sqrt{\varepsilon})^d} \; \exp[\,\frac{i\;P.\Delta}{\sqrt{\varepsilon}}\,] U_{AS}(P/\sqrt{\varepsilon},q;\varepsilon) \qquad (5.11)$$

and in the path integral $U_{AS}(P/\sqrt{\varepsilon},q;\varepsilon)$ will contribute only with terms up to $O(\varepsilon)$. The refinement we make here is that $U_{AS}$ is not approximated by $\exp[-i\varepsilon H_{AS}]$ but instead $U_{AS}$ is computed up to $O(\varepsilon)$ from (5.10). The anti-standard symbol of $\hat{U}$ is obtained writing $\hat{U}$ in anti-standard ordered way ($\hat{P}$' s at the left hand side) and replacing then $\hat{P}$ by $P$ and $\hat{q}$ by $q$. To order (5.11) we develop the exponential and reorder the operators in each term of the series (keeping only termps up to $O(\varepsilon)$, from (5.9) one has that the commutator between $\hat{q}$ and $\hat{P}$ is proportional to $\sqrt{\varepsilon}$), finally we resume the series. This calculation is done explicitly in the Appendix of Graham's work [G15] , so we don't give the details here. The result is

$$U_{AS}(P/\sqrt{\varepsilon},q;\varepsilon) = \exp[-\frac{c}{2} P_\mu P_\nu G^{\mu\nu}(q,\frac{\varepsilon}{2})].[\sum_{k=0}^{6} P_{\nu_1}\ldots P_{\nu_k} B^{\nu_1\ldots\nu_k}(q,\bar{t}) + O(\varepsilon^{3/2})] \tag{5.12}$$

where the term with k=0 is independent of P and the quantities $B, B^{\nu_1}, \ldots, B^{\nu_1\ldots\nu_6}$ are given by

$$B(q,t) = 1 - \frac{\varepsilon}{c} V \tag{5.13a}$$

$$B^{\nu}(q,t) = \sqrt{\varepsilon}[i A^{\nu}] \tag{5.13b}$$

$$B^{\mu\nu}(q,t) = \varepsilon[-\frac{1}{2} A^{\mu}A^{\nu} + \frac{c}{2} G^{\nu\lambda}\partial_\lambda A^{\mu} - \frac{c^2}{8} G^{\kappa\lambda}\partial_\kappa\partial_\lambda G^{\mu\nu} + \frac{c}{4} A^{\lambda}\partial_\lambda G^{\mu\nu}] \tag{5.13c}$$

$$B^{\mu\nu\kappa}(q,t) = \sqrt{\varepsilon}[+\frac{ic^2}{4} G^{\mu\lambda}(\partial_\lambda G^{\nu\kappa})] \tag{5.13d}$$

$$B^{\mu\nu\kappa\lambda}(q,t) = \varepsilon[-\frac{c^2}{4} A^{\mu}G^{\nu\sigma}\partial_\sigma G^{\kappa\lambda} + \frac{1}{12} c^3 G^{\nu\rho}\partial_\rho(G^{\mu\sigma}\partial_\sigma G^{\kappa\lambda}) + \frac{1}{24} c^3 G^{\rho\sigma}(\partial_\rho G^{\mu\nu})(\partial_\sigma G^{\kappa\lambda})] \tag{5.13e}$$

$$B^{\mu\nu\kappa\lambda\sigma}(q,t) = 0$$

$$B^{\mu\nu\kappa\lambda\rho\sigma}(q,t) = \varepsilon[-\frac{1}{32} c^4 G^{\mu\alpha}(\partial_\alpha G^{\kappa\lambda})G^{\nu\beta}(\partial_\beta G^{\rho\sigma})]. \tag{5.13f}$$

We can compare (5.12) with $\exp[-i\varepsilon H_{AS}(p,q)]$ by reexpressing P in terms of p. We see then that the terms which were neglected in the approach of chapter II are the ones with $k \geq 2$ in (5.12). So the point to show is that these terms do not contribute in the limit $N \to \infty$ and can be dropped.

We shall give the proof first in the case of a one-dimensional configuration space (d = 1) , the extension to higher dimensions being essentially a question of notations. Formula (5.12) is then of the form

$$U_{AS}(P/\sqrt{\varepsilon},q;\varepsilon) = \exp[-\frac{c}{2} P^2 G(q,\frac{\varepsilon}{2})] (\sum_{k=0}^{6} P^k B_{(k)}(q,t) + O(\varepsilon^{3/2})) \tag{5.14}$$

with ( ' denotes $\partial/\partial q$)

$$B_{(0)} = 1 - \frac{\varepsilon}{c} V \tag{5.15a}$$

$$B_{(1)} = i \sqrt{\varepsilon}\, A \tag{5.15b}$$

$$B_{(2)} = \varepsilon[- \frac{1}{2} A^2 + \frac{c}{2} GA' - \frac{c^2}{8} GG'' + \frac{c}{4} AG' \tag{5.15c}$$

$$B_{(3)} = i \sqrt{\varepsilon}\, \frac{c^2}{4} GG' \tag{5.15d}$$

$$B_{(4)} = \varepsilon[- \frac{c^2}{4} A\, GG' + \frac{c^3}{12} G(GG')' + \frac{c^3}{24} G(G')^2] \tag{5.15e}$$

$$B_{(5)} = 0 \tag{5.15f}$$

$$B_{(6)} = \varepsilon[- \frac{c^4}{32} (GG')^2 ]. \tag{5.15g}$$

In (5.11) with (5.14)-(5.15) the following integrals appear:

$$I_k = \frac{1}{2\pi\sqrt{\varepsilon}} \int dP \exp[ \frac{iP\Delta}{\sqrt{\varepsilon}} - \frac{1}{2} c\, P^2 G]\, P^k \ , \ k = 0,1,2,3,4,6. \tag{5.16}$$

(G stands for $G(q, \frac{\varepsilon}{2} ))$. They satisfy the recursion relation:

$$I_k = \frac{i}{cG} \frac{\Delta}{\sqrt{\varepsilon}} I_{k-1} + (k-1) \frac{1}{cG} I_{k-2} \ . \tag{5.17}$$

This follows directly from

$$I_k = \frac{\partial^k}{\partial J^k} Z(J)\,|_{J=0} \tag{5.18}$$

with

$$Z(J) = \frac{1}{2\pi\sqrt{\varepsilon}} \int d\, P \exp [ \frac{iP\Delta}{\sqrt{\varepsilon}} - \frac{1}{2} c\, P^2 G + JP]$$

$$= I_0 \exp [ \frac{i}{cG} \frac{\Delta}{\sqrt{\varepsilon}} J + \frac{1}{2cG} J^2 ] \tag{5.19}$$

$$I_0 = (2\pi c\varepsilon)^{-1/2} \exp [- \frac{\Delta^2}{2\varepsilon c G} ] \ . \tag{5.20}$$

From (5.17) - (5.19) one obtains that

$$I_k = I_o(cG)^{-k} \sum_{\ell=o}^{[k/2]} ( \frac{i\Delta}{\sqrt{\varepsilon}} )^{k-2\ell} (c\, G)^{\ell} \binom{k}{2\ell} (2\ell- 1)!! \tag{5.21}$$

with $[k/2]$ the highest entire number smaller then or equal to $k/2$ and $(2\ell-1)!! = (2\ell- 1)(2\ell-3) \ldots 1, ((-1)!! = 1 )$

Define

$$a = \frac{1}{\sqrt{c\, G}} \frac{\Delta}{\sqrt{\varepsilon}} , \ J_k = I_k / I_o \tag{5.22}$$

then

$$J_1 = a \tag{5.23a}$$

$$J_2 = a^2 + 1 \tag{5.23b}$$

$$J_3 = a^3 + 3a \tag{5.23c}$$

$$J_4 = a^4 + 6a^2 + 3 \tag{5.23d}$$

$$J_6 = a^6 + 15a^4 + 45a^2 + 15. \tag{5.23e}$$

Using this in (5.11) with (5.14)-(5.15) one finds (setting c=1, afterwards c can be recuperated replacing G by cG and V by $\frac{1}{c}$ V)

$$<q'|\hat{U}(\hat{P}/\sqrt{\varepsilon},\hat{q};\varepsilon)\ |q> = I_\circ \sum_{k=0}^{6} J_k\ B_k \tag{5.24}$$

$$= [2\pi\varepsilon G(q,\tfrac{1}{2}\varepsilon)]^{-1/2} \exp\left[-\frac{\Delta^2}{2\varepsilon G(q,\frac{1}{2}\varepsilon)}\right] (1 + x\Delta + y\varepsilon + z\Delta^2 + u\Delta^3/\varepsilon + v\Delta^4/\varepsilon + w\,\Delta^6/\varepsilon^2 + \mathcal{O}(\varepsilon^{3/2})) \tag{5.25}$$

where the coefficients x,y,z,u,v,w, are evaluated at $(q,\bar{t})$ ,$\bar{t}\in[0,\varepsilon]$ and are given by.

$$x = -\frac{A}{G} - \frac{3}{4}\frac{G'}{G} \tag{5.26a}$$

$$y = -V - \frac{1}{2}\frac{A^2}{G} + \frac{1}{2}A' + \frac{1}{8}G'' - \frac{1}{2}\frac{AG'}{G} - \frac{3}{32}\frac{G''}{G} \tag{5.26b}$$

$$z = \frac{1}{2}\frac{A^2}{G^2} - \frac{1}{2}\frac{A'}{G} - \frac{3}{8}\frac{G''}{G} + \frac{5}{4}\frac{AG'}{G^2} + \frac{21}{32}\frac{G'^2}{G^2} \tag{5.26c}$$

$$u = \frac{1}{4}\frac{G'}{G^2} \tag{5.26d}$$

$$v = -\frac{1}{4}\frac{AG'}{G^3} + \frac{1}{32}\frac{G''}{G^2} - \frac{11}{32}\frac{G'^2}{G^3} \tag{5.26e}$$

$$w = \frac{1}{32}\frac{G'^2}{G^4}. \tag{5.26f}$$

This improved approximation to the short time propagator is at first sight not related to one of the approximations used in the previous chapters. However, when an expression as (5.25) is used as short time propagator in the functional integral the following replacements can be made in the factor multiplying the Gaussian $I_\circ$ (See (5.20)) without changing the value of the functional integral:

$$\Delta^2 \doteq \varepsilon\, c\, G \tag{5.27a}$$

$$\Delta^3/\varepsilon \doteq 3\varepsilon c G \Delta \tag{5.27b}$$

$$\Delta^4/\varepsilon \doteq 3\,\varepsilon\,(cG)^2 \tag{5.27c}$$

$$\Delta^6/\varepsilon^2 \doteq 15\varepsilon(cG)^3. \tag{5.27d}$$

This can be considered as an averaging procedure with respect to the Gaussian weight factor $\exp[-\Delta^2/2\varepsilon G]$ and expresses the stochastic nature of the paths. The symbol $\doteq$ means equal as far as use in the functional integral is concerned. (See also the next section and the references given after (5.36) )

In terms of the quantity a defined in (5.22) the relations (5.27) read

$$a^2 \doteq -1,\ a^3 \doteq -3a,\ a^4 \doteq 3,\ a^6 \doteq -15 \tag{5.28}$$

and from (5.22- 23) we see that this means

$$I_k \doteq 0\ ,\quad k = 2,3,4,6. \tag{5.29}$$

This shows that the terms which were neglected in the approach of the previous chapters (the ones coming from the integrals $I_k$, $k \geqslant 2$) can be neglected since they are effectively zero in the limit $N \to \infty$ .

Nevertheless the knowledge of the improved approximation (5.25 - 5.26) to the short time propagator will turn out to be necessary to relate all the definitions of the path integral that will be studied in the next chapter. To avoid confusion with the many short time propagators that can be used in the path integral we shall call the result (5.25- 5.26) the short time propagator obtained by Graham's scaling method. Its calculation in an arbitrary dimension is given in section 5.5.

Let us now comment on the generalization to higher dimensions. In (5.11) - (5.12) appear the integrals

$$I_k^{\{\mu_i\}} = \int dP/(2\pi\sqrt{\varepsilon})^d \exp\left[\frac{i\,P.\Delta}{\sqrt{\varepsilon}} - \frac{c}{2} P_\mu P_\nu G^{\mu\nu}(q, \tfrac{\varepsilon}{2})\right] P_{\mu_1} \cdots P_{\mu_k} \tag{5.30}$$

which have the value

$$I_k^{\{\mu_i\}} = I_\circ\, c^{-k}\, G_{\mu_1\alpha_1} \cdots G_{\mu_k\alpha_k} \sum_{\ell=o}^{[\frac{k}{2}]} \psi^{\alpha_1 \ldots \hat{\alpha}_{r_1} \ldots \alpha_k}\, \phi_{2\ell}^{\alpha_{r_1} \ldots \alpha_{r_{2\ell}}} \tag{5.31}$$

where

$$I_\circ = [(2\pi c\varepsilon)^d \det G^{\mu\nu}(q, \tfrac{\varepsilon}{2})]^{-1/2} \exp\left[-\frac{1}{2\varepsilon c} G_{\mu\nu}(q, \tfrac{\varepsilon}{2})\, \Delta^\mu \Delta^\nu\right], \tag{5.32}$$

$$\psi^{\lambda_1 \ldots \lambda_n} \equiv \left(\frac{i}{\sqrt{\varepsilon}}\right)^n \Delta^{\lambda_1} \ldots \Delta^{\lambda_n} , \tag{5.33}$$

$\hat{\alpha}_j$ is a missing index and

$$\phi_{2p}^{\lambda_1 \ldots \lambda_{2p}} = \sum_{i=1}^{2p} cG^{\lambda_1 \lambda_i} \phi_{2p-2}^{\lambda_1 \ldots \hat{\lambda}_i \ldots \lambda_{2p}} , \; p = 2,3,\ldots, \tag{5.34a}$$

$$\phi_2^{\lambda_1 \lambda_2} = c\, G^{\lambda_1 \lambda_2} . \tag{5.34b}$$

In the same way as in the one-dimensional case it follows from the relations

$$\Delta^\mu \Delta^\nu \doteq \varepsilon c G^{\mu\nu} \tag{5.35a}$$

$$\Delta^\mu \Delta^\nu \Delta^\kappa / \varepsilon \doteq \varepsilon c \, [\Delta^\mu G^{\nu\rho} + \Delta^\nu G^{\mu\rho} + \Delta^\kappa G^{\mu\nu}] \tag{5.35b}$$

$$\Delta^\mu \Delta^\nu \Delta^\kappa \Delta^\lambda / \varepsilon \doteq \varepsilon c^2 [G^{\mu\nu} G^{\kappa\lambda} + G^{\mu\kappa} G^{\nu\lambda} + G^{\mu\lambda} G^{\nu\kappa}] \tag{5.35c}$$

$$\Delta^\mu \Delta^\nu \Delta^\kappa \Delta^\lambda \Delta^\rho \Delta^\sigma / \varepsilon \doteq \varepsilon c^3 \phi_6^{\mu\nu\kappa\lambda\rho\sigma} \tag{5.35d}$$

where, expressed in terms of $G^{\mu\nu}$, $\phi_6$ is a sum of 15 terms, that

$$I_k^{\{\mu_i\}} \doteq 0 \qquad k = 2,3,4,6.$$

The relation (5.35a) was used first by B. DeWitt in [D9] . (There the notation $\doteq$ was introduced). Mayes and Dowker [M10] and Mc.Laughlin and Schulman [M1] proved (5.35a) and (5.35c) by an averaging procedure. (5.35b) can be handled in the same way (See Appendix 5.1) The factors $G^{\mu\nu} \equiv G^{\mu\nu}(q, \frac{\varepsilon}{2})$ in (5.35) are a consequence of the $G_{\mu\nu}(q, \frac{\varepsilon}{2})$ in (5.32) .In the next section we discuss generalizations.

Remark

There is of course nothing special about the use of the antistandard symbol of $\hat{U}(\hat{P}/\sqrt{\varepsilon},\hat{q})$ in (5.11) Consider for example a correspondence rule characterised by an $\Omega$-function $\Omega(uv)$. According to formula (2.44) the $\theta_\Omega$-symbol of the operator $\hat{U}$ is given by

$$U^{\Theta_\Omega}(P/\sqrt{\varepsilon},q) = \overleftrightarrow{\Omega}\left(-\sqrt{\varepsilon}\,\frac{\partial^2}{\partial q \partial P}\right) U_{AS}(P/\sqrt{\varepsilon},q;\varepsilon) \tag{5.36}$$

which is of the form

$$U^{\Theta_\Omega}(P/\sqrt{\varepsilon},q;\varepsilon) = \exp[-\frac{c}{2} P^2 G(q, \frac{\varepsilon}{2})](B_{(o)} - \sqrt{\varepsilon}\,\bar{\Omega}'(o)B'_{(1)} - \varepsilon \frac{c}{2}\,\bar{\Omega}''(0)G'' +$$

$$+ P(B_{(1)} + \sqrt{\varepsilon}\, c \quad \bar{\Omega}'(0)G') +$$

$$+ \sum_{\ell=2}^{6} P^k \overset{\Theta_\Omega}{B}_{(\ell)} (q,t) + O(\varepsilon^{3/2}) \} \tag{5.39}$$

where the quantities $\overset{\Theta_\Omega}{B}_{(k)}$, $k \geq 2$ can be expressed in terms of the quantities $B_{(k)}$ of (5.15) and G.

In the functional integral there appears (compare with (2.31))

$$u^{\Theta_\Omega}(P_j/\sqrt{\varepsilon},q_j,q_{j-1}) = \Omega(-\Delta_j \frac{\partial}{\partial \bar{q}_j})\, U^{\Theta_\Omega}(P_j/\sqrt{\varepsilon};\, \bar{q}_j;\varepsilon) \tag{5.40}$$

and as before it can be shown that the terms in $P^k$, $k \geq 2$ in (5.28) don't contribute in the limit $N \to \infty$. The approximation that contributes equals up to equivalence $\exp[-i\varepsilon h^\theta(p_j,q_j,q_{j-1})]$ as can be checked from (5.39) and (2.43-44). On the other hand the short time propagator obtained by Graham's scaling method is unique and independent of the ordering chosen for $\hat{U}(\hat{P}/\sqrt{\varepsilon},\hat{q};\varepsilon)$ (See Weiss [W3] ).

## 5.3. Relations between short time propagators

In this section we investigate the relations between the different short time propagators that can be used in a functional integral representation. In chapter 2 we have shown in detail how different short time propagators can be obtained e.g. by reordering of the non-commuting factors in the Hamiltonian operator. The question arises how to go in a path integral representation from one discretization to another (inequivalent) one without making explicit reference to the underlying reordering in the operator formalism. The answer lies in the replacements "$\doteq$" which we have already introduced both in phase space functional integrals (section 2.4., relations (2.104)) and in configuration space functional integrals (previous section, relations (5.27) and ( 5.35)). The equality in the sense $\doteq$ has been studied in the case of configuration space functional integrals by several authors ([M10],[M1],[C5]) and we shall not repeat their arguments here. Instead of this we shall study first the phase space functional integral and draw conclusions for the configuration space path integral afterwards.

It is very illustrative to go back first to the more or less complicated case that we have met in chapter 2 .

We had there the functional integral (2.101)and we had developed $h^\gamma(p,q',q,\tau)$ given by (2.102) using (2.103) . To prove the replacements (2.104) i.e.

$$\varepsilon p_\mu \Delta^\kappa \doteq i\varepsilon\delta_\mu{}^\kappa \tag{2.104a}$$

$$\varepsilon p_\mu p_\nu \Delta^\kappa \doteq i\varepsilon p_\mu \delta_\nu{}^\kappa + i\varepsilon p_\nu \delta_\mu{}^\kappa \tag{2.104b}$$

$$\varepsilon p_\mu p_\nu \Delta^\kappa \Delta^\lambda \doteq -\varepsilon\delta_\mu{}^\kappa\delta_\nu{}^\lambda - \varepsilon\delta_\mu{}^\lambda\delta_\nu{}^\kappa \tag{2.104c}$$

we write the resulting expression as

$$I(q,t;q_\circ,t_\circ) = \int\Big(\prod_{i=1}^{N} dq_i \prod_{j=1}^{N+1}\frac{dp_j}{(2\pi)^d}\Big)\exp\{-i\varepsilon\sum_{j=1}^{N+1}[-\frac{ic}{2}p_{j\mu}p_{j\nu}(\Delta^\kappa\partial'_\kappa g^{\mu\nu}_{[\gamma]} + \frac{1}{2}\Delta^\kappa\Delta^\lambda\partial'_\kappa\partial'_\lambda g^{\mu\nu}_{[\gamma]}) - p_{j\mu}\Delta^\kappa\partial'_\kappa a^\mu_{[\gamma]}]\}\exp\{+i\varepsilon\sum_{j=1}^{N+1}[p_{j\mu}\frac{\Delta j^\mu}{\varepsilon} - H(p_j,q_{j-1},\bar{t}_j)]\} \tag{5.41}$$

which is a functional integral of the type we have introduced in chapter 3 , the first exponential being the discretized functional that is averaged, the second being a weight factor of the usual form.

The properly time ordered operator whose matrix element is represented by (5.41) is

$$P\exp\{-i\int_{t_\circ}^{t} d\tau\Big(-\frac{ic}{2}[\hat{q}^\kappa(\tau),\hat{p}_\mu\hat{p}_\nu(\tau)]\partial'_\kappa g^{\mu\nu}_{[\gamma]}\Big|_{\hat{q}(\tau)} - \frac{ic}{4}[\hat{q}^\kappa(\tau),[\hat{q}^\lambda(\tau),\hat{p}_\mu(\tau)\hat{p}_\nu(\tau)]\,.\,\partial'_\kappa\partial'_\lambda g^{\mu\nu}_{[\gamma]}\Big|_{\hat{q}(\tau)} - [\hat{q}^\kappa(\tau),\hat{p}_\mu(\tau)]\,\partial'_\kappa a^\mu_{[\gamma]}\Big|_{\hat{q}(\tau)}\Big)\} \tag{5.42}$$

which because of the commutation relations is also

$$P\exp\{-i\int_{t_\circ}^{t} d\tau\Big(-\frac{ic}{2}(i\,\hat{p}_\mu\delta_\nu{}^\kappa + i\,\hat{p}_\nu\delta_\mu{}^\kappa)\partial'_\kappa g^{\mu\nu}_{[\gamma]} - \frac{ic}{4}(-\delta_\mu{}^\kappa\delta_\nu{}^\lambda - \delta_\mu{}^\lambda\delta_\nu{}^\kappa)\,.\,\partial'_\kappa\partial'_\lambda g^{\mu\nu}_{[\gamma]} - i\delta_\mu{}^\kappa\partial'_\kappa a^\mu_{[\gamma]}\Big)\}\Big|_{\hat{q}(\tau)} \tag{5.43}$$

Reintroducing this in a functional integral of the form (5.41) one finishes the proof of (2.104). This proof shows clearly that the use of $\doteq$ is linked to the reordering of noncommuting factors in the operator formalism and illustrates the relation between the "uncertainty" expressed by commutation relations and the stochastic nature of the paths.

As illustrated in section 2.4. the relations (2.104) provide a straightforward method to change from one discretization to another inequivalent one. (Inequivalent in the sense that $H^\gamma$ will change when $\doteq$ is used, the value of the functional integral does not change.)

We shall derive now the $\doteq$ relations in the configuration space functional integral. Consider the following phase space functional integral

$$I(q,t;q_0,t_0) = \lim_{N\to\infty} \int_{q_0}^{q} \Big( \prod_{i=1}^{N} dq_i \prod_{j=1}^{N+1} \frac{dp_j}{2\pi} \Big) \exp\{i\varepsilon \sum_{j=1}^{N+1} [p_j \frac{\Delta_j}{\varepsilon} + \frac{ic}{2} p_j^2 g_{[\gamma]}(q_j,q_{j-1}) + (p_j\Delta_j)^n]\} \tag{5.44}$$

In the same way as above it can be shown that it is equal to

$$I(q,t;q_0,t_0) = \lim_{N\to\infty} \int_{q_0}^{q} \Big( \prod_{i=1}^{N} dq_i \prod_{j=1}^{N+1} \frac{dp_j}{2\pi}\Big) \exp\{i\varepsilon \sum_{j=1}^{N+1} [p_j \frac{\Delta_j}{\varepsilon} + \frac{ic}{2} p_j^2 g_{[\gamma]}(q_j,q_{j-1}) + i^n n!\ ]\}. \tag{5.45}$$

Indeed, the term $(p_j\Delta_j)^n$ in (5.44) can be considered as coming from the operator

$$\underbrace{[\hat{q},[\hat{q},[\ \dots\ [\hat{q},\hat{p}^n]\ ]\ ]..\ ]}_{n \text{ times}}\ . \tag{5.46}$$

For example

$$\langle q_j|[\hat{q},\hat{p}]|q_{j-1}\rangle \Leftrightarrow p_j(q_j-q_{j-1}) = p_j\Delta_j \tag{5.47a}$$

$$\langle q_j|[\hat{q},[\hat{q},\hat{p}^2]]|q_{j-1}\rangle \Leftrightarrow p_j^2(q_j^2 - 2q_jq_{j-1} - q_{j-1}^2\ ) = p_j^2\Delta_j^2 \tag{5.47b}$$

$$\langle q_j|[\hat{q},[\hat{q},[\hat{q},\hat{p}^3]]]\ |q_{j-1}\rangle \Leftrightarrow p_j^3(q_j^3 - 3q_j^2q_{j-1} + 3q_jq_{j-1}^2-q_{j-1}^3) = p_j^3\Delta_j^3 \tag{5.47c}$$

and the operator (5.46) can be written as $i^n n!$.

Integration over the p-variables in (5.44) and (5.45) will give us two configuration space functional integrals with short time propagators related by the replacements inthe sense $\doteq$ .

We write (5.44) as

$$I(q,t;q_0,t_0)= \lim_{N\to\infty} \int_{q_0}^{q} \prod_{i=1}^{N} dq_i \Big( \prod_{j=1}^{N+1} \frac{dp_j}{2\pi} \exp\{i\varepsilon[p_j \frac{\Delta_j}{\varepsilon} + \frac{ic}{2} p_j^2 g_{[\gamma]}(q_j,q_{j-1})]\} \times$$

$$\times(1 + i\varepsilon\,(p_j\Delta_j)^n + O(\varepsilon^2))) \tag{5.48}$$

(Remember that $p_j$ and $\Delta_j$ are quantities of order $\varepsilon^{-1/2}$ and $\varepsilon^{1/2}$ respectively) and using (cf. (5.16-21))

$$\bar{I}_k = \int \frac{dp}{2\pi} \exp\{ip\Delta - \frac{\varepsilon c}{2} p^2 g_{[\gamma]}\} p^k \tag{5.49}$$

$$= \bar{I}_o \sum_{\ell=0}^{[\frac{k}{2}]} \binom{k}{2\ell} (2\ell-1)!!(i\Delta)^{k-2\ell} (\varepsilon\ c\ g_{[\gamma]})^{\ell-k} \tag{5.50}$$

$$\bar{I}_o = [2\pi\varepsilon c\ g_{[\gamma]}]^{-1/2} \exp[-\frac{\Delta^2}{2\varepsilon c\ g_{[\gamma]}}] \tag{5.51}$$

we obtain

$$I(q,t;q_o,t_o) = \lim_{N\to\infty} \int_{q_o}^{q} \prod_{i=1}^{N} dq_i \,( \prod_{j=1}^{N+1} [2\pi\varepsilon c g_{[\gamma]}(q_j,q_{j-1})]^{-1/2} \exp[-\frac{\Delta_j^2}{2\varepsilon c g_{[\gamma]}(q_j,q_{j-1})}] .$$

$$.[1 + i\varepsilon \sum_{\ell=0}^{[\frac{n}{2}]} \binom{n}{2\ell} (2\ell-1)!! i^{n-2\ell} (\frac{\Delta_j^2}{\varepsilon c\ g_{\gamma}})^{n-\ell} ] ). \tag{5.52}$$

On the other hand we obtain from (5.45)

$$I(q,t;q_o,t_o) = \lim_{N\to\infty} \int_{q_o}^{q} \prod_{i=1}^{N} dq_i ( \prod_{j=1}^{N+1} [2\pi\varepsilon c g_{[\gamma]}]^{-1/2} \exp[-\frac{\Delta_j^2}{2\varepsilon c g_{[\gamma]}}][1+i\varepsilon(i^n n!)]). \tag{5.53}$$

Comparing (5.52) and (5.53) for n=1,2,3 we conclude that

$$i\Delta_j^2/\varepsilon c\ g_{[\gamma]} \doteq i\ , \tag{5.54a}$$

$$-\Delta_j^{\ 4}/(\varepsilon c\ g_{[\gamma]})^2 + \Delta_j^2/(\varepsilon c g_{[\gamma]}) \doteq -2, \tag{5.54b}$$

$$-i\Delta_j^6/(\varepsilon c g_{[\gamma]})^3 + i\ 3\Delta_j^{\ 4}/(\varepsilon c g_{[\gamma]}) \doteq -i6\ , \tag{5.54c}$$

or also

$$\Delta_j^2 \doteq \varepsilon c\ g_{[\gamma]}(q_j,q_{j-1})\ , \tag{5.55a}$$

$$\Delta_j^{\ 4}/\varepsilon \doteq 3\ \varepsilon(c\ g_{[\gamma]}(q_j,q_{j-1}))^2, \tag{5.55b}$$

$$\Delta_j^{\ 6}/\varepsilon^2 \doteq 15\varepsilon\ (c\ g_{[\gamma]}(q_j,q_{j-1}))^3. \tag{5.55c}$$

These are the relations (5.27a,c,d) generalized to the case of an arbitrary dominating exponential. We now proceed to the derivation of the generalization of (5.27b), the relation for $\Delta_j^3$, and consider the phase space functional integral

$$I(q,t;q_\circ,t_\circ) = \lim_{N\to\infty}\int_{q_\circ}^{q_N} \prod_{i=1}^{N} dq_i \Big( \prod_{j=1}^{N+1} \frac{dp_j}{2\pi} \exp\{i\varepsilon\, [p_j \frac{\Delta_j}{\varepsilon} + \frac{ic}{2} p_j^2\, g_{[\gamma]}(q_j,q_{j-1}) + p_j^2\Delta_j]\}\Big) \tag{5.56}$$

$$= \lim_{N\to\infty}\int_{q_\circ}^{q} \prod_{i=1}^{N} dq_i \Big( \prod_{j=1}^{N+1} \frac{dp_j}{2\pi} \exp\{i\varepsilon[p_j \frac{\Delta_j}{\varepsilon} + i\,\frac{c}{2}\, p_j^2 g_{[\gamma]}(q_j,q_{j-1})]\}\ (1+ i\varepsilon p_j^2\Delta_j - \frac{1}{2}\,\varepsilon^2 p_j^4\Delta_j^2 + O(\varepsilon^{3/2}))\Big). \tag{5.57}$$

Remark that $p_j^2\Delta_j$ is a quantity of order $\sqrt{\varepsilon}$ and we have to develop the exponential in series up to second order. The quantity $p_j^2\Delta_j$ can be considered as coming from the operator $[\hat{q},\hat{p}^2] = 2i\hat{p}$ and can therefore be replaced by $2ip_j$, the quantity $p_j^4\,\Delta_j^2$ as coming from $[\hat{q},[\hat{q},\hat{p}^4]] = -12\hat{p}^2$.

Consequently $I(q,t,;q_\circ,t_\circ)$ also equals

$$I(q,t;q_\circ,t_\circ) = \lim_{N\to\infty}\int_{q_\circ}^{q} \prod_{i=1}^{N} dq_i \Big( \prod_{j=1}^{N+1} \frac{dp_j}{2\pi} \exp\{i\varepsilon[p_j \frac{\Delta_j}{\varepsilon} + \frac{ic}{2}\, p_j^2 g_{[\gamma]}(q_j,q_{j-1})]\ \}\ (1 + i\,\varepsilon(2ip_j) - \frac{1}{2}\,\varepsilon^2(-12p_j^2) + O(\varepsilon^{3/2})\Big). \tag{5.58}$$

Integrating over the p-variables using (5.49-5.51) one obtains from (5.57) and (5.58) respectively ($C$ denotes $cg_{[\gamma]}$)

$$I(q,t;q_\circ,t_\circ) = \lim_{N\to\infty}\int_{q_\circ}^{q} \prod_{i=1}^{N} dq_i \Big( \prod_{j=1}^{N+1} [2\pi\varepsilon C]^{-1/2} \exp[-\frac{\Delta_j^2}{2\varepsilon C}].\ [1 + i\varepsilon\,(-\frac{\Delta_j^2}{(\varepsilon C)^2} + \frac{1}{\varepsilon C})\,\Delta_j + - \frac{1}{2}\varepsilon^2\,(+\frac{\Delta_j^4}{(\varepsilon C)^4} - 6\,\frac{\Delta_j^2}{(\varepsilon C)^3} + \frac{3}{(\varepsilon C)^2})\,\Delta_j^2\,]\Big) \tag{5.59}$$

and

$$I(q,t;q_\circ,t_\circ) = \lim_{N\to\infty}\int_{q_\circ}^{q} \prod_{i=1}^{N} dq_i \Big( \prod_{j=1}^{N+1} [2\pi\varepsilon C]^{-1/2} \exp[-\frac{\Delta_j^2}{2\varepsilon C}]\ [1 + i\varepsilon\,(2i\,\frac{i\Delta_j}{\varepsilon C}) + 6\varepsilon^2\ (-\frac{\Delta_j^2}{(\varepsilon C)^2} + \frac{1}{\varepsilon C})]\Big). \tag{5.60}$$

Comparing (5.59) and (5.60) gives

$$- i\, \Delta_j^3/\varepsilon C^2 + i\, \Delta_j/C - \tfrac{1}{2}\, \Delta_j{}^6/\varepsilon^2 C^4 + 3\Delta_j{}^4/\varepsilon C^3 - \tfrac{3}{2}\, \Delta_j^2/C^2$$

$$\doteq -2i\, \Delta_j/C - 6\Delta_j^2/C^2 + 6\varepsilon/C \tag{5.61}$$

or using (5.54) and $C = cg_{[\gamma]}$

$$\Delta_j^3 /\varepsilon \doteq 3\, c\, g_{[\gamma]}(q_j,q_{j-1})\Delta_j\ . \tag{5.62}$$

which is the generalization of (5.27b).

The generalization to higher dimensions (relations (5.35)) can be treated in the same way.

It should be stressed that we have derived the allowed replacements for products of $\Delta$'s appearing multiplied with a Gaussian exp $(- \Delta_j^2/2\varepsilon c g_{[\gamma]})$. When powers of $\Delta$ appear in the argument of an exponential, this exponential has in principle to be developed in series before (5.55)(5.62) can be used. For terms of order $\varepsilon$ (e.g. $\Delta^2$) the replacement can also be made directly in the argument of the exponential (we have used this in (2.108)).

In general the replacements "$\doteq$" provide an easy method to change from one discretization to another inequivalent one. Take for example the propagator for the Hamiltonian $H(\hat{p},\hat{q}) = - \frac{ic}{2}\, \hat{p}^2 - \hat{p}A(\hat{q})$, its configuration space functional integral representation in prepoint discretization is

$$K(q,t;q_\circ,t_\circ) = \lim_{N\to\infty} \int \prod_{i=1}^{N} dq_i \Big( \prod_{j=1}^{N+1} [\, 2\pi\varepsilon c\, ]^{-1/2} \exp \{- \frac{\varepsilon}{2c} \Big( \frac{\Delta_j}{\varepsilon} + A(q_{j-1})\Big)^2\} \Big). \tag{5.63}$$

Developing $A(q_{j-1})$ around $\bar{q}_j = \frac{1}{2}(q_{j-1}+q_j)$ one obtains

$$K(q,t;q_\circ,t_\circ) = \lim_{N\to\infty} \int \prod_{i=1}^{N} dq_i \Big( \prod_{j=1}^{N+1} [\, 2\pi\varepsilon c]^{-1/2} \exp\{ - \frac{\varepsilon}{2c} \Big(\frac{\Delta_j}{\varepsilon} + A(\bar{q}_j)\Big)^2 + \frac{1}{2c}\, \Delta_j^2\, A'(\bar{q}_j)\} \tag{5.64}$$

and using $\Delta_j^2 \doteq \varepsilon c$

$$K(q,t;q_\circ,t_\circ) = \lim_{N\to\infty} \int \prod_{i=1}^{N} dq_i \Big( \prod_{j=1}^{N+1} [\, 2\pi\varepsilon c\, ]^{-1/2} \exp\{- \frac{\varepsilon}{2c} \Big(\frac{\Delta_j}{\varepsilon} + A(\bar{q}_j)\Big)^2 + \frac{\varepsilon}{2c}\, A'(\bar{q}_j)\}\Big) \tag{5.65}$$

the functional integral representation of K in midpoint discretization.

In other words it is precisely because of the stochastic properties of the paths ($\Delta \sim \sqrt{\varepsilon}$) which have a quantitive manifestation in the relations involving $\doteq$ that many short time propagators can be used in the path integral, they all being related by these same relations to the short time propagator obtained by

Graham's scaling method (5.25-26).

## 5.4. Covariant formulation, curvature and normal coordinates.

Consider again equation (1.1) with $\mathcal{L}(q,\partial,t)$ given by (1.3) but take $G^{\mu\nu}, A^{\mu}$ and V time independent:

$$\phi(q,t) = [\frac{1}{2} c\ \partial_\mu \partial_\nu\ G^{\mu\nu}(q) + \partial_\mu A^\mu(q) + \frac{1}{c} V(q)\ ]\ \ \phi(q,t). \tag{5.66}$$

Until now we have always been working in one given system of coordinates $q = (q^1,\ldots,q^d)$ . Inthe following we shall study functional integral representations of the propagator of (5.66) which are defined in terms of covariant quantities (i.e. scalars, vectors...). Therefore, it is necessary to study first the transformation properties under changes of variables.

The transformation properties of (5.66) under general (time independent coordinate transformations $q = q(\bar q)$ are fixed if we assume that $\phi(q,t)$ is a scalar density (or relative scalar) of weight one.[W2] This means that in the new coordinate system $\bar q, \bar\phi(\bar q,t)$ is related to $\phi(q,t)$ by

$$\phi(q,t) = \bar\phi(\bar q(q),t)\ \left|\det \frac{\partial \bar q^\alpha}{\partial q^\beta}\right| \equiv \bar\phi(\bar q(q),t)\ \left|\frac{\partial \bar q^\alpha}{\partial q^\beta}\right| \tag{5.67}$$

From (5.66) and (5.67) one can derive the transformation properties of $G^{\mu\nu}$, $A^\mu$, and V by writing the equation for $\phi(q,t)$ in the form (5.66) with derivatives $\bar\partial \equiv \partial/\partial\bar q$ and transformed quantities $\bar G^{\mu\nu}$, $\bar A^\mu$, $\bar V$. Let us do this explicitly. We first prove the result

$$\partial_\mu\, \ell n\left|\frac{\partial \bar q^\alpha}{\partial q^\beta}\right| = \bar\partial_\rho\ \left(\frac{\partial \bar q}{\partial q^\mu}\right)$$

which is an immediate consequence of the formula for differentiating determinants[W2]

$$\frac{\partial}{\partial x}\ (\det a^j{}_i(x)) = A^i{}_j\ \frac{\partial}{\partial x}\ (a^j{}_i)\ ,$$

with $A^i{}_j$ the cofactor of the matrix element $a^j{}_i(x)$. In our case we take $a^j{}_i = \partial\bar q^j/\partial q^i$ which has the cofactor $A^i{}_j = \partial q^i/\partial\bar q^j\ \det \partial\bar q^\mu/\partial q^\nu$ .

Hence

$$\partial_\mu \ln \left|\frac{\partial \bar{q}^\alpha}{\partial q^\beta}\right| = \frac{\partial q^i}{\partial \bar{q}^j}\,\partial_\mu\left(\frac{\partial \bar{q}^j}{\partial q^i}\right) = \frac{\partial}{\partial \bar{q}^j}\left(\frac{\partial \bar{q}^j}{\partial q^\mu}\right). \tag{5.68}$$

If X(q) is an arbitrary function of q we obtain from (5.68) the useful formula

$$\partial_\mu\left(\left|\frac{\partial \bar{q}^\alpha}{\partial q^\beta}\right| X\right) = \frac{\partial \bar{q}^\rho}{\partial q^\mu}\,\bar{\partial}_\rho\left(\left|\frac{\partial \bar{q}^\alpha}{\partial q^\beta}\right| X\right)$$

$$= \left|\frac{\partial \bar{q}^\alpha}{\partial q^\beta}\right| \bar{\partial}_\rho\left(\frac{\partial \bar{q}^\rho}{\partial q^\mu} X\right) + \left(\partial_\mu \ln \left|\frac{\partial \bar{q}^\alpha}{\partial q^\beta}\right| - \bar{\partial}_\rho \frac{\partial \bar{q}^\rho}{\partial q^\mu}\right)\left|\frac{\partial \bar{q}^\kappa}{\partial q^\lambda}\right| X$$

$$= \left|\frac{\partial \bar{q}^\alpha}{\partial q^\beta}\right| \bar{\partial}_\rho\left(\frac{\partial \bar{q}^\rho}{\partial q^\mu} X\right). \tag{5.69}$$

Transforming equation (5.66) and using (5.69) we find

$$\left|\frac{\partial \bar{q}^\alpha}{\partial q^\beta}\right| \frac{\partial \bar{\phi}}{\partial t} = \frac{1}{2} c\, \partial_\mu \partial_\nu G^{\mu\nu} \left|\frac{\partial \bar{q}^\alpha}{\partial q^\beta}\right| \bar{\phi} + \partial_\mu A^\mu \left|\frac{\partial \bar{q}^\alpha}{\partial q^\beta}\right| \bar{\phi} + V \left|\frac{\partial \bar{q}^\alpha}{\partial q^\beta}\right| \bar{\phi}$$

$$= \left|\frac{\partial \bar{q}^\alpha}{\partial q^\beta}\right| \left[\frac{1}{2} c \bar{\partial}_\rho \frac{\partial \bar{q}^\rho}{\partial q^\mu} \bar{\partial}_\sigma \frac{\partial \bar{q}^\sigma}{\partial q^\nu} G^{\mu\nu} \bar{\phi} + \partial_\rho \frac{\partial \bar{q}^\rho}{\partial q^\mu} A^\mu \bar{\phi} + V \bar{\phi}\right]$$

or

$$\frac{\partial \bar{\phi}}{\partial t} = \left[\frac{1}{2} c\, \bar{\partial}_\rho \bar{\partial}_\sigma \frac{\partial \bar{q}^\rho}{\partial q^\mu} \frac{\partial \bar{q}^\sigma}{\partial q^\nu} G^{\mu\nu} + \bar{\partial}_\rho\left(\frac{\partial \bar{q}^\rho}{\partial q^\mu} A^\mu - \frac{1}{2} c \frac{\partial^2 \bar{q}^\rho}{\partial q^\mu \partial q^\nu} G^{\mu\nu}\right) + V\right] \bar{\phi}.$$

We conclude

$$\bar{G}^{\rho\sigma} = G^{\mu\nu} \frac{\partial \bar{q}^\rho}{\partial q^\mu} \frac{\partial \bar{q}^\sigma}{\partial q^\nu}$$

$$\bar{A}^\rho = \frac{\partial \bar{q}^\rho}{\partial q^\mu} A^\mu - \frac{1}{2} c \frac{\partial^2 \bar{q}^\rho}{\partial q^\mu \partial q^\nu} G^{\mu\nu} \tag{5.70}$$

$$\bar{V} = V$$

and see that $D^{\mu\nu}$ transforms as a contravariant tensor. If we assume that it is invertible (which is what we have been doing all the time) we can interpret it as the metric tensor of a Riemannian manifold and use the notations and results of differential geometry.[W2,V3] It is not necessary to do so, but it provides us with the possibility of an elegant presentation. The covariant metric tensor $G_{\mu\nu}(q)$ is the inverse matrix of $G^{\mu\nu}(q)$.

If there exists a coordinate system such that the metric is constant (i.e. independent of q) it defines a Euclidean space, if such a coordinate system cannot be found the manifold has nonvanishing curvature. In general the curvature is given by the Riemann-Christoffel tensor[W2]

$$R^{\mu}{}_{\nu\rho\sigma} = \partial_\rho \Gamma_\nu{}^\mu{}_\sigma - \partial_\sigma \Gamma_\nu{}^\mu{}_\rho + \Gamma_\lambda{}^\mu{}_\rho \Gamma_\nu{}^\lambda{}_\sigma - \Gamma_\lambda{}^\mu{}_\sigma \Gamma_\nu{}^\lambda{}_\rho \tag{5.71}$$

where the Christoffel symbols $\Gamma_\nu{}^\mu{}_\sigma$ are defined by

$$\Gamma_\nu{}^\mu{}_\sigma = \frac{1}{2} G^{\mu\lambda} [\partial_\nu G_{\sigma\lambda} + \partial_\sigma G_{\nu\lambda} - \partial_\lambda G_{\nu\sigma}]$$

In the following we shall also need the Ricci tensor

$$R_{\nu\sigma} = R^{\mu}{}_{\nu\sigma\mu} \tag{5.72}$$

and the scalar curvature

$$R = G^{\nu\rho} R_{\nu\sigma} \tag{5.73}$$

It has to be remarked that the definitions of curvature used in the literature sometimes differ in sign. The convention used here is the one of [W2].

The determinant of the covariant metric tensor, $G(q) = \det G_{\mu\nu}(q)$, transforms as a scalar density of weight two, ($dq\sqrt{G(q)}$ is the invariant measure on the manifold), consequently

$$[G(q)]^{1/2}\phi(q,t) \equiv S(q,t) \tag{5.74}$$

is a scalar. Using $\nabla_\mu$ for covariant differentiation, the covariant derivatives of a scalar V and the divergence of a vector $V^\mu$ are given by

$$\nabla_\mu V = \partial_\mu V$$

$$\nabla_\mu V^\mu = \frac{1}{\sqrt{G(q)}} \partial_\mu \sqrt{G(q)}\, V^\mu(q).$$

The equation for S(q,t) which follows from (5.66) and (5.74) can then be written in the following covariant form

$$\dot{S}(q,t) = [\frac{1}{2} c\, \Delta_2 + \nabla_\mu h^\mu + \frac{1}{c} V]\, S(q,t) \tag{5.75}$$

In this equation $\Delta_2$ stands for the Laplace-Beltrami operator

$$\Delta_2 \equiv \frac{1}{\sqrt{G(q)}} \partial_\mu \sqrt{G(q)} \; G^{\mu\nu}(q) \, \partial_\nu \tag{5.76}$$

and $h^\mu$ for

$$h^\mu(q) \equiv A^\mu(q) + \frac{1}{2} c \frac{1}{\sqrt{G(q)}} \partial_\nu \sqrt{G(q)} \; G^{\mu\nu}(q) \tag{5.77}$$

which is a vector (contrary to $A^\mu$ ) as can be seen from (5.70) and the following calculation

$$\frac{1}{\sqrt{\bar{G}}} \bar{\partial}_\mu (\bar{G}^{\mu\nu} \sqrt{\bar{G}}) = \bar{\partial}_\mu (\ln \sqrt{\bar{G}}\,) \bar{G}^{\mu\nu} + \bar{\partial}_\mu \left(G^{\lambda\rho} \frac{\partial \bar{q}^\mu}{\partial q^\lambda} \frac{\partial \bar{q}^\nu}{\partial q^\rho}\right)$$

$$= \bar{\partial}_\mu \left[\ln \sqrt{G} - \ln \left|\frac{\partial \bar{q}^\alpha}{\partial q^\beta}\right|\right] G^{\lambda\rho} \frac{\partial \bar{q}^\mu}{\partial q^\lambda} \frac{\partial \bar{q}^\nu}{\partial q^\rho} + \partial_\lambda (G^{\lambda\rho}) \frac{\partial \bar{q}^\nu}{\partial q^\rho} +$$

$$+ G^{\lambda\rho} \bar{\partial}_\mu \left( \frac{\partial \bar{q}^\mu}{\partial q^\lambda} \frac{\partial \bar{q}^\nu}{\partial q^\rho} \right)$$

$$= \frac{\partial \bar{q}^\nu}{\partial q^\rho} [G^{\lambda\rho} \partial_\lambda (\ln \sqrt{G}) + \partial_\lambda G^{\lambda\rho} ] - G^{\lambda\rho} \frac{\partial \bar{q}^\nu}{\partial q^\rho} \left[\partial_\lambda \ln \left| \frac{\partial \bar{q}^\alpha}{\partial q^\beta} \right| - \bar{\partial}_\mu \left(\frac{\partial \bar{q}^\mu}{\partial q^\lambda}\right)\right] +$$

$$+ G^{\lambda\rho} \frac{\partial^2 \bar{q}^\nu}{\partial q^\lambda \partial q^\rho}$$

$$= \frac{\partial \bar{q}^\nu}{\partial q^\rho} \left[\frac{1}{\sqrt{G}} \partial_\lambda \sqrt{G} \, G^{\lambda\rho}\right] + G^{\lambda\rho} \frac{\partial^2 \bar{q}}{\partial q^\lambda \partial q^\rho} .$$

In (5.75) the equation for S is then given in manifestly covariant form. S being a scalar all terms on the right hand side are scalars : the Laplace-Beltrami operator $\Delta_2 = \nabla_\mu G^{\mu\nu} \nabla_\nu$ acting on S, the divergence of the vector $h^\mu S$ and the scalar VS.

The propagator $I(q,t;q_\circ,t_\circ)$ for the equation (5.75) is related to the propagator $K(q,t;q_\circ,t_\circ)$ for the equation (5.66) by

$$I(q,t;q_\circ,t_\circ) = G(q)^{-1/2} K(q,t;q_\circ,t_\circ) G^{+1/2}(q_\circ). \tag{5.78}$$

The factor $G(q_\circ)^{1/2}$ is present to assure that the initial condition $I(q,t;q_\circ,t_\circ) = \delta(q-q_\circ)$ is satisfied.

Very useful in practical calculations are the so-called normal coordinates. They are introduced in the following way [V3]: the geodesics on the manifold with metric $G_{\mu\nu}(q)$ are the solutions of the equations

$$\ddot{q}^{\mu}(t) = -\Gamma_{\kappa}{}^{\mu}{}_{\lambda}(q(t))\,\dot{q}^{\kappa}(t)\dot{q}^{\lambda}(t) \tag{5.79}$$

which are the Euler-Lagrange equations following from the Lagrangian

$$L_{o}(\dot{q},q) = \frac{1}{2}\, G_{\mu\nu}(q)\dot{q}^{\mu}\dot{q}^{\nu}. \tag{5.80}$$

The geodesics for the initial conditions

$$q(t=0)=q_{o}\ ;\ \dot{q}(t=0) = a$$

can be written in terms of a series in $a^{\mu}t$ which starts as follows (more details are given in [V3])

$$\begin{aligned} q^{\mu}(t) &= q^{\mu}_{o} + a^{\mu}t - \frac{1}{2}\,\Gamma_{\kappa}{}^{\mu}{}_{\lambda}(q_{o})a^{\kappa}a^{\lambda}t^{2} \\ &- \frac{1}{3!}[\partial_{\lambda}\Gamma_{\nu}{}^{\mu}{}_{\kappa} - 2\,\Gamma_{\nu}{}^{\mu}{}_{\alpha}\Gamma_{\kappa}{}^{\alpha}{}_{\lambda}](q_{o})\ a^{\kappa}a^{\lambda}a^{\nu}t^{3} \ldots \ . \end{aligned} \tag{5.81}$$

If one now defines the transformation of coordinates q to coordinates y by

$$\begin{aligned} q^{\mu}(t) &= q^{\mu}_{o} + y^{\mu}(t) - y^{\mu}_{o} - \frac{1}{2}\,\Gamma_{\kappa}{}^{\mu}{}_{\lambda}(q_{o})(y^{\kappa}(t) - y^{\kappa}_{o})(y^{\lambda}(t) - y^{\lambda}_{o}) - \\ &- \frac{1}{3!}[\partial_{\lambda}\Gamma_{\nu}{}^{\mu}{}_{\kappa} - 2\Gamma_{\nu}{}^{\mu}{}_{\alpha}\Gamma_{\kappa}{}^{\alpha}{}_{\lambda}]\ (q_{o})(y^{\kappa}(t)-y^{\kappa}_{o})(y^{\lambda}(t)- y^{\lambda}_{o})(y^{\nu}(t) -y^{\nu}_{o}) + \ldots \end{aligned} \tag{5.82}$$

where the series in (5.82) is obtained from the series in (5.81) by replacing

$$y(t) - y_{o} = a\ t \tag{5.83}$$

one has that in coordinates y the geodesics through the point $y_{o}$ are given by (5.83). These coordinates y are called normal coordinates with respect to the origin $y_{o}$.

We shall mark all quantities in normal coordinates by a star. Normal coordinates offer the advantage that all first order derivatives of the metric tensor vanish at $y = y_{o}$ :

$$\star(\partial_{\lambda}G_{\mu\nu})(y\ ) = 0 \quad \text{at } y = y_{o}. \tag{5.84}$$

Consequently also the Christoffel symbols vanish at the origin and results written in terms of normal coordinates and evaluated at the origin can be transformed covariantly to an arbitrary coordinate system by replacing all starred tensors by their unstarred counterparts and ordinary derivatives by covariant derivatives. Furthermore, at the origin of the normal coordinate system the Riemann-Christoffel tensor is related to the metric in a very simple way:

$$\overset{\star}{R}_{\mu\nu\rho\sigma} = \overset{\star}{(}\partial_\nu\partial_\rho G_{\mu\sigma} - \partial_\nu\partial_\sigma G_{\mu\rho}) \text{ at } y = y_\circ$$

or also

$$\overset{\star}{(}\partial_\kappa\partial_\lambda G_{\mu\nu}) = -\frac{1}{3}\overset{\star}{(}R_{\mu\kappa\nu\lambda} + R_{\mu\lambda\nu\kappa}) \text{ at } y = y_\circ. \qquad (5.85)$$

And using (5.84-85) in the Taylor expansion of $\overset{\star}{G}_{\mu\nu}(y)$ around $y_\circ$ one obtains ($\Delta \equiv y-y_\circ$)

$$\overset{\star}{G}_{\mu\nu}(y) = \overset{\star}{G}_{\mu\nu}(y_\circ) - \frac{1}{3}\overset{\star}{R}_{\mu\kappa\nu\lambda}(y_\circ)\,\Delta^\kappa\Delta^\lambda + \mathcal{O}(\Delta^3) \qquad (5.86)$$

and also

$$\overset{\star}{G}(y) = \overset{\star}{G}(y_\circ)\,(1 + \frac{1}{3}\overset{\star}{R}_{\mu\nu}(y_\circ)\Delta^\mu\Delta^\nu + \mathcal{O}(\Delta^3)). \qquad (5.87)$$

These are all the formulae in normal coordinates that we shall need here.

## 5.5. Short time propagator from a WKB-type expansion

### 5.5.1. Introduction

The original idea of path integration is that the propagator is given by a sum of contributions of all possible paths, and this independent of the time interval considered. In Feynman's formulation of quantum mechanics [F3] it was shown that for simple systems the paths $q(\tau)$ contribute with weight $\exp[\frac{i}{\hbar}A[q(\tau)]]$ where $A$ is the classical action functional. On the basis of a stationary phase argument one can argue that in first approximation only the path that makes the action stationary (the classical path) contributes. Feynman therefore postulated that on short time scales this is the only path that contributes to the propagation and that the short time propagator is of the form $\mathcal{N}\exp[\frac{i}{\hbar}A[q_{cl}(\tau)]]$ where $A[q_{cl}(\tau)]$ is the classical action evaluated at the classical trajectory between fixed initial and final points and $\mathcal{N}$ is a normalization factor, which has to be determined by other means. Later on [M21] it became clear however that this normalization factor is build up by the contributions from "quadratic fluctuations" around the classical path. (In a semi-classical expansion an arbitrary path is expanded around the classical path. Terms which are quadratic in the deviation of a path from the classical path are called quadratic fluctuations). In Pauli's treatment of Feynman's theory [P3] the short time propagator was therefore taken to be equal to the semi-classical or WKB-approximation to the propagator. B. De Witt [D 9] showed that on a curved Riemannian manifold the semiclassical propagator fails to be the exact short time propagator by a factor involving the scalar curvature R. This indicates that here also higher order terms in the deviation of an arbitrary path from the classical path contribute to the short time propagator.

Let us investigate this in more detail. First we introduce in the covariant equation (5.75) the "Planck's constant" which we call $\eta$ in the following way

$$\dot{S}(q,t) = \frac{1}{2}\eta c\, \Delta_2\, S + \nabla_\mu h^\mu\, S + \frac{1}{\eta c}\, VS\ . \tag{5.88}$$

The propagator $I(q,t;q_\circ,t_\circ;\eta)$ of (5.88) has an asymptotic series expansion in $\eta$ (See chapter 9)

$$I(q,t;q_\circ,t_\circ;\eta) = I_{WKB}(1 + \sum_{k=1}^{\infty} \eta^k\, I_k) \tag{5.89}$$

where $I_{WKB}$ is the semi-classical approximation satisfying (5.88) up to terms of order $\eta$. $I_{WKB}$ is proportional to $\exp[\frac{i}{\eta} A[q_{cl}]]$ where $A[q_{cl}]$ is the classical action of the system with Hamiltonian

$$H_{cl}(p,q) = -\frac{1}{2}\, ic\, p_\mu p_\nu G^{\mu\nu}(q) - p_\mu h^\mu(q) + \frac{i}{c}\, V(q) \tag{5.90}$$

and Lagrangian

$$L_{cl}(\dot{q},q) = \frac{i}{c}\, [\frac{1}{2}\, G_{\mu\nu}(q)\, (\dot{q}^\mu + h^\mu(q))(\dot{q}^\nu + h^\nu(q)) - V(q)]\ . \tag{5.91}$$

The quantities $I_k$ in (5.89) are the higher order corrections to the WKB-approximations. The series (5.89) taken at $\eta=1$ is the propagator of (5.75)
The WKB-approximation $I_{WKB}$ can be obtained from a phase space functional integral representation by expanding the paths around the classical path and integrating over the quadratic fluctuations around this path. (More details are given in chapter 9 ) It is given by

$$I_{WKB}(q,t;q_\circ,t_\circ,\eta) = [2\pi i\eta]^{-d/2}\, [G(q)/G(q_\circ)]^{-1/4}\, [\det - \frac{\partial^2 A_{cl}}{\partial q^\mu \partial q_\circ^\nu}]^{1/2}$$

$$\cdot\ \exp[\int_{t_\circ}^{t} \frac{1}{2}\nabla_\mu h^\mu(q_{cl}(\tau))d\tau\,]\ .\ \exp[\frac{i}{\eta}\, A_{cl}\,] \tag{5.92}$$

where $A_{cl}$ is the classical action for the Lagrangian $L_{cl}$ and boundary conditions $q(t) = q$, $q(t_\circ) = q_\circ$.

The quantity $G(q)^{-1/4}\, [\det - \frac{\partial^2 A_{cl}}{\partial q^\mu \partial q_\circ^\nu}]^{1/2}\, G(q_\circ)^{-1/4}$ is a scalar, and so are the arguments of the exponentials in (5.92). We therefore find, as it should be since we have introduced $\eta$ in a covariant way, that $I_{WKB}(q,t;q_o,t_o;\eta)$ has the same transformation properties as $I(q,t;q_o,t_o;\eta)$ .

The aim of this section is to obtain from the expansion (5.89) the short time propagator up to order $\varepsilon$ and considering from the beginning $\Delta q$ as a quantity of order $\sqrt{\varepsilon}$ . We expect to find the same result as the one of Graham's scaling method (but now in an arbitrary dimension).

We set from now on $t_o=0$, $t = \varepsilon$ .

First of all we remark that we shall in general pick up contributions not only from $I_{WKB}$, but also from the first order correction $I_1$. Indeed, consider the equation

$$\dot{S} = \frac{1}{2} \eta c \, \Delta_2 \, S \tag{5.93}$$

which describes free diffusion (c=1) or free quantummechanical motion (c=i) on the manifold with metric $G_{\mu\nu}(q)$. The solution $I(q,\varepsilon;q_o,0;\eta)$ of this equation will depend on $\varepsilon$ and $\eta$ through the combination $\varepsilon\eta$ only. Therefore the first order correction $\eta I_1$, which picks up contributions from cubic and quartic fluctuations around the classical path, will be of order $\varepsilon$ . Hence, in order to determine an approximation up to $\mathcal{O}(\varepsilon)$ one has to approximate $I_{WKB}(1+\eta I_1)$ up to $\mathcal{O}(\varepsilon)$. In fact, in chapter 10 we derive from the general expression for the first order correction in the case of equations of the form (5.88) its approximation up to order $\varepsilon$ :

$$I_1^{\varepsilon} = - \frac{1}{12} \varepsilon c R \, . \tag{5.94}$$

In a flat space R=0 and the WKB-approximation provides a good short time propagator in that case (e.g. Pauli's calculation in [P3] )
(De Witt's method [D9 ] to find the contribution from $I_1$ was to consider the equation satisfied by $I_{WKB}$ , the approximation up to $\mathcal{O}(\varepsilon)$ of $I_{WKB}$ and to see by what factor it misses to be a solution up to order $\varepsilon$ .)

### 5.5.2. $I_{WKB}$ up to order $\varepsilon$

We shall calculate here $I_{WKB}^{\varepsilon}$ , the approximation up to order $\varepsilon$ (with $\Delta$ of order $\sqrt{\varepsilon}$ ) of $I_{WKB}$. The algebra will be simplified a lot if we use normal coordinates y with as origin the prepoint of the propagator. ${}^{\star}I_{WKB}(y,\varepsilon,y_o,0;\eta)$, the WKB-approximation to the propagator is given by (5.92) in terms of starred quantities. From (5.87) one has that

$$[{}^{\star}G(y)/{}^{\star}G(y_o)]^{-1/4} = 1 - \frac{1}{12} {}^{\star}R_{\mu\nu}(y_o)\Delta^{\mu}\Delta^{\nu} + \mathcal{O}(\Delta^3). \tag{5.95}$$

Furthermore , since in normal coordinates first order derivatives of the metric vanish at the origin:

$$\star h^{\mu}(y_\circ) = \star A^{\mu}\ (y_\circ), \quad \star(\nabla_{\mu} h^{\mu})(y_\circ) = \star(\partial_{\mu} A^{\mu})\ (y_\circ) \tag{5.96}$$

and

$$\exp \frac{1}{2} \int_0^{\varepsilon} {}^{\star}(\nabla_{\mu} h^{\mu})\, d\tau = 1 + \varepsilon\, \frac{1}{2}\star(\partial_{\mu} A^{\mu})(y_\circ) + \mathcal{O}(\varepsilon^2). \tag{5.97}$$

The calculation of the expansion of the classical action needs a little more effort but is also quite easy in normal coordinates.

First of all we split ${}^{\star}L_{cl}(\dot{y},y)$ in two parts:

$$\star L_{cl}(\dot{y},y) = \frac{i}{c}\left[\ \star L_\circ(\dot{y},y) + \star L_1(\dot{y},y)\ \right] \tag{5.98}$$

with $L_\circ$ given by (5.80) and

$$\star L_1\ (\dot{y},y) = \star(G_{\mu\nu} h^{\mu})(y)\dot{y}^{\nu} + \frac{1}{2}\star(G_{\mu\nu} h^{\mu} h^{\nu})(y) \overset{\star}{-} V(y) \tag{5.99}$$

and remark that the contribution of $\star L_1$ to the expansion of the action up to $\mathcal{O}(\varepsilon)$ is independent of the path $y(\tau)$ and given by

$$\begin{aligned}\star A^{1}_{c\ell} = \frac{i}{c} \int_0^{\varepsilon} \star L_1(\dot{y},y)d\tau &= \frac{i}{c}\star(G_{\mu\nu} h^{\mu})\Delta^{\nu} + \frac{1}{2}\star(\partial_{\lambda} G_{\mu\nu} h^{\mu})\Delta^{\lambda}\Delta^{\nu} + \\ &+ \varepsilon\, {}^{\star}(\frac{1}{2}\, G_{\mu\nu} h^{\mu} h^{\nu} - V)\ \Big|_{y_\circ} + \mathcal{O}(\varepsilon^{3/2})\end{aligned} \tag{5.100}$$

where $|_{y_\circ}$ indicates that y-dependent functions are evaluated at $y_\circ$. Eq.(5.100) readily follows by expanding in the integrand ${}^{\star}L_1$

$$\star[G_{\mu\nu} h^{\mu}\ (y(\tau))] = \star[G_{\mu\nu} h^{\mu}](y_\circ) + (y^{\lambda}(\tau) - y^{\lambda}_\circ)\star[\partial_{\lambda} G_{\mu\nu} h^{\mu}] + \ldots. \tag{5.101a}$$

$$y(\tau) = y_\circ + \tau\ \dot{y}(0)\ \ldots \tag{5.101b}$$

and using that

$$\dot{y}(\tau) = \frac{\Delta}{\varepsilon} + \mathcal{O}(\varepsilon^{\circ}). \tag{5.102}$$

Written as in (5.100) this expansion of $A^{1}_{cl}$ is valid in any coordinate system, in normal coordinates we can use (5.96) and obtain

$$\star\ A^{1}_{c\ell} = \frac{i}{c}\left[\ {}^{\star}(G_{\mu\nu} A^{\mu})\Delta^{\nu} + \frac{1}{2}\star(G_{\mu\nu}\partial_{\lambda} A^{\mu})\Delta^{\lambda}\Delta^{\nu} + \varepsilon\star(\frac{1}{2}\, G_{\mu\nu} A^{\mu} A^{\nu} - V)\right]\Big|_{y_\circ} + \mathcal{O}(\varepsilon^{3/2})\ . \tag{5.103}$$

On the other hand the contribution of $\star L_\circ$ to the expansion of the action is up to order $\varepsilon$ the same for the classical path $y_{cl}(\tau)$ of $\star L_{cl}$ and for the classical path of $\star L_\circ$ which is the geodesic $y^\circ_{cl}(\tau) = y_\circ + \Delta \frac{\tau}{\varepsilon}$. The action of $\star L_\circ$ along $y^\circ_{cl}(\tau)$ is simply $\varepsilon \frac{1}{2}\star G_{\mu\nu}(y_\circ)\frac{\Delta^\mu}{\varepsilon}\frac{\Delta^\nu}{\varepsilon}$ since as can be checked using the equation of motion (5.79) $L_\circ$ is constant on the geodesic and $\dot{y}^\circ_{cl}(0) = \Delta/\varepsilon$. Next it can be shown easily using the expansion of $y(\varepsilon)$ around $y(0)$ and the equations of motion for $y_{cl}(\tau)$ and $y^\circ_{cl}(\tau)$ that $\dot{y}^\circ_{cl}(0) = \dot{y}_{cl}(0) + O(\sqrt{\varepsilon})$. We find then that

$$A_{cl} = \frac{i}{c}\left[\frac{1}{2}\varepsilon \star G_{\mu\nu}\left(\frac{\Delta^\mu}{\varepsilon} + \star A^\mu\right)\left(\frac{\Delta^\nu}{\varepsilon} + \star A^\nu\right) - \star V + \frac{1}{2}\star G_{\mu\nu}\star(\partial_\lambda A^\mu)\Delta^\nu\Delta^\lambda\right]\Big|_{y_\circ} + O(\varepsilon^{3/2})$$

and (5.104)

$$-\frac{\partial^2 A_{cl}}{\partial y^\mu \partial y^\nu_o} = \frac{i}{c\varepsilon}\star G_{\mu\nu} - \frac{1}{2}\star(\partial_\mu A_\nu - \partial_\nu A_\mu) + O(\varepsilon^{1/2})\Big|_{y_\circ}$$

$$= \frac{i}{c\varepsilon}\star G_{\mu\lambda}\left(\delta^\lambda{}_\nu + \frac{1}{2}\, ic\varepsilon G^{\lambda\kappa}(\partial_\kappa A_\nu - \partial_\nu A_\kappa) + O(\varepsilon^{3/2})\right)\Big|_{y_\circ}. \tag{5.105}$$

The determinant of this matrix is computed using det AB= det A det B and det B = exp Tr ℓn B and one obtains

$$\left[\det -\frac{\partial^2 A_{cl}}{\partial y^\mu \partial y^\nu_o}\right]^{1/2} = \left(\frac{i}{c\varepsilon}\right)^{d/2}\star G(y_\circ)^{1/2}\,(1 + O(\varepsilon^{3/2}))\,. \tag{5.106}$$

Putting all these approximations together we obtain for the WKB-propagator up to order $\varepsilon$ :

$$\star I_{WKB}(y,\varepsilon;y_\circ,0;\eta) = (2\pi\varepsilon c)^{-d/2}\star G(y_\circ)^{+1/2}\Big(1 - \frac{1}{12}\star R_{\mu\nu}\Delta^\mu\Delta^\nu + \varepsilon\,\frac{1}{2}\star(\partial_\mu A^\mu) +$$

$$+ O(\varepsilon^{3/2})\Big)\exp\Big\{-\frac{1}{\eta c}\,\varepsilon\,\frac{1}{2}\star G_{\mu\nu}\left(\frac{\Delta^\mu}{\varepsilon} + \star A^\mu\right)\left(\frac{\Delta^\nu}{\varepsilon} + \star A^\nu\right) - \star V +$$

$$+ \eta\,\frac{1}{2}\star(G_{\mu\nu}\partial_\lambda A^\mu)\Delta^\nu\Delta^\lambda\Big|_{y_\circ}\Big\} \tag{5.107}$$

$$= (2\pi\varepsilon c)^{-d/2}\star G(y_\circ)^{1/2}\exp\left[-\frac{1}{2\varepsilon c\eta}G_{\mu\nu}(y_\circ)\Delta^\mu\Delta^\nu\right].\Big\{1 +$$

$$-\frac{1}{\eta c}\star(G_{\mu\nu}A^\mu)\Delta^\nu + \varepsilon\left(\frac{1}{2}\star(\partial_\mu A^\mu) - \frac{1}{\eta c}\star\left(\frac{1}{2}G_{\mu\nu}A^\mu A^\nu - V\right)\right) + \tag{5.108}$$

$$+\Delta^\mu\Delta^\nu\left(\frac{1}{2(\eta c)^2}\star(G_{\mu\alpha}A^\alpha G_{\nu\beta}A^\beta) - \frac{1}{2c}\star(G_{\nu\lambda}\partial_\mu A^\lambda) - \frac{1}{12}\star R_{\mu\nu}\right) + O(\varepsilon^{3/2})\Big\}\Big|_{y_\circ}$$

Formula (5.108) is the expression we call $\star I^{\varepsilon}_{WKB}$ .

5.5.3. Short time propagator on a Riemannian manifold.

Following our arguments of paragraph 5.5.1 that the propagator of (5.88) up to order $\varepsilon$ is given by $\star I = \star I^{\varepsilon}_{WKB}\ (1+\eta I^{\varepsilon}_{1})$ and using the relation (5.78) between the propagators of equations (5.75) and (5.66) we obtain for the propagator K up to order $\varepsilon$ (set c=1)

$$\star K(y,\varepsilon;y_\circ,0) := \star[G(y)/G(y_\circ)]^{1/2}\ \star I^{\varepsilon}_{WKB}(y,\varepsilon;y_\circ,0;\eta = 1).(1+I^{\varepsilon}_{1}) \tag{5.109}$$

$$= (2\pi\varepsilon)^{-d/2}\ \star G(y)^{1/2}\ \exp\,[-\tfrac{1}{2\varepsilon}\star G_{\mu\nu}(y_\circ)\Delta^{\mu}\Delta^{\nu}\ ]\ .$$

$$[1 - \star(G_{\mu\nu}A^{\mu})\Delta^{\nu} + \varepsilon\star(\tfrac{1}{2}\partial_{\mu}A^{\mu} - \tfrac{1}{2}\,G_{\mu\nu}A^{\mu}A^{\nu} + V - \tfrac{1}{12}\,R)$$

$$+ \star(\tfrac{1}{2}\,G_{\mu\alpha}A^{\alpha}G_{\nu\beta}A^{\beta} - \tfrac{1}{2}\,G_{\mu\lambda}\partial_{\nu}A^{\lambda} - \tfrac{1}{12}\,R_{\mu\nu})\Delta^{\mu}\Delta^{\nu} + O(\varepsilon^{3/2})]\Big|_{y_\circ}\ . \tag{5.110}$$

We shall now compare this result with that obtained by the method of section 5.1. According to (5.11-13) and (5.30) the latter is given by

$$\star K(y,\varepsilon;y_\circ,0) = \sum_{k=0}^{6}\ I_k^{\{\mu_i\}}\ \star B^{\{\mu_i\}}\ (y_\circ)\ . \tag{5.111}$$

In normal coordinates and evaluated at the origin the coefficients $\star B^{\{\mu_i\}}$ are

$$\star B\ \ = 1 - \varepsilon\star V \tag{5.112a}$$

$$\star B^{\mu}\ = i\sqrt{\varepsilon}\ \star A^{\mu} \tag{5.112b}$$

$$\star B^{\mu\nu} = \varepsilon\star[-\tfrac{1}{2}\,A^{\mu}A^{\nu} + \tfrac{1}{2}\,G^{\nu\lambda}\partial_{\lambda}A^{\mu} - \tfrac{1}{8}\,G^{\kappa\lambda}(\partial_{\kappa}\partial_{\lambda}G^{\mu\nu})]$$

$$= \varepsilon\star[-\tfrac{1}{2}\,A^{\mu}A^{\nu} + \tfrac{1}{2}\,G^{\nu\lambda}\partial_{\lambda}A^{\mu} - \tfrac{1}{12}\,R^{\mu\nu}] \tag{5.112c}$$

$$\star B^{\mu\nu\rho} = \star B^{\mu\nu\kappa\lambda} = \star B^{\mu\nu\kappa\lambda\rho\sigma}\ = 0 \tag{5.112d}$$

and we need in fact only $I_k^{\{\mu_i\}}$ ,k= 0,1,2 which greatly facilitates the calculation. Using (5.30-34) we find that

$$\star K(y,\varepsilon;y_\circ,0) = (2\pi\varepsilon)^{-d/2}\star G(y_\circ)^{1/2}\ \exp[\ -\tfrac{1}{2\varepsilon}\ G_{\mu\nu}(y_\circ)\Delta^{\mu}\Delta^{\nu}]\ \times$$

$$\cdot(1 - \star(G_{\mu\nu}A^\mu)\Delta^\nu + \varepsilon(\tfrac{1}{2}\star(\partial_\mu A^\mu) - \star(\tfrac{1}{2}G_{\mu\nu}A^\mu A^\nu - V)) - \varepsilon\tfrac{1}{12}\star R$$

$$+ \Delta^\mu\Delta^\nu(\tfrac{1}{2}\star A_\mu A_\nu - \tfrac{1}{2}\star(\partial_\mu A_\nu) + \tfrac{1}{12}\star R_{\mu\nu}))\Big|_{y_\circ} . \quad (5.113)$$

Formula (5.110) can be made to coincide with (5.113) by expanding $\star G(y)^{1/2} = \star G(y_\circ)^{1/2}(1 + \frac{1}{6}\star R_{\mu\nu}\Delta^\mu\Delta^\nu + O(\Delta^3))$. The fact that we find this same result by two completely different methods confirms the validity of both. One can say that the derivation based on a WKB-expansion gives more insight into the meaning of the different terms (especially those involving the curvature, see also the next chapter) and that Graham's scaling method that we have used in section 5.2. is more straightforward from the technical point of view.

From the results in normal coordinates one can obtain the result in an <u>arbitrary coordinate system</u> by replacing all starred tensors by their unstarred counterparts, replacing derivatives $\partial$ (at the origin!) by covariant derivatives $\nabla$ and by transforming $\Delta y = y - y_\circ$ into $\Delta q = q - q_\circ$ by using the inverse of (5.82) :

$$\Delta y^\mu = \Delta q^\mu + \tfrac{1}{2}\Gamma_\kappa{}^\mu{}_\lambda \Delta q^\kappa \Delta q^\lambda + \tfrac{1}{3!}[\partial_\lambda\Gamma_\nu{}^\mu{}_\kappa + \Gamma_\nu{}^\mu{}_\alpha\Gamma_\kappa{}^\alpha{}_\lambda]\cdot$$

$$\cdot\Delta q^\kappa\Delta q^\lambda\Delta q^\nu + O(\Delta q)^4). \quad (5.114)$$

One obtains

$$K^\varepsilon(q,\varepsilon;q_\circ,0) = (2\pi\varepsilon)^{-d/2}G(q)^{1/2}\exp[-\tfrac{1}{2\varepsilon}G_{\mu\nu}(q_\circ)\Delta q^\mu\Delta q^\nu] \ .$$

$$.\ [1 - G_{\mu\nu}h^\mu\Delta q^\nu + \varepsilon[-\tfrac{1}{2}G_{\mu\nu}h^\mu h^\nu + V - \tfrac{1}{12}R + \tfrac{1}{2}\nabla_\mu h^\mu]$$

$$+ [-\tfrac{1}{2}(\partial_\mu G_{\lambda\nu}h^\lambda) - \tfrac{1}{12}R_{\mu\nu}]\Delta q^\mu\Delta q^\nu +$$

$$+ [-\tfrac{1}{4}(\partial_\mu G_{\nu\kappa})]\ \Delta q^\mu\Delta q^\nu\Delta q^\kappa +$$

$$+ [-\tfrac{1}{12}\partial_\mu\partial_\nu G_{\kappa\lambda} - \tfrac{1}{24}\Gamma_{\mu\alpha\nu}\Gamma_\kappa{}^\alpha{}_\lambda]\ \Delta q^\mu\Delta q^\nu\Delta q^\kappa\Delta q^\lambda/\varepsilon +$$

$$+ \tfrac{1}{2}\{G_{\mu\nu}h^\mu\Delta q^\nu + \tfrac{1}{4}(\partial_\mu G_{\nu\kappa})\Delta q^\mu\Delta q^\nu\Delta q^\kappa/\varepsilon\}^2 + O(\varepsilon^{3/2})] . \quad (5.115)$$

The expression (5.115) gives the propagator up to order $\varepsilon$ taking into account that $\Delta q = O(\sqrt{\varepsilon})$, and this in an arbitrary coordinate system. We shall study its special properties in the next chapter.

In <u>one dimension</u> the contravariant metric was denoted by $G(q)$ , then the covariant metric and its determinant are $G(q)^{-1}$ , the vector $h^\mu$ equals $A + \frac{1}{4}G'$ and the Christoffel symbol $\Gamma_1{}^1{}_1$ equals $-\frac{1}{2}\frac{G'}{G}$. One has then

$$K^{\varepsilon}(q,\varepsilon;q_{\circ},0) = (2\pi\varepsilon G(q))^{-1/2} \exp\left[-\frac{1}{2\varepsilon}\frac{\Delta q^2}{G(q_\circ)}\right]$$

$$\cdot\Big[ 1 - \frac{1}{G}( A+ \frac{1}{4} G')\Delta q + \varepsilon\Big[-\frac{1}{2}\frac{1}{G}( A+ \frac{1}{4} G')^2 + V +$$

$$+ \frac{1}{2}\sqrt{G}\left(\frac{A + \frac{1}{4}G'}{\sqrt{G}}\right)'\Big] +$$

$$+ \Big[-\frac{1}{2}\left(\frac{A+\frac{1}{4}G'}{G}\right)'\Big]\Delta q^2$$

$$+ \frac{1}{4}\frac{G'}{G^2}\Delta q^3/\varepsilon$$

$$+ \Big[-\frac{1}{12}\left(\frac{1}{G}\right)'' + \frac{1}{96}\frac{G'^2}{G^3}\Big]\Delta q^4/\varepsilon$$

$$+ \frac{1}{2}\Big\{\frac{1}{G}( A+ \frac{1}{4} G')\Delta q - \frac{1}{4}\frac{G'}{G^2}\Delta q^3/\varepsilon\Big\}^2 + O(\varepsilon^{3/2})\Big]\Big|_{q_\circ} . \tag{5.116}$$

One can check that by developing in (5.116) $G(q)^{1/2}$ around $q_\circ$ one obtains (5.25-26) in the case of time-independent functions G,A and V.

After having finished the calculation we stress once more that without the use of normal coordinates the calculations would have been much more tedious. In an arbitrary coordinate system the quantities $B^{\mu\nu\kappa}$ , $B^{\mu\nu\kappa\lambda}$ and $B^{\mu\nu\kappa\lambda\rho\sigma}$ do not vanish and we need in the calculation by the method of section (5.2) the integrals $I_k^{\{\mu_i\}}$ up to k=6. Consequently by the integration over P powers of $\Delta$ up to order $\Delta^6$ are generated (cf. our final result (5.115)). On the other hand, the expansion of the classical action in an arbitrary coordinate system contains terms proportional to $\Delta^3/\varepsilon$ and $\Delta^4/\varepsilon$ and the expansion of the exponential of the classical action generates the term $\Delta^6/\varepsilon^2$. [CS] The utmost attainable simplification (apart from the use of normal coordinates) is in the first method the use of $\hat{U}(\hat{P}/\sqrt{\varepsilon},\hat{q};\varepsilon)$ in Weyl ordered form since $B^{\mu\nu\kappa}_{\Theta_W}$ and $B^{\mu\nu\kappa\lambda\rho\sigma}_{\Theta_W}$ vanish or correspondingly in the second method the expansion of the classical action around the midpoint since then the coefficient of the term $\sim \Delta^3/\varepsilon$ vanishes and no term in $\Delta^6/\varepsilon^2$ is generated. (The latter has been exploited by Mc Laughlin and Schulman [M1]). The most complete calculations in this context have been performed by Weiss in [W3].

Remark

Suppose that we did not know the contribution to the short time propagator of the first order correction in the WKB-expansion. Then the comparison of (5.110) and (5.113) would have been a method to determine this contribution.

But a much simpler method is the following. It is possible to relate (5.110) to one of the short time propagators associated with a simple discretization using the replacement (5.35a) (cf. an analogous method used in the phase space functional integral used in section 2.4) It is obvious that the prepoint discretization is advocated: the short time propagator corresponding to prepoint discretization is simply

$$\int \frac{dp}{(2\pi)^d} \exp\,[ip.\Delta - i\varepsilon H_{AS}(p,y_\circ)] =$$

$$= (2\pi\varepsilon)^{-d/2}\,{\star}G(y_\circ)^{1/2} \exp\{-\varepsilon[\tfrac{1}{2}{\star}G_{\mu\nu}(y_\circ)(\tfrac{\Delta^\mu}{\varepsilon} + {\star}A^\mu(y_\circ))(\tfrac{\Delta^\nu}{\varepsilon} + {\star}A^\nu(y_\circ)) - {\star}V(y_\circ)]\}$$

$$\doteq (2\pi\varepsilon)^{-d/2}{\star}G(y_\circ)^{1/2} \exp[-\tfrac{1}{2\varepsilon}{\star}G_{\mu\nu}(y_\circ)\Delta^\mu\Delta^\nu](1 - i{\star}(G_{\mu\nu}A^\mu)\Delta^\nu + \varepsilon{\star}V) \tag{5.117}$$

On the other hand expanding again ${\star}G(y)^{1/2}$ in (5.110) around the prepoint and using (5.35a) one obtains (supposing that $I_1^\varepsilon$ was still unknown)

$${\star}K(y,\varepsilon;y_\circ,0) \doteq (2\pi\varepsilon)^{-d/2}\,{\star}G(y_\circ)^{1/2} \exp[-\tfrac{1}{2\varepsilon}{\star}G_{\mu\nu}(y_\circ)\Delta^\mu\Delta^\nu]\,\cdot$$

$$\cdot(1 - i{\star}(G_{\mu\nu}A^\mu)\Delta^\nu + \varepsilon{\star}V + \tfrac{1}{12}\varepsilon{\star}R + I_1^\varepsilon) \tag{5.118}$$

which gives (5.94).

### 5.5.4. Time dependent metric

In sections 5.2. and 5.3. we allowed for an explicit time dependence in the metric and the potentials but we restricted ourselves to the one dimensional case. In section 5.4- 5.5. we studied the multidimensional case but did not consider explicit time dependence of the metric and the potentials. Here we will complete our work and study the multidimensional case with time dependent metric. Again the essentials can be explained using normal coordinates. Graham's scaling method leads in the case that $G_{\mu\nu}$, $A^\mu$ and $V$ depend on time explicitly, to the short time propagator (5.113) but with the factor $G^{1/2}$ and $G_{\mu\nu}$ in the dominating exponential evaluated at $(y_\circ,\frac{\varepsilon}{2})$ and the other functions in $(y_\circ,\bar{t})$, $\bar{t}$ arbitrary. (Compare with (5.25)).

Let us now also repeat the derivation from a WKB-expansion.

Some results have to be reformulated. We go back to the beginning of section (5.4). Let $K(q,t;q_\circ,t_\circ)$ be the propagator for the equation

$$\dot{\phi}(q,t) = [\tfrac{1}{2}c\,\partial_\mu\partial_\nu G^{\mu\nu}(q,t) + \partial_\mu A^\mu(q,t) + \tfrac{1}{c}V(q,t)]\,\phi(q,t). \tag{5.119}$$

The quantity

$$I(q,t;q_\circ,t_\circ) = \left[ \frac{G(q_\circ,t_\circ)}{G(q,t)} \right]^{1/2} K(q,t;q_\circ,t_\circ) \tag{5.120}$$

is then the propagator for the equation

$$\dot{S}(q,t) = \frac{1}{2} c\, \Delta_2 S + \nabla_\mu h^\mu S + \frac{1}{c} V S - \frac{1}{2} \left( \frac{\partial}{\partial t} \ln G \right) S. \tag{5.121}$$

Introducing $\eta$ as before, i.e. replacing c by $\eta c$ in (5.121) one obtains an equation with associated classical Lagrangian

$$L_{cl}(\dot{q}(\tau),q(\tau),\tau) = \frac{i}{c} \{ \frac{1}{2} G_{\mu\nu}(q,\tau)(\dot{q}^\mu + h^\mu(q,\tau))(\dot{q}^\nu + h^\nu(q,\tau)) - V(q,\tau) \}. \tag{5.122}$$

The WKB-approximation to $I(q,t;q_\circ,t_\circ;\eta)$ is now given by

$$I_{WKB}(q,t;q_\circ,t_\circ;\eta) = [2\pi i\eta]^{-d/2} \left[\det - \frac{\partial^2 \Lambda_{cl}}{\partial q^\mu \partial q_\circ^\nu}\right]^{1/2} \exp\left[\frac{i}{\eta} \Lambda_{cl}\right]$$

$$\cdot \exp\{ \int_{t_\circ}^{t} d\tau [ -\frac{1}{4} \dot{x}^\mu \partial_\mu \ln G(x(\tau),\tau) +$$

$$+ \frac{1}{2} \nabla_\mu h^\mu (x(\tau),\tau) - \frac{1}{2} \frac{\partial}{\partial \tau} \ln G(x(\tau),\tau)]\} \tag{5.123}$$

and up to order $\varepsilon$ the first order correction is again (5.94). (We remark that for time independent metric (5.120) and (5.123) equal (5.78) and (5.92) respectively.)

Using

$$\left[\frac{G(q,t)}{G(q_\circ,t_\circ)}\right]^{+1/2} = \exp \{ + \frac{1}{2} \int_{t_\circ}^{t} d\tau [ \dot{x}^\mu \partial_\mu \ln G(x(\tau),\tau) + \frac{\partial}{\partial\tau} \ln G(x(\tau),\tau)] \} \tag{5.124}$$

with $x(\tau)$ the classical path between $(q_\circ,t_\circ)$ and $(q,t)$ for the Lagrangian (5.122) one obtains that the short time propagator $K^\varepsilon$ is given by the expansion up to order $(t-t_\circ) = \varepsilon$ of

$$[2\pi i]^{-d/2} \left[\det - \frac{\partial^2 A_{cl}}{\partial q^\mu \partial q_\circ^\nu}\right]^{1/2} \exp[i\, A_{cl}] \exp\{ \int_{t_\circ}^{t} d\tau [\frac{1}{4} \dot{x}^\mu \partial_\mu \ln G(x(\tau),\tau)$$

$$+ \frac{1}{2} \nabla_\mu h^\mu (x(\tau),\tau) \}(1 - \frac{1}{12} \varepsilon c R)]\} \tag{5.125}$$

We take now $t_\circ=0$, $t = \varepsilon$, introduce normal coordinates belonging to the Lagrangian

$$L_\circ(\dot{q},q) = \frac{1}{2} G_{\mu\nu}(q,0)\dot{q}^\mu \dot{q}^\nu \tag{5.126}$$

and with origin $q_\circ$ and consider the expansion of (5.125) in these coordinates. To find the expansion of the classical action we develop $\star G_{\mu\nu}(y(\tau),\tau) = \star G_{\mu\nu}(y(\tau),0) + \tau \frac{\partial}{\partial\tau} {}^\star G_{\mu\nu}(y,\tau)\Big|_{\tau=0} + \mathcal{O}(\tau^2)$ and write $\star L_{cl} = \frac{i}{c}[\star L_\circ + \frac{1}{4}\tau\frac{\partial}{\partial\tau}\star G^{\mu\nu}(y,\tau)\big|_{\tau=0}\dot{y}^\mu\dot{y}^\nu + \star L_1 + \mathcal{O}(\tau^2)]$ with $\star L_\circ$ given by (5.126). The contribution up to $\mathcal{O}(\varepsilon)$ of $\star L_\circ$ and $\star L_1$ is as before, the new term gives $\frac{1}{4}\frac{\partial}{\partial\tau}{}^\star G_{\mu\nu}(y_\circ,\tau)\big|_{\tau=0}\Delta^\mu\Delta^\nu$. Combining this with the $\frac{1}{2\varepsilon}\star G_{\mu\nu}(y_\circ,0)\Delta^\mu\Delta^\nu$ coming from $\star L_\circ$ gives then

$$\star A_{cl} = \frac{i}{c}\,\frac{1}{2\varepsilon}\star G_{\mu\nu}(y_\circ,\tfrac{1}{2}\,\varepsilon)\Delta^\mu\Delta^\nu + \star A^1_{c\ell} + \mathcal{O}(\varepsilon^{3/2}) \tag{5.127}$$

and

$$\det\left[-\frac{\partial^2 A_{cl}}{\partial y^\mu \partial y^\nu_\circ}\right]^{1/2} = \left(\frac{c\varepsilon}{i}\right)^{-d/2} \star G(y_\circ,\tfrac{1}{2}\,\varepsilon)^{1/2}(1+\mathcal{O}(\varepsilon^{3/2})). \tag{5.128}$$

The approximation up to order $\varepsilon$ of the second exponential in (5.125) is

$$1 + \frac{1}{12}\star R_{\mu\nu}\Delta^\mu\Delta^\nu + \varepsilon\star(\partial_\mu A^\mu)$$ as can be checked computing it as $A^1_{c\ell}$ in (5.100) and using (5.87)

Putting all approximations together we find again (5.113) with the factor $\star G^{1/2}$ and $\star G_{\mu\nu}$ in the exponential evaluated at $(y_\circ, \frac{\varepsilon}{2})$. We see that the derivation of the propagator in the limit $\Delta = \mathcal{O}(\sqrt{\varepsilon})$ up to order $\varepsilon$ from a WKB-expansion is in complete agreement with the result obtained by the operator method of section 5.2.

## Appendix 5.1 : Alternative proof of (5.62)

The proof of the validity of the step $\langle q'|U(\varepsilon,0)|q\rangle \approx \langle q'|1-i\varepsilon\hat{H}(\hat{p},\hat{q},\bar{t})|q\rangle$ given in section 5.2 uses the relations in terms of $\doteq$ (equal as far as use in the functional integral is concerned). Our proof of these relations given in the text uses however the formalism of chapter 3 which is based on the approximation $\hat{U}=1-i\varepsilon\hat{H}$ ! To complete our arguments it is therefore necessary to stress that the relations in terms of $\doteq$ can be proved independently of the approximation $\hat{U}=1-i\varepsilon\hat{H}$. (See the works of Mayes and Dowker [M10] and Mc Laughlin and Schulman [M1]).

To illustrate this we shall give here a proof of the relation

$$\Delta_j^\kappa\Delta_j^\lambda\Delta_j^\rho/\varepsilon \doteq c\, g^{\kappa\lambda}_{[\gamma]}(q_j,q_{j-1})\Delta_j^\rho+g^{\kappa\rho}_{[\gamma]}(q_j,q_{j-1})\Delta_j^\lambda+g^{\rho\lambda}_{[\gamma]}(q_j,q_{j-1})\Delta_j^\kappa \qquad \text{(A5.1)}$$

following the methods of [M6] and [M1].

We consider the integral

$$\int dq'(2\pi c\varepsilon)^{-d/2}g(q',q)^{1/2}\exp\{-\frac{1}{2c\varepsilon}g_{\mu\nu}(q',q)\Delta^\mu\Delta^\nu\}\,\frac{1}{\varepsilon}\Delta^\kappa\Delta^\lambda\Delta^\rho\, f(q',q) \qquad \text{(A5.2)}$$

representing one of the many integrations in the functional integral ($q'$-dependence of other short time propagators is contained in $f(q',q)$ which is not singular at $q'=q$). Developing $q'$-dependent functions around $q$ one obtains

$$\int dq'(2\pi c\varepsilon)^{-d/2}g(q,q)^{1/2}\exp\{-\frac{1}{2c\varepsilon}g_{\mu\nu}(q,q)\Delta^\mu\Delta^\nu\}[\frac{1}{\varepsilon}\Delta^\kappa\Delta^\lambda\Delta^\rho f(q,q)\,.$$
$$(1+\frac{1}{2}\Delta^\mu\frac{\partial'_\mu g}{g}\Big|_{q'=q}-\frac{1}{2\varepsilon}\Delta^\mu\Delta^\nu\Delta^\sigma\partial'_\sigma g_{\mu\nu}\Big|_{q'=q}-\Delta^\sigma\frac{\partial'_\sigma f}{f}\Big|_{q'=q})+\mathcal{O}(\varepsilon^{3/2})]. \qquad \text{(A5.3)}$$

We use that the integrand is only needed up to $\mathcal{O}(\varepsilon)=\mathcal{O}(\Delta^2)$. In (A5.3) only appear simple moments of a Gaussian integral and it equals

$$\int dq'(2\pi c\varepsilon)^{-d/2}g(q,q)^{1/2}\exp\{-\frac{1}{2c\varepsilon}g_{\mu\nu}(q,q)\Delta^\mu\Delta^\nu\}f(q,q)[c(\Delta^\kappa g^{\lambda\rho}(q,q)+\Delta^\lambda g^{\kappa\rho}(q,q)$$
$$+\Delta^\rho g^{\kappa\lambda}(q,q))(1+\frac{1}{2}\Delta^\mu\frac{\partial'_\mu g}{g}\Big|_{q'=q}-\frac{1}{2\varepsilon}\Delta^\mu\Delta^\nu\Delta^\sigma\,\partial'_\sigma g_{\mu\nu}\Big|_{q'=q}-\Delta^\sigma\frac{\partial'_\sigma f}{f}\Big|_{q'=q})$$
$$-\frac{1}{2}\varepsilon c^3(g^{\kappa\mu}g^{\lambda\nu}g^{\rho\sigma}+g^{\kappa\nu}g^{\lambda\sigma}g^{\rho\mu}+g^{\kappa\sigma}g^{\lambda\mu}g^{\rho\nu}+g^{\kappa\nu}g^{\lambda\mu}g^{\rho\sigma}+g^{\kappa\mu}g^{\lambda\sigma}g^{\rho\nu}+g^{\kappa\sigma}g^{\lambda\nu}g^{\rho\mu})\partial'_\sigma g_{\mu\nu}\Big|_{q'=q}$$
$$+\mathcal{O}(\varepsilon^{3/2})] \qquad \text{(A5.4)}$$

The last six terms arise from the pairing of the indices of $(\kappa,\lambda,\rho)$ and $(\mu,\nu,\sigma)$. They cancell against the terms that arise when we develop in $(\Delta^\kappa g^{\lambda\rho}(q,q)+\Delta^\lambda g^{\kappa\rho}(q,q)+\Delta^\rho g^{\kappa\lambda}(q,q))$ the factor $g^{\lambda\rho}(q,q)$ around $g^{\lambda\rho}(q',q)$ etc. We obtain then

$$\int dq'(2\pi c\varepsilon)^{-d/2}g(q,q)^{1/2}\exp\{-\frac{1}{2c\varepsilon}g_{\mu\nu}(q,q)\Delta^\mu\Delta^\mu\}f(q,q)[c(\Delta^\kappa g^{\lambda\rho}(q',q)+\Delta^\lambda g^{\kappa\rho}(q',q)$$
$$+\Delta^\rho g^{\kappa\lambda}(q',q))(1+\frac{1}{2}\Delta^\mu\frac{\partial'_\sigma g}{g}\Big|_{q'=q}-\frac{1}{2\varepsilon}\Delta^\mu\Delta^\nu\Delta^\sigma\partial'_\sigma g_{\mu\nu}\Big|_{q'=q}-\Delta^\sigma\frac{\partial'_\sigma f}{f}\Big|_{q'=q})+\mathcal{O}(\varepsilon^{3/2})]$$
$$=\int dq'(2\pi c\varepsilon)^{-d/2}g(q',q)^{1/2}\exp\{-\frac{1}{2\varepsilon}g_{\mu\nu}(q',q)\Delta^\mu\Delta^\nu\}f(q',q)\,.$$
$$[c(\Delta^\kappa g^{\lambda\rho}(q',q)+\Delta^\lambda g^{\kappa\rho}(q',q)+\Delta^\rho g^{\kappa\lambda}(q',q))+\mathcal{O}(\varepsilon^{3/2})] \qquad \text{(A5.5)}$$

which proofs (A5.1). We see here also that $\doteq$ means that when integration is performed with the Gaussian weight factor the results differ only by terms of $\mathcal{O}(\varepsilon^{3/2})$ that do not contribute in the functional integral.

## CHAPTER VI : COVARIANT DEFINITIONS OF FUNCTIONAL INTEGRALS

### 6.1. Introduction

Nonlinear point transformations play an important role in the applications of functional integrals in various branches of modern physics and many papers have been addressed to the subject. The first observation that there are problems with changes of variables in the path integral was made by Edwards and Gulyaev [E1] when studying the change from cartesian to polar coordinates in the quantum-mechanical free particle propagator. More recently Gervais and Jevicki [G9] have investigated the problem for configuration space functional integrals in the mid-point discretization. Their conclusion was that a formal change of variables in the continuum expression of the functional integral leads to erroneous results. A correct change of variables can however be done in the discretized expression taking into account the stochastic properties of the paths i.e. $\Delta q \sim \sqrt{\varepsilon}$ . This is shown in appendix 6.1.

That a formal change of variables in the functional integral defined in terms of a correspondence rule (e.g. Weyl's correspondence in [G9]) is related to the following well-known result from the theory of quantization: the relation between an operator theory and a c-number theory cannot be given by one and the same correspondence rule before and after a nonlinear canonical transformation is made. Translated into path integral language this means: the discretization associated with a correspondence rule cannot be the same before and after a formal change of variables in the path integral. There are at least two possible attitudes towards this fact. Either one insists on using the same correspondence rule, then the transformation of the integrand in the continuum expression of the functional integral is not according to the usual rules of calculus. Or one tries to find a discretization which is compatible with formal change of variables which is what we shall do in section 6.4.

Another approach (not based on correspondence rules) to functional integrals in arbitrary coordinates dates back to the works of Pauli [P3] and De Witt [D12] that we have already quoted in the previous chapter. It consists in defining the configuration space functional integral by iteration of a short time propagator which is the WKB-approximation. This was in fact a generalization of Feynman's postulate [F3] that the short time propagator is proportional to the exponential of the classical action. In section 6.2. we study these definitions and show how they lead to covariant path integral representations. Very useful is the knowledge of the special short time propagator that we have computed in the previous chapter.

A covariant definition of path integrals based directly on the properties of this special short time propagator was given by Deininghaus and Graham [D ] . In section 6.3. we investigate in detail their method and generalise it to the case of our equation 5.3. i.e. including explicit time-dependence
In section 6.4. we elaborate on covariant discretization procedures and section 6.5 is devoted to Dekker's covariant definition using Fourier series.

## 6.2. Feynman's definitionand its generalizations

It was postulated by Feynman[F3] on the basis of Dirac's [D19] stationary phase argument that in quantum mechanics the short time propagator $K(q_j,t_j;q_{j-1},t_{j-1})$ is proportional to $\exp[i/\hbar\, A_{cl}]$ where $A_{cl}$ is the classical action evaluated at the classical path between $(q_{j-1},t_{j-1})$ and $(q_j,t_j)$. In section 5.5.1. we have already mention that this postulate has been refined to the statement that the short time propagator is a WKB-approximation and in section 5.5. we have shown that if one takes for classical Lagrangian $L_{cl}$, given by (5.91) , then one has to take into account also the first order correction to the WKB-approximation It is however perfectly possible to postulate following Feynman that the short time propagator is of the form $N \exp[i\, A^F_{cl}]$ where $A^F_{cl}$ is now the classical action for a Lagrangian $L^F$ which has to be determined and $N$ is a normalization factor. This short time propagator should be related by the relations $\doteq$ to the short time propagator obtained (in the previous chapter) by Graham's scaling method and from a WKB-expansion and also by the same relations to any of the other short time propagators used in the first four chapters. This provides us with an easy method to determine $L^F$ , the Feynman Lagrangian.

Let us first consider the case of a metric with vanishing curvature. From the results of section 5.5.4. one has that a good short time propagator $K(q,\varepsilon;q_\circ,0)$ is in this case the expression

$$K(q,\varepsilon;q_\circ,0) = [2\pi c\eta\varepsilon]^{-d/2}\ [G(q,\tfrac{1}{2}\varepsilon)]^{1/2}\exp[\varepsilon\ \tfrac{1}{2}\nabla_\mu h^\mu + \tfrac{i}{\eta}\ A_{cl}]\,\Big|_{\eta=1} \tag{6.1}$$

with $A_{cl}$ the classical action for the Lagrangian

$$L_{cl}\ (\dot{q},q,t) = \frac{i}{c}[(\tfrac{1}{2}\ G_{\mu\nu}(\dot{q}^\mu + h^\mu\ )\ (\dot{q}^\nu + h^\nu) - V]. \tag{6.2}$$

The expansion up to $\mathcal{O}(\varepsilon)$ of (6.1) is precisely the same as the expansion up to $\mathcal{O}(\varepsilon)$ of

$$K^F(q,\varepsilon;q_\circ,0) = [2\pi c\eta\varepsilon]^{-d/2}\ [G(q,\ \tfrac{1}{2}\ \varepsilon)]^{1/2}\ \exp[\tfrac{i}{\eta}\ A^F_{cl}]\Big|_{\eta=1} \tag{6.3}$$

where $A^F_{cl}$ is the classical action for the Feynman Lagrangian

$$L^F(\dot{q},q) = \frac{i}{c}\,[\frac{1}{2}\,G_{\mu\nu}(\dot{q}^\mu + h^\mu)\,(\dot{q}^\nu + h^\nu\,) - V - \eta c\,\frac{1}{2}\,\nabla_\mu h^\mu]\ . \tag{6.4}$$

Indeed, remember that the contribution of potential terms to the expansion of the classical action is independent of the path considered and given by (5.100) and that the contribution of the kinetic term is independent of the potential terms. Therefore one has $\frac{i}{\eta}\,A^F_{cl} = \varepsilon\,\frac{1}{2}\,\nabla_\mu h^\mu + \frac{i}{\eta}\,A_{cl} + O(\varepsilon^{3/2})$ and it also follows that in a flat space Feynman's propagator gives when expanded up to order $\varepsilon$ precisely the propagator obtained by Graham's scaling method (I.e. without using the relation $\doteq$ ) The Lagrangian determined by the postulate is (6.4), the normalization factor $N$ is $[2\pi c\varepsilon]^{-d/2}\ [G(q,\,\frac{1}{2}\varepsilon)\,]^{\,1/2}$.

Now we turn to the general case when $G^{\mu\nu}$ defines a curved space. A priori one has the choice to include the effects of curvature either in the normalization $N$ or in the Lagrangian $L^F(\dot{q},q,t)$.[U1,P2]
Following most authors[C5,D25,H9] we include them in the Lagrangian and put forward

$$K^F(q,\varepsilon;q_\circ,0) = [2\pi c\eta\varepsilon]^{-d/2}\ G(q,\,\frac{1}{2}\varepsilon\,)^{\,1/2}\ \exp\,[\frac{i}{\eta}\,A^F_{cl}]\,|_{\eta=1}\ . \tag{6.5}$$

From the short time propagator of section 5.5.4. $K^F$ can be obtained by interpreting in (5.125) both the factor

$$\exp\{\int d\tau\,[\frac{1}{4}\,\dot{x}^\mu\partial_\mu\ \ell n\ G(x(\tau),\tau)\ +\ \frac{1}{2}\ \ \nabla_\mu h^\mu\ (x(\tau),\tau)\,]\,\}\ \simeq$$

$$\simeq(1+\ \varepsilon\,\frac{1}{2}\,\nabla_\mu h^\mu - \frac{1}{12}\,R_{\mu\nu}\ \Delta^\mu\Delta^\nu)\ \doteq\ (1\,+\,\varepsilon(\,\frac{1}{2}\ \ \nabla_\mu h^\mu\ \ -\frac{1}{12}\,\eta c\ R))$$

coming from the normalization of the WKB-approximation and the contribution $(1 - \frac{1}{12}\varepsilon\eta\ cR)$ coming from the first order correction as terms coming from a potential. One finds

$$L^F(\dot{q},q,t) = \frac{i}{c}\,[\,\frac{1}{2}\,G_{\mu\nu}(\dot{q}^\mu + h^\mu)(\dot{q}^\nu + h^\nu)\ -\ V\ -\eta c\,\frac{1}{2}\,\nabla_\mu h^\mu\ +\,\frac{1}{6}\,(\eta c)^2 R\,]. \tag{6.6}$$

The short time propagator (6.5) leads to the functional integral representation

$$K(q,t;q_\circ,t_\circ) = \lim_{N\to\infty}\ \int\prod_{i=1}^{N} dq_i\ \prod_{j=1}^{N+1}\{\,[\ \frac{G(q_j,\,\frac{1}{2}(t_{j-1}+t_i)^{1/2}}{(2\pi c\varepsilon)^d}\,]\,\exp[\,i\,A^F_{cl}(q_{j-1},t_j,q_{j-1},t_{j-1})]\} \tag{6.7}$$

with $A^F_{cl}(q_j,t_j;q_{j-1},t_{j-1})$ the classical action for the Lagrangian $L^F$ (with $\eta = 1$)

evaluated at the trajectory $q(\tau)$ that makes this action stationary for $q(t_{j-1}) = q_{j-1}$ , $q(t_j) = q_j$.

Formally (5.114) is denoted by

$$K(q,t;q_0,t_0) = \int_F \mathcal{D}q \left(\frac{\sqrt{G}}{c^{d/2}}\right)\exp\left[ i \int_{t_0}^{t} d\tau\, L^F(\dot{q},q,\tau)\right] \delta(q(t)-q)\ \delta(q(t_0)-q_0). \tag{6.8}$$

We note that $L^F$ is a scalar and that this definition of the functional integral is formally covariant in the sense that if a change of variables $q(\bar{q})$ is made then the new path integral representation of the propagator can be obtained from the old one by formally changing variables in (6.8) by the usual rules of calculus.

In a curved space the expansion up to order $\varepsilon$ of (6.8) does not coincide with Graham's short time propagator but is related to it by the relation (5.35a) ($\doteq$). Therefore we do not expect the Lagrangian $L^F$ to be of particular importance. It is simply the Lagrangian such that iteration of $K^F$ gives in the limit the propagator of a given differential equation. Why this Lagrangian is different from the classical Lagrangian $L_{cl}$ which can be associated to the differential equation is clear in our approach: it exhibits the contributions from the complete WKB-approximation and the first order correction to it.

As in chapter 2 one can put the inverse problem and ask for the differential equation satisfied by the functional integral with short time propagator $[2\pi c\eta\varepsilon]^{-d/2}[G(q_j, \frac{1}{2}(t_{j-1} + t_j))]^{1/2}$ $\exp[\frac{i}{\eta} A]$ where $A$ is now the classical action for a given Lagrangian L. Especially one can choose $L_{cl}$ for L and find that the propagator satisfies (5.88) with V replaced by $V + \eta c\, \frac{1}{2}\nabla_\mu h^\mu + \frac{1}{6}(\eta c)^2 R.$ [C5]

Above we have determined $L^F$ by relating (6.5) to the short time propagator obtained by Graham's scaling method. This was also done by Weiss in [W3] . But there are much shorter ways to find $L^F$. One can compute the differential equation satisfied by the path integral [C5] , this calculation is short and elegant when normal coordinates are used.[H9,D25,R1]

Using the relations (5.35) one can also relate the expansion of (6.5) to a short time propagator in a simple discretization, this method was used by us in [L13] (See also [H4]). Also then the use of normal coordinates would still simplify the calculations.

## 6.3. Graham's covariant interpretation of path integrals

In Ref. [G15] Graham derived a new path integral representation for the fundamental solution of a general Fokker-Planck equation. The motivation for his calculation stemmed from the need for a generalization of the so called Onsager-Machlup Lagrangian for general nonlinear diffusion processes. We will come back to this in chapter 7. Here we focus our attention to the definition of the functional integral, generalizing Graham's method to the case of our equation 5.3. (i.e. including explicit time dependence). In [G15] a covariant path integral representation was derived by disregarding the fact that in the path integral the paths that contribute most are the nondifferentiable ones with $\Delta q$ of order $\sqrt{\varepsilon}$. The question how such a path integral has to be interpreted in doing actual calculations has been answered by Deininghaus and Graham in [D2] (A short earlier communication is [G17]).

In this section we explain the methods of these papers applying them to our equation 5.3. The relation to the definitions that we have given before is made clear.

### 6.3.1. Derivation of covariant path integral

In section 5.2. we have derived the expansion of $K(q_j,t_j;q_{j-1},t_{j-1})$ following Graham in [G1]. There we have proved that owing to the stochastic nature of the paths $(\Delta = \mathcal{O}(\sqrt{\varepsilon}))$ this expansion can effectively be replaced in the functional integral by the result obtained using the approximation $\hat{U} = 1 - i\varepsilon\hat{H}$ for the short time evolution operator. Graham however puts emphasis on obtaining in the continuum expression of the functional integral a Lagrangian that is a scalar. This can be achieved by looking to what expression the short time propagator (5.25) (5.115) reduces for a differentiable path, which is a path such that $\Delta/\varepsilon \underset{\varepsilon\to 0}{\to} \dot{q}$ is a finite quantity:

$$K^{\varepsilon}(q,\varepsilon;q_{\circ},0)\Big|_{\substack{\Delta/\varepsilon=\dot{q}\ \text{finite}\\ \varepsilon\to 0}} = (2\pi\varepsilon)^{-\frac{d}{2}} G(q,\tfrac{\varepsilon}{2})^{+\frac{1}{2}} \exp[-\varepsilon\tfrac{1}{2}G_{\mu\nu}(q_{\circ},\tfrac{\varepsilon}{2})\dot{q}^{\mu}\dot{q}^{\nu}\,.$$

$$(1-\varepsilon G_{\mu\nu}h^{\mu}\dot{q}^{\nu}+\varepsilon[-\tfrac{1}{2}G_{\mu\nu}h^{\mu}h^{\nu}+V+\tfrac{1}{12}R+\tfrac{1}{2}\nabla_{\mu}h^{\mu}]+\mathcal{O}(\varepsilon^{2}))\Big|_{q_{\circ}}$$

$$= (2\pi\varepsilon)^{-\frac{d}{2}} G(q,\tfrac{\varepsilon}{2})^{+\frac{1}{2}} \exp\{-\varepsilon[\tfrac{1}{2}G_{\mu\nu}(\dot{q}^{\mu}+h^{\mu})(\dot{q}^{\nu}+h^{\nu})-V-\tfrac{1}{2}\nabla_{\mu}h^{\mu}+\tfrac{1}{12}R]\}\Big|_{q_{o},\frac{\varepsilon}{2}} (1+\mathcal{O}(\varepsilon^{2}))\,. \tag{6.9}$$

This limit of the short time propagator is used to introduce the formally covariant configuration space path integral

$$K(q,t;q_o,t_o) = \int_G \mathcal{D}(\sqrt{G}\,q)\ \exp\left[-\int_{t_o}^{t} d\tau\ L^G\,(\dot q,q,\tau)\delta(q(t)-q)\quad \delta(q(t_o)-q_o)\right. \tag{6.10}$$

with

$$L^G(\dot q,q) = \frac{1}{2}\,G_{\mu\nu}(\dot q^\mu+h^\mu)(\dot q^\nu+h^\nu) - V - \frac{1}{2}\nabla_\mu h^\mu + \frac{1}{12}R \tag{6.11}$$

and formally

$$\mathcal{D}(\sqrt{G}\ q) = \lim_{N\to\infty}\ \frac{G(Q,t-\frac{\varepsilon}{2})^{1/2}}{(2\pi\varepsilon)^{d/2}}\ \prod_{i=1}^{N} dq_i\frac{G(q_i,t_i-\frac{\varepsilon}{2})^{1/2}}{(2\pi\varepsilon)^{d/2}}\ . \tag{6.12}$$

(Notice the appearance of the invariant volume element $dq\,\sqrt{G}$ .
The shortest way to obtain expression (6.9) is to do the scaling $p=P/\sqrt{\varepsilon}$ as in section 5.2., to do the integration over P-variables using (5.30)-(5.34) but keeping only the term with $\ell = [\frac{k}{2}]$ . This approximation consists precisely in dropping the terms proportional to $\Delta^n, n \geq 2$ which vanish in the limit $\varepsilon \to 0$, $\frac{\Delta}{\varepsilon}$ finite. This is the procedure followed by Graham in [G15].

It should be clear that the terms that are dropped have to be recuperated when doing actual calculations since they contribute when the functional integral is computed as the limit for $N\to\infty$ of the multidimensional integral. This is what we study now.

### 6.3.2. Interpretation of covariant path integral

From the calculations that led us to (6.10) one knows that it is in fact just a convenient (because formally covariant) short hand notation for the limit

$$K(q,t;q_o,t_o) = \lim_{N\to\infty} \int \prod_{i=1}^{N} dq_i \prod_{j=1}^{N+1} K^\varepsilon\,(q_j,t_j;q_{j-1},t_{j-1}) \tag{6.13}$$

with $K^\varepsilon(q_j,t_j;q_{j-1},t_{j-1})$ the short time propagator (5.115). From the notation (6.10-6.12) we only know the "rough approximation" [D2] (6.9) and (suppose we don't know $K^\varepsilon$ yet) in order to have a complete definition (6.10-6.12) has to be supplemented by a method to obtain $K^\varepsilon$ from the knowledge of $L^G$ . In other words we have to determine the function $Z(q_{j-1},\Delta_j,\varepsilon)$ such that

$$K^\varepsilon(q_j,t_j;q_{j-1},t_{j-1}) = (2\pi\varepsilon)^{-d/2}G(q_j,t_j-\tfrac{\varepsilon}{2})^{\frac{1}{2}}\exp[-\,L^G(\tfrac{\Delta_j}{\varepsilon},q_{j-1},t_j-\tfrac{\varepsilon}{2})]Z(q_{j-1},\Delta_j,\varepsilon). \tag{6.14}$$

In Ref. [D2] it is shown that Z must satisfy the conditions

$$Z(q,\varepsilon\dot{q},\varepsilon)\Big|_{\varepsilon\to 0} = 1 \tag{6.15a}$$

$$\frac{1}{\varepsilon}[Z(q,\varepsilon\dot{q},\varepsilon) - 1]\Big|_{\varepsilon\to 0} = 0 \tag{6.15b}$$

$$Z(q,\Delta,\varepsilon)\Big|_{\varepsilon\to 0,\Delta^2/\varepsilon\ \text{constant}} = 1. \tag{6.15c}$$

The first two follow from (6.9), the third one from the fact that in the limit $\varepsilon \to 0$, $\Delta^2/\varepsilon$ constant $K^\varepsilon$ goes to $(2\pi\varepsilon)^{-d/2}G(q_{j-1},t_j-\frac{\varepsilon}{2})^{1/2}\exp[-\frac{1}{2\varepsilon}G_{\mu\nu}(q_{j-1},t_j-\frac{\varepsilon}{2})\Delta^\mu\Delta^\nu]$.

Taking into account (6.15) one has that Z is of the form (in an obvious symbolic notation)

$$Z(q,\Delta,\varepsilon) = 1 + \sum_{k=1}^{\infty} a_k(q,\bar{t})\Delta^{2k}/\varepsilon^{k-1} + \sum_{k=1}^{\infty} b_k(q,\bar{t})\,\Delta^{2k+1}/\varepsilon^k\ . \tag{6.16}$$

The series starts with 1 because of (6.15a) ; terms proportional to $\varepsilon$ and $\Delta$ are missing because of (6.15a,b). Equation (6.15a) rules out all terms $\sim\Delta^k/\varepsilon^\ell$ for $k \leq 2\ell$ . Terms $\sim \Delta^k/\varepsilon^\ell$ for $k > 2\ell + 2$ are at least or order $\varepsilon^{3/2}$ when $\Delta = O(\sqrt{\varepsilon})$ is taken into account and are also absent in $K^\varepsilon$ and Z.
In order to determine the unknown coefficient functions $a_k(q)$ and $b_k(q)$ we impose that K as given by (6.14) satisfies

$$K^{2\varepsilon}(q_{j+1},t_{j+1};q_{j-1},t_{j-1}) = \int dq_j\ K^\varepsilon\ (q_{j+1},t_{j+1},q_j,t_j)K^\varepsilon(q_j,t_j;q_{j-1},t_{j-1}). \tag{6.17}$$

Relation (6.17) is just the basic semi-group law (1.4) and the determination of the coefficients $a_k$ and $b_k$ can be interpreted [D3] as the determination of a fixed point of the nonlinear transformation
$\phi^\varepsilon(\{a_k,b_k\}) \to \phi^{2\varepsilon}(\{\bar{a}_k,\bar{b}_k\})$

$$\phi^{2\varepsilon}(q_{j+1},t_{j+1};q_{j-1},t_{j-1}) = \int dq_j\phi^\varepsilon(q_{j+1},t_{j+1};q_j,t_j)\phi^\varepsilon(q_j,t_j;q_{j-1},t_{j-1})\ . \tag{6.18}$$

Let us denote by $\phi^{\varepsilon\star}$ the function $\phi^\varepsilon$ evaluated at the fixed point $\bar{a}_k = a_k$, $\bar{b}_k = b_k$ . We want to investigate whether $\phi^{\varepsilon\star} \equiv K^\varepsilon$ and whether the fixed point is unique.

The equality sign in (6.17) is important. Any of the approximations to the short time propagator related to it in the sense $\doteq$ will satisfy (6.17) only in the sense $\doteq$ .

In [D2] Deininghaus and Graham determined the fixed point value of the coefficients first in the example of a free particle in polar coordinates and second, using normal coordinates, for the case of equation (5.66), i.e. time independent metric and potential $h^{\nu}$ (and no potential V). Here we present the calculation in the one dimensional case and for time-dependent metric and potentials.

We want to construct from the path integral representation (6.10-6.12) the short time propagator $K^{\varepsilon}$ by identifying it with the fixed point of (6.18) for

$$\phi^{\varepsilon}(q_j,t_j;q_{j-1},t_{j-1}) = (2\pi\varepsilon)^{-1/2}G(q_j,t_j-\tfrac{\varepsilon}{2})^{-1/2}\exp[-\varepsilon L^{G}(\tfrac{\Delta_j}{\varepsilon},\, q_{j-1}^{(\alpha)},t_{j-1}^{(\mu)})]. \cdot Z\,(q_{j-1}^{(\alpha)},\, \Delta_j,\varepsilon\,) \tag{6.19}$$

with $Z(q,\Delta,\varepsilon)$ of the form (6.16) (Remember again that in one dimension we denote by G the contravariant metric). Several remarks are in order here. The term $\varepsilon L^{G}(\frac{\Delta_j}{\varepsilon}, q_{j-1}^{(\alpha)}, t_{j-1}^{(\mu)})$ can be interpreted as a discrete approximation to $\int_{t_{j-1}}^{t_j} d\tau\;\; L^{G}(\dot q(\tau), q(\tau), \tau)$. It can be shown by explicit calculation that the method of imposing (6.18) works irrespective of the points

$$q_{j-1}^{(\alpha)} = (1-\alpha)q_{j-1} + \alpha q_j \quad , \quad t_{j-1}^{(u)} = (1-u)\, t_{j-1} + u t_j \tag{6.20}$$

in which the q and t dependence of $L^{G}$ is made discrete. For every $\alpha$ and u the calculation of Z leads to a different value of this quantity in such a way that $\phi^{\varepsilon}$ remains the same. It is in this sense that the path integral (6.10) is independent of the discretization. We shall put here $\alpha = 0$ and keep u arbitrary. On the other hand the prefactor $1/\sqrt{G}$ has to be taken in the point $(q_j,\frac{1}{2}(t_{j-1}+t_j))$ (See (6.9)), if not the conditions (6.15a,b) or the Lagrangian have to be modified.

The coefficients $a_k$ and $b_k$ in (6.16) depend also on time $\bar t$ but one remarks immediately that here $\bar t$ can be taken anywhere in the interval of length $\varepsilon$ , the difference in Z for different choices being at least of order $\varepsilon^{3/2}$.

The integrand in (6.18) is

$$(2\pi\varepsilon)^{-\frac{1}{2}}[G(q_{j+1},t_j^{(1/2)})G(q_j,t_{j-1}^{(1/2)})]^{-\frac{1}{2}} \exp[-\varepsilon L^G(\frac{\Delta_{j+1}}{\varepsilon},q_j,t_j^{(u)})-$$

$$-\varepsilon L^G(\frac{\Delta_j}{\varepsilon},q_{j-1},t_{j-1}^{(u)}]Z(q_j,\Delta_{j+1},\varepsilon)\,Z(q_{j-1},\Delta_j,\varepsilon) \tag{6.21a}$$

with

$$L^G(\dot q,q,\tau)=\frac{1}{2G}\dot q^2+\frac{(A+\frac{1}{4}G')}{G}\dot q+[\frac{1}{2G}(A+\frac{1}{4}G')^2-V-\frac{1}{2}\sqrt{G}\ (\frac{A+\frac{1}{4}G'}{\sqrt{G}})']$$

$$=\frac{1}{2G(q,\tau)}\dot q^2+C(q,\tau)\,\dot q+D(q,\tau). \tag{6.21b}$$

In order to perform up to the required order the $q_j$- integration in (6.18) the $q_j$-dependent functions in (6.21) are developed around $q_{j-1}$.
We also develop the functions depending on $t_{j-1}^{(u)}$ and $t_j^{(u)}$ around $\bar t^{(u)} \equiv \bar t$ given by

$$\bar t=(1-u)t_{j-1}+u\,t_{j+1}=\bar t_j+u\varepsilon\quad=\bar t_{j+1}-(1-u)\varepsilon\quad. \tag{6.22}$$

and $G(q_j,t_{j-1}^{(1/2)})$ around $(q_{j-1},\bar t)$, $G(q_{j+1},t_j^{(1/2)})$ around $(q_{j+1},\bar t_j)$
In this way one obtains for (6.21)

$$(2\pi\varepsilon)^{-1}\ [G(q_{j+1},\bar t_j)G(q_{j-1},\bar t)]^{-1/2}\exp[-\frac{1}{2\varepsilon G(q_{j-1},\bar t)}(\Delta_{j+1}^2+\Delta_j^2)]\ \cdot$$

$$\cdot(1+\frac{1}{4}\varepsilon\frac{\partial_t G}{G})(1-\frac{1}{2}\Delta_j\frac{G'}{G}-\frac{1}{4}\Delta_j^2\frac{G''}{G}+\frac{3}{8}\Delta_j^2\frac{G'^2}{G^2}+\frac{1}{2}(2u-\frac{1}{2})\varepsilon\frac{\partial_t G}{G})\cdot$$

$$\cdot\exp[-\frac{1}{2\varepsilon}\Delta_{j+1}^2(-\Delta_j\frac{G'}{G^2}-\frac{1}{2}\Delta_j^2\frac{G''}{G^2}+\Delta_j^2\frac{G'^2}{G^3}-(1-u)\varepsilon\frac{\partial_t G}{G})+\frac{1}{2\varepsilon}\Delta_j^2\,u\,\varepsilon\frac{\partial_t G}{G}$$

$$-(\Delta_{j+1}+\Delta_j)C(q_{j-1},\bar t)-\Delta_{j+1}\Delta_j\ C'(q_{j-1},\bar t)-2\varepsilon D(q_{j-1},\bar t)]\cdot$$

$$\cdot(1+\sum_{k=1}^{\infty}(a_k(q_{j-1})\frac{\Delta_{j+1}^{2k}}{\varepsilon^{k-1}}+b_k(q_{j-1})\frac{\Delta_{j+1}^{2k+1}}{\varepsilon^k}+b_k'(q_{j-1})\frac{\Delta_j\,\Delta_{j+1}^{2k+1}}{\varepsilon^k}))\quad\cdot$$

$$\cdot(1+\sum_{k=1}^{\infty}(a_k(q_{j-1})\frac{\Delta_j^{2k}}{\varepsilon^{k-1}}+b_k(q_{j-1})\frac{\Delta_j^{2k+1}}{\varepsilon^k}))\ . \tag{6.23}$$

Next the second exponential in (6.23) is developed and the products are worked out up to the required order. Then we can define $\Delta$ and $\nabla$ by

$$\Delta\equiv\Delta_{j+1}+\Delta_j\ =q_{j+1}-q_{j-1} \tag{6.24a}$$

$$\nabla = -\frac{1}{2}(\Delta_{j+1} - \Delta_j) = q_j - \frac{1}{2}(q_{j+1} + q_{j-1}) \tag{6.24b}$$

so that $\Delta_{j+1}$ and $\Delta_j$ are expressed in terms of $\Delta$ and $\nabla$ by

$$\Delta_{j+1} = \frac{1}{2}\Delta - \nabla , \quad \Delta_j = \frac{1}{2}\Delta + \nabla . \tag{6.24c}$$

The first exponential in (6.23) can be written as

$$\exp\left[-\frac{1}{G(q_{j-1},\bar{t})}\left(\frac{1}{2(2\varepsilon)}\Delta^2 + \frac{1}{\varepsilon}\nabla^2\right)\right] \tag{6.25}$$

and the integration over $q_j$ is performed using

$$\int \frac{dq_j}{[\pi\varepsilon G(q_{j-1},\bar{t})]^{1/2}} \exp[-\nabla^2/(\varepsilon G(q_{j-1},\bar{t})]\nabla^p = \begin{cases} 1 & ,p=0 \\ [\frac{1}{2}\varepsilon G(q_{j-1},\bar{t})]^{p/2}(p-1)!!, & 0 < p \text{ even} \\ 0 & ,0 < p \text{ odd} \end{cases} \tag{6.26}$$

The result can then be written in the form

$$[2\pi(2\varepsilon)]^{-1/2}\, G(q_{j+1},t_j)^{-1/2} \exp\{-2\varepsilon\, L^G(\frac{\Delta}{2\varepsilon},q_{j-1},\bar{t})\}\; \bar{\bar{Z}}(\Delta,q_{j-1},2\varepsilon) \tag{6.27}$$

with

$$\bar{Z}(\Delta,q_{j-1},2\varepsilon) = 1 + \bar{c}(q_{j-1},\bar{t})\Delta + \bar{d}(q_{j-1},\bar{t})2\varepsilon + \sum_{k=1}^{\infty} \bar{a}_k(q_{j-1},\bar{t})\,\Delta^{2k}/(2\varepsilon)^{k-1} +$$

$$+ \sum_{k=1}^{\infty} \bar{b}_k(q_{j-1},\bar{t})\Delta^{2k+1}/(2\varepsilon)^k . \tag{6.28}$$

Remark that the renormalization group transformation has generated terms proportional to $\Delta$ and $2\varepsilon$ . They should be zero when $a_k$ and $b_k$ take their fixed point values. If not, new terms in the Lagrangian are generated.

To investigate in more detail now the transformation $\{a_k,b_k\}\to\{\bar{a}_k,\bar{b}_k,\bar{c},\bar{d}\}$ it is rewarding to look first at the coefficients $\bar{b}_k$ for $k \geq 2$ .

In the formula that one obtains from (6.23) all kinds of products $\Delta_j^m\Delta_{j+1}^n = (\frac{1}{2}\Delta+\nabla)^m (\frac{1}{2}\Delta-\nabla)^n$ appear. When this is worked out each term is proportional to $\Delta^p\Delta^{m+n-p}$ for some p, $0 \leq p \leq m+n$. An odd power of $\nabla$ gives no contribution (See (6.26)). Therefore if m+n is even only even powers of $\Delta$ remain and odd powers of $\Delta$ if m+n is odd. To find $\bar{b}_k$ $k \geq 2$ it is enough to determine the contribution coming from

$$\sum_{k=1}^{\infty} b_k(q_{j-1})\left(\frac{\Delta_{j+1}^{2k+1}}{\varepsilon^k}+\frac{\Delta_j^{2k+1}}{\varepsilon^k}\right)=\sum_{k=1}^{\infty} b_k(q_{j-1})\sum_{\substack{m=1\\ m\ \text{odd}}}^{2k+1} 2\binom{2k+1}{m}\left(\frac{\Delta}{2}\right)^m \nabla^{2k+1-m}. \tag{6.29}$$

Using (6.26) one obtains

$$\bar{b}_n(q_{j-1},\bar{t}) = 2^{-n}\Big(b_n(q_{j-1},\bar{t}) + \sum_{k=n+1}^{\infty}\binom{2k+1}{2n+1}[2(k-n)-1]!!\,[\tfrac{1}{2}G(q_{j-1},\bar{t})]^{k-n}\cdot$$

$$\cdot b_k(q_{j-1},\bar{t}),\ n\geq 2 \tag{6.30}$$

Since this is a linear transformation with eigenvalues $2^{-n}$ $n=2,3,\ldots$, the only finite fixed point $\bar{b} = b$ is

$$b_n^\star(q_{j-1},\bar{t}) = 0 \quad , n = 2,3,\ldots \quad . \tag{6.31}$$

Inserting these fixed point values for $b_n^\star$ , $n \geq 2$ in the expression (6.23) greatly simplifies the transformation equations for $a_k$. The coefficients $\bar{a}_n, n\geq 4$ are given by

$$\bar{a}_n(q_{j-1},\bar{t}) = 2^{-n}\Big(a_n(q_{j-1},\bar{t}) + \sum_{k=n+1}^{\infty}\binom{2k}{2n}[2(k-n)-1]!!\,[\tfrac{1}{2}G(q_{j-1},\bar{t})]^{k-n} a_k(q_{j-1},\bar{t}) \tag{6.32}$$

and we find by the same argument as above

$$a_n^\star(q_{j-1},\bar{t}) = 0, \quad n=4,5,\ldots \quad . \tag{6.33}$$

It should be remarked that for arbitrary $b_k, \bar{a}_n$ contains contributions proportional to $b_k b_\ell$ ,$k,\ell \geq n$, to $b_k'$ $k \geq n$, and to $b_k, k \geq n-2$ so that at first sight we had a complicated nonlinear transformation.

Now we turn to the determination of the fixed point values of $a_1,a_2,a_3$ and $b_1$. The transformation equations they satisfy are

$$\bar{a}_1 = \frac{1}{2}a_1 - \frac{1}{4}C' + \frac{3}{8}C\frac{G'}{G} + \frac{3}{2}(a_2 - Cb_1)G +$$

$$+\frac{45}{8}a_3G^2 + \frac{9}{16}b_1^{\,2}G^2 - \frac{15}{16}b_1G' - \frac{1}{8}\frac{G''}{G} + \frac{33}{128}\frac{G'^2}{G^2}$$

$$-\frac{1}{4}\left(u-\frac{1}{2}\right)\frac{\partial_t G}{G^2}, \tag{6.34a}$$

$$\bar{b}_1 = \frac{1}{2}b_1 + \frac{1}{8}\frac{G'^2}{G^2}, \tag{6.34b}$$

$$\bar{a}_2 = \frac{1}{4} a_2 + \frac{15}{8} a_3 G - \frac{3}{16} b_1{}^2 G + \frac{1}{32} \frac{G''}{G^2} - \frac{13}{128} \frac{G'^2}{G^3} + \frac{1}{8} b_1' \tag{6.34c}$$

$$\bar{a}_3 = \frac{1}{8} a_3 + \frac{1}{16} b_1{}^2 + \frac{1}{128} \frac{G'^2}{G^4} + \frac{1}{16} b_1 \frac{G'}{G^2} . \tag{6.34d}$$

The conditions that express that $\bar{c}$ and $\bar{d}$ in (6.28) are zero are

$$- \frac{3}{8} \frac{G'}{G} + \frac{3}{2} b_1 G = 0, \tag{6.35a}$$

$$\frac{1}{4} C'G + \frac{1}{2} a_1 G + \frac{3}{4} a_2 G^2 + \frac{15}{8} a_3 G^3 - \frac{15}{16} b_1{}^2 G^3 +$$

$$+ \frac{1}{32} G'' - \frac{9}{128} \frac{G'^2}{G} - \frac{3}{8} b_1' G^2 + \frac{1}{4} (u - \frac{1}{2}) \frac{\partial_t G}{G} = 0. \tag{6.35b}$$

From (6.34) one obtains the fixed point values

$$a_1^\star(q_{j-1},\bar{t}) = - \frac{1}{2} C' - \frac{1}{2} (u - \frac{1}{2}) \frac{\partial_t G}{G^2} \tag{6.36a}$$

with $C$ given by (6.21b) ,

$$b_1^\star(q_{j-1},\bar{t}) = \frac{1}{4} \frac{G'}{G^2} \tag{6.36b}$$

$$a_2^\star (q_{j-1},\bar{t}) = \frac{1}{12} \frac{G''}{G^2} - \frac{5}{32} \frac{G'^2}{G^3} , \tag{6.36c}$$

$$a_3^\star (q_{j-1},\bar{t}) = \frac{1}{32} \frac{G'^2}{G^4} \tag{6.36d}$$

and one can check easily that the conditions (6.35) are satisfied.

Putting this solution (6.31) (6.33) (6.36) into (6.19)(6.16) we find the following fixed point of the transformation (6.18)

$$\phi^{\varepsilon\star}(q_j,t_j;q_{j-1},t_{j-1}) = (2\pi\varepsilon)^{-1/2} G(q_j,t_j - \frac{\varepsilon}{2})^{-1/2} \exp [- \varepsilon L^G (\frac{\Delta_j}{\varepsilon}, q_{j-1},\bar{t}_j)]$$

$$[ 1 + (-\frac{1}{2} ( \frac{A + \frac{1}{4}G'}{G} )' - \frac{1}{2} (u - \frac{1}{2}) \frac{\partial_t G}{G^2}) \Delta_j^2 + \frac{1}{4} \frac{G'^2}{G^2} \Delta_j^3/\varepsilon +$$

$$+ ( \frac{1}{12} \frac{G''}{G^2} - \frac{5}{32} \frac{G'^2}{G^3}) \Delta_j^4/\varepsilon + \frac{1}{32} \frac{G'^2}{G^4} \Delta_j^6/\varepsilon^2 ] \Big|_{q_{j-1},\bar{t}_j} \tag{6.37}$$

which is up to $\mathcal{O}(\varepsilon^{3/2})$ the same as (remember $\bar{t}_j = \frac{1}{2}(t_{j-1}+t_j) + (u - \frac{1}{2})\varepsilon$)

$$\phi^{\varepsilon\star}(q_j,t_j;q_{j-1},t_{j-1}) = (2\pi\varepsilon)^{-1/2} G(q_j,\tfrac{1}{2}(t_{j-1}+t_j))^{-\frac{1}{2}} \exp[-\varepsilon L^{G}(\tfrac{\Delta_j}{\varepsilon},q_{j-1},\tfrac{1}{2}(t_{j-1}+t_j))]$$

$$[1 - \tfrac{1}{2}(\frac{A+\frac{1}{4}G'}{G})'\Delta_j^2 + \tfrac{1}{4}\frac{G'^2}{G^2}\Delta_j^3/\varepsilon + (\tfrac{1}{12}\frac{G''}{G^2} - \tfrac{5}{32}\frac{G'^2}{G^3})\Delta_j^4/\varepsilon +$$

$$+ \tfrac{1}{32}\frac{G'^2}{G^4}\Delta_j^6/\varepsilon^2]\Big|_{(q_{j-1},\bar{t}_j)} . \tag{6.38}$$

Now we ask whether this fixed point of the transformation (6.18) is indeed a good short time propagator. We do this by comparing it with (5.116). Using (6.21b), developing

$$\exp[-L^{G}(\tfrac{\Delta_j}{\varepsilon}, q_{j-1}, \tfrac{1}{2}(t_{j-1}+t_j)] =$$

$$\exp[-\tfrac{1}{2}\frac{\Delta_j^2}{2\varepsilon G(q_{j-1},1/2(t_{j-1}+t_j))}](1 - \frac{(A+\frac{1}{4}G')}{G}\Delta_j + \tfrac{1}{2}\frac{(A+\frac{1}{4}G')^2}{G^2}\Delta_j^2 -$$

$$- \varepsilon[\tfrac{1}{2G}(A+\tfrac{1}{4}G')^2 - V - \tfrac{1}{2}\sqrt{G}(\frac{A+\frac{1}{4}G'}{\sqrt{G}})'] + O(\varepsilon^{3/2}))_{q_{j-1}} \tag{6.39}$$

and working out the product with $Z(q_{j-1},\Delta_j,\varepsilon)$ up to order $\varepsilon$ we obtain $\phi^{\varepsilon\star}$ in the same form as (5.116). We see then that $\phi^{\varepsilon\star}$ coincides with $K^{\varepsilon}$.

## 6.4 Covariant discretizations

The covariant configuration space path integral representations studied in the previous sections were all basedon short time propagators which can be defined in terms of covariant notions such as classical action, WKB-approximation. Now we ask the question : do there exist discretizations as defined in chapter I such that the resulting continuuum expression of the path integral is formally covariant. More specifically, do there exist discretizations $\gamma_{c\ell}$, $\gamma^F$ and $\gamma^G$ such that $L^{\gamma_{c\ell}} \equiv L_{c\ell}$, $L^{\gamma_F} \equiv L^F$, $L^{\gamma^G} \equiv L^G$, irrespective of the coordinate system used. Such discretizations cannot be defined in terms of a correspondence rule since this is not a covariant concept. But using the generalizations introduced in section 2.2.3 we shall now show by explicit construction that covariant discretizations can be constructed. We start from the covariant equation (5.75) which is explicitly

$$\dot{S}(q,t) = [\tfrac{1}{2}c\frac{1}{\sqrt{G(q)}}\partial_\mu\sqrt{G(q)}G^{\mu\nu}(q)\partial_\nu + \frac{1}{\sqrt{G(q)}}\partial_\mu\sqrt{G(q)}h^\mu(q) + \tfrac{1}{c}V(q)]\, S(q,t). \tag{6.40}$$

In the same way as in section 1.4 we introduce an operator formalism. The propagator $I(q,t;q_o,t_o)$ of (6.40) is given by

$$I(q,t;q_o,t_o) = \langle q|U(t,t_o)|q_o\rangle \tag{6.41}$$

where $U(t,t_o)$ is the evolution operator associated with the operator

$$\hat{H}(\hat{p},\hat{q}) = -\frac{ic}{2}\frac{1}{\sqrt{G(\hat{q})}}\hat{p}_\mu\sqrt{G(\hat{q})}G^{\mu\nu}(\hat{q})\hat{p}_\nu - \frac{1}{\sqrt{G(\hat{q})}}\hat{p}_\mu\sqrt{G(\hat{q})}h^\mu(\hat{q}) + \frac{i}{c}V(\hat{q}). \tag{6.42}$$

In order to obtain a representation of the form

$$I(q,t;q_o,t_o) = \int_{\gamma_{c\ell}} \mathcal{D}q\mathcal{D}p \; \exp\{i\int_{t_o}^{t} d\tau[p.\dot{q}-H_{c\ell}(p,q)]\}\; \delta(q(t)-q)\delta(q(t_o)-q_o) \tag{6.43}$$

with $H_{c\ell}(p,q)$ given by (5.90) i.e.

$$H_{c\ell}(p,q) = -\frac{ic}{2}p_\mu p_\nu G^{\mu\nu}(q) - p_\mu h^\mu(q) + \frac{i}{c}V(q) \tag{6.44}$$

we have to construct a function $h^{\gamma_{c\ell}}(p,q',q)$ such that

$$\langle q'|\hat{H}(\hat{p},\hat{q})|q\rangle = \int\frac{dp}{(2\pi)^d}\exp[ip.\Delta]\; h^{\gamma_{c\ell}}(p,q',q) \tag{6.45}$$

and

$$\lim_{q'\to q} h^{\gamma_{c\ell}}(p,q',q) = H_{c\ell}(p,q). \tag{6.46}$$

Using (6.42) one obtains that

$$\langle q'|\hat{H}(\hat{p},\hat{q})|q\rangle = -\frac{ic}{2}\frac{1}{\sqrt{G(q')}}\langle q'|\hat{p}_\mu\sqrt{G(\hat{q})}G^{\mu\nu}(\hat{q})\hat{p}_\nu|q\rangle +$$

$$+\int\frac{dp}{(2\pi)^d}\exp[ip.\Delta][-\frac{1}{\sqrt{G(q')}}p_\mu\sqrt{G(q)}h^\mu(q) + \frac{i}{c}V(q)] \tag{6.47}$$

and the problem reduces to finding the function $b(p,q',q)$ such that

$$\langle q'|\hat{p}_\mu\sqrt{G(\hat{q})}G^{\mu\nu}(\hat{q})\hat{p}_\nu|q\rangle = \int\frac{dp}{(2\pi)^d}\exp[ip.\Delta]\; b(p,q',q) \tag{6.48}$$

and

$$\lim_{q'\to q} b(p,q',q) = p_\mu p_\nu\sqrt{G(q)}G^{\mu\nu}(q). \tag{6.49}$$

This problem has partially been solved in section 2.5 (examples 4 and 5). There it was shown that the correspondence rule relating $\hat{p}_\mu\hat{B}(\hat{q})\hat{p}_\nu$ to $p_\mu p_\nu B(q)$ is determined by the $\Omega$-function (2.122). In the case under consideration here the operator $\hat{B}(\hat{q})$ is a symmetric tensor in $\mu$ and $\nu$ and summation over $\mu$ and $\nu$ is performed. Therefore we can use the more simple correspondence rule determined by (2.123).

According to (2.36) the function $b(p,q',q)$ we are looking for is ($\bar{q}=(q'+q)/2$)

$$b(p,q',q) = \Omega(i\frac{\partial}{\partial\bar{q}},-\Delta)p_\mu p_\nu\sqrt{G(\bar{q})}G^{\mu\nu}(\bar{q})$$

$$= p_\mu p_\nu[\sqrt{G(\bar{q})}G^{\mu\nu}(\bar{q}) - \frac{1}{8}\Delta^\kappa\Delta^\lambda\bar{\partial}_\kappa\bar{\partial}_\lambda\sqrt{G(\bar{q})}G^{\mu\nu}(\bar{q})] \tag{6.50}$$

and the function $h^{\gamma_{c\ell}}(p,q',q)$ defining the discretization $\gamma_{c\ell}$ is

$$h^{\gamma_{c\ell}}(p,q',q) = -\frac{1}{2}icp_\mu p_\nu g^{\mu\nu}_{c\ell}(q',q) - p_\mu a^\mu_{c\ell}(q',q) + \frac{i}{c}V(q) \tag{6.51}$$

with

$$g^{\mu\nu}_{c\ell}(q',q) = \frac{1}{\sqrt{G(q')}}[\sqrt{G(\bar{q})}G^{\mu\nu}(\bar{q}) - \frac{1}{8}\Delta^\kappa\Delta^\lambda\bar{\partial}_\kappa\bar{\partial}_\lambda\sqrt{G(\bar{q})}G^{\mu\nu}(\bar{q})] \tag{6.52}$$

$$a^\mu_{c\ell}(q',q) = \sqrt{\frac{G(q)}{G(q')}}\,h^\mu(q). \tag{6.53}$$

We remark that the mapping from $\hat{H}(\hat{p},\hat{q})$ into $H_{c\ell}(p,q)$ constructed here is an example of the generalizations studied in section 2.2.3.

From the phase space functional integral (6.4 ) one obtains the configuration space functional integral

$$I(q,t;q_o,t_o) = {}_{\gamma_{c\ell}}\!\int\mathcal{D}q\left(\frac{\sqrt{G}}{c^{d/2}}\right)\exp\{i\int_{t_o}^{t}d\tau\; L_{c\ell}(\dot{q},q)\}\;\delta(q(t)-q)\delta(q(t_o)-q_o) \tag{6.54}$$

with $L_{c\ell}(\dot{q},q)$ the classical Lagrangian (5.91) and the discretization determined by (6.51-6.53) as (See section 2.2.4)

$$I(q,t;q_o,t_o) = \lim_{N\to\infty}\int\Big(\prod_{i=1}^{N}dq_i\prod_{j=1}^{N+1}(2\pi c\varepsilon)^{-d/2}g_{c\ell}(q_j,q_{j-1})\Big)^{1/2}.$$

$$\exp\{-\frac{\varepsilon}{c}\sum_{j=1}^{N+1}[\frac{1}{2}g^{c\ell}_{\mu\nu}(\frac{\Delta^\mu_j}{\varepsilon}+a^\mu_{c\ell})(\frac{\Delta^\nu_j}{\varepsilon}+a^\nu_{c\ell}) - V]\}. \tag{6.55}$$

Remark however that

$$g^{c\ell}_{\mu\nu}(q',q)a^\mu_{c\ell}(q',q)a^\nu_{c\ell}(q',q) = G_{\mu\nu}(q)h^\mu(q)h^\nu(q) + \mathcal{O}(\Delta) \tag{6.56}$$

so that using equivalence of discretization (6.5 ) can be simplified.

The result that there exists this discretization $\gamma_{c\ell}$ such that the Hamiltonian in the phase space functional integral representation of the propagator is the classical one has been used in the calculation of higher order corrections in the semi-classical expansion. In [L11,L12,L14] we have used a discretization $\gamma_3$ equivalent to $\gamma_{c\ell}$. The discretization $\bar{\gamma}$ used in[L17](WKB-Type expansion for Langevin equations) and [L20] (On quantization and point canonical transformations in phase space functional integrals) is a more complicated version of the $\gamma_{c\ell}$ that we have constructed here.

Now we ask for discretizations that lead to the Lagrangians of Feynman and Graham. We have only been able to construct one in a flat space. Then both $L^F$ and $L^G$ are given by

$$L^G_{R=0}(\dot{q},q) = \frac{i}{c}\left[\frac{1}{2}G_{\mu\nu}(\dot{q}^\mu + h^\mu)(\dot{q}^\nu + h^\nu) - V - \frac{1}{2}c\nabla_\mu h^\mu\right]. \tag{6.57}$$

We proceed as follows : the term linear in $\hat{p}$ in $\hat{H}(\hat{p},\hat{q})$ can be written as

$$-\frac{1}{\sqrt{G(q)}}\hat{p}_\mu\sqrt{G(\hat{q})}h^\mu(\hat{q}) = -\frac{1}{\sqrt{G(\hat{q})}}\frac{1}{2}[\hat{p}_\mu\sqrt{G(\hat{q})}h^\mu(\hat{q}) + \sqrt{G(\hat{q})}h^\mu(\hat{q})\hat{p}_\mu] + \\ + \frac{i}{2}\frac{1}{\sqrt{G(\hat{q})}}\partial_\mu\sqrt{G(\hat{q})}h^\mu(\hat{q}). \tag{6.58}$$

This leads to the discretization $\tilde{\gamma}$ defined by

$$h^{\tilde{\gamma}}(p,q',q) = -\frac{ic}{2}\,p_\mu p_\nu g^{\mu\nu}_{[\tilde{\gamma}]}(q',q) - p_\mu a^\mu_{[\tilde{\gamma}]}(q',q) + \frac{i}{c}v_{[\tilde{\gamma}]}(q',q) \tag{6.59}$$

with

$$g^{\mu\nu}_{[\tilde{\gamma}]}(q',q) = g^{\mu\nu}_{c\ell}(q',q) \tag{6.60}$$

$$a^\mu_{[\tilde{\gamma}]}(q',q) = \frac{1}{\sqrt{G(q')}}\frac{1}{2}[\sqrt{G(q)}h^\mu(q) + \sqrt{G(q')}h^\mu(q')] \tag{6.61}$$

$$v_{[\tilde{\gamma}]}(q',q) = V(q) + \frac{c}{2}\nabla_\mu h^\mu(q). \tag{6.62}$$

In the configuration space path integral the discretization that follows from (6.59) -(6.62) can be simplified using equivalence of discretizations :

$$g^{[\tilde{\gamma}]}_{\mu\nu}(q',q)a^\mu_{[\tilde{\gamma}]}(q',q) = G_{\mu\nu}(\bar{q})h^\mu(\bar{q}) + \mathcal{O}(\Delta^2)\ , \tag{6.63}$$

$$g^{[\tilde{\gamma}]}_{\mu\nu}(q',q)a^\mu_{[\tilde{\gamma}]}(q',q)a^\nu_{[\tilde{\gamma}]}(q',q) = G_{\mu\nu}(q)h^\mu(q)h^\nu(q) + \mathcal{O}(\Delta) \tag{6.64}$$

and we finally arrive at

$$I(q,t;q_o,t_o) = \lim_{N\to\infty}\int\left(\prod_{i=1}^{N}dq_i\prod_{j=1}^{N+1}(2\pi c\varepsilon)^{-d/2}g_{[\tilde{\gamma}]}(q_j,q_{j-1})^{1/2}\right). \tag{6.65}$$

$$\exp\left\{-\frac{\varepsilon}{c}\sum_{j=1}^{N+1}\left[\frac{1}{2}g^{[\tilde{\gamma}]}_{\mu\nu}(q_j,q_{j-1})\frac{\Delta^\mu_j}{\varepsilon}\frac{\Delta^\nu_j}{\varepsilon} + \bar{G}_{\mu\nu}h^\mu(\bar{q}_j)\Delta_j + \left(\frac{1}{2}G_{\mu\nu}h^\mu h^\nu - V + \frac{c}{2}\nabla_\mu h^\mu\right)(q_{j-1})\right]\right\}$$

$$= \int_{\tilde{\gamma}'}\mathcal{D}q\left(\frac{\sqrt{G}}{c^{d/2}}\right)\exp\left\{i\int_{t_o}^{t}d\tau L^G_{R=0}(\dot{q},q)\right\}\delta(q(t)-q)\delta(q(t_o)-q_o) \tag{6.66}$$

## 6.5 On the use of Fourier series in path integrals

In chapter 5 we have computed the propagator up to order $\varepsilon$ in the limit $\Delta\sim\sqrt{\varepsilon}$. The calculation from a WKB-expansion given in section 5.5 shows clearly that not only the classical path contributes to this approximation but also terms up to fourth order in the deviations from this path.

Another method to obtain a short time propagator taking into account that the paths must remain in principle fully arbitrary, no matter how short the time interval, was developed by Dekker and uses a Fourier series analysis of the paths (A technique used already in [F5]). The Fokker-Planck equation in one dimension was treated in [D3] diffusion and quantization in curved spaces in [D4-D7] and[D30]. The methods used in [D4] and[D5] have been criticized by us in [L15]. We shall repeat here the arguments of [L15] as far as they are still relevant for the improved versions [D6,D7,D30] .

We consider again equation (5.66) :

$$\dot{\phi}(q,t) = [\tfrac{1}{2}c\partial_\mu\partial_\nu G^{\mu\nu}(q) + \partial_\mu A^\mu(q) + \tfrac{1}{c}V(q)\ ]\phi(q,t). \tag{6.67}$$

The method proposed in [D5]to determine a path integral representation for the propagator $K(q,t;q_o,t_o)$ is to write the representation as

$$K(q,t;q_o,t_o) = \lim_{N\to\infty} [(2\pi\varepsilon c)^d\beta(q)]^{-1/2} \int(\prod_{i=1}^{N} dq_i[(2\pi c\varepsilon)^d\beta(q_i)]^{-1/2}). \\ .\exp -\tfrac{1}{c}A[q(t')] \tag{6.68}$$

where

$$A[q(t')] = \int_{t_o}^{t} dt'\ L(\dot{q}(t'),q(t')) \tag{6.69}$$

with

$$L(\dot{q},q) = \tfrac{1}{2}\beta_{\mu\nu}(q)(\dot{q}^\mu+\alpha^\mu(q))(\dot{q}^\nu+\alpha^\nu(q)) + \gamma(q)\ . \tag{6.70}$$

The functions $\beta_{\mu\nu}(q)$, $\alpha^\mu(q)$ and $\gamma(q)$ are to be determined in such a way that (6.68) satisfies (6.67). In order to carry out this program one has either to specify how $A[q(t')]$ in (6.68) is expressed in terms of the $q_i$, or give an equivalent prescription. The proposal of [D5] is to consider only differentiable paths. Peactically the short time propagator $K(q,\tau;q_o,0)$ is computed using (6.68) up to $\mathcal{O}(\tau)$ which is all one needs to identify the equation satisfied by the path integral. In [D5] the calculation is done going first to locally Euclidean coordinates $\xi^\mu(q)$ with origin at $q_o$,i.e. $\xi(q_o)=0$ and then writing the paths $\xi(t')$ as Fourier series in the form

$$\xi^\mu(t') = \xi^\mu_f\ t'/\tau + y^\mu(t') \tag{6.71}$$

with

$$y^{\mu}(t') = \lim_{N\to\infty} \sum_{k=1}^{N} a_k^{\mu} \sin(\frac{k\pi t'}{\tau}) \tag{6.72}$$

and to interprete (6.68) as a multiple integral over $da_k^{\mu}$ after performing a change of variables $:\{\xi_j^{\mu}\} \to \{a_k^{\mu}\}$ $1\leq j,k\leq N$ , $\xi(0)=\xi_o=0$, $\xi(\tau)=\xi_{N+1}=\xi_f$. In [L1 ] we have made two remarks concerning this approach.

The first remark concerned the choice of a locally Cartesian coordinate system. In the case of a curved space, the metric tensor can be taken locally Euclidean in one point (or, at most, in all points of a fixed curve if one uses Fermi coordinates) but not in all the curves [W2]. In fact, in points in the neighborhood of the point (or fixed curve) considered , the metric tensor differs by terms of the order of the distance squared which contribute to the path integral. This remark invalidates the calculation of [D5] in the case of a curved space. In [D6-D7] Dekker followed our suggestion of [L15] to use normal Riemannian coordinates these bringing (as we know from the previous chapter) the utmost attainable simplification. We come back to this below.

In flat space one can go to Cartesian coordinates as done in Ref. [D5] but then the rest of the computation can be carried out without the use of Fourier series representation since the final result is as we have shown in [L15] essentially a consequence of considering only differentiable paths.

Our second remark concerned the use of Fourier series and the interpretation of the transformation $\{\xi_j^{\mu}\}\to\{a_k^{\mu}\}$ . This change of variables cannot be done as in [D5] and to see this we follow a treatment of Chang in [C4] and consider the Gaussian integral

$$I = \lim_{N\to\infty} F_N(t) = \lim_{N\to\infty} (2\pi\epsilon c)^{-1/2} \int_{q(0)=0}^{q(t)=0} \prod_{i=1}^{N} dq_i (2\pi\epsilon c)^{-1/2} \exp-\int_0^t d\tau \frac{1}{2c}(\dot{q}(\tau)^2+\omega^2 q(\tau)^2) \tag{6.73}$$

Putting $q(\tau) = \sum_k a_k \sin(\frac{k\pi\tau}{t})$ and proceeding as in [D5] one obtains

$$F_N(t) = (2\pi\epsilon c)^{-(N+1)/2} J_N \int \prod_{k=1}^{N} da_k \exp[-\frac{t}{4c} \sum_{k=1}^{N} a_k^2 (\frac{k^2\pi^2}{t^2} - \omega^2)] \tag{6.74}$$

where the Jacobian $J_N = \det \partial q(t_j)/\partial a_k = \det \sin(kj\pi/N)$. Doing the Gaussian integrals in (6.74) one has

$$F_N(t) = J_N\sqrt{N+1} \prod_{k=1}^{N} \frac{\sqrt{2(N+1)}}{k\pi} \cdot \frac{1}{\sqrt{2\pi ct}} \prod_{k=1}^{N} (1-\frac{\omega^2 t^2}{k^2\pi^2})^{-1/2} \tag{6.75}$$

and using $(\sin\ \omega t/\omega t) = \prod_{k=1}^{\infty} (1-\frac{\omega^2 t^2}{k^2\pi^2})$ we have $I = K\ (\omega/2\pi c \sin \omega t)^{1/2} \equiv K\ \tilde{I}$ with

$$K = \lim_{N\to\infty} J_N \sqrt{N+1} \prod_{k=1}^{N} \frac{\sqrt{2(N+1)}}{k\pi} \tag{6.76}$$

The value of I for $\omega = 0$ is $(2\pi ct)^{-1/2}$ and consequently the limit in (6.76) should be one and $I = \tilde{I}$ which is in fact the correct result obtained by direct calculation of $\lim I_N, N\to\infty$, with

$$I_N = (2\pi\varepsilon c)^{-1/2} \int_{q_0=0}^{q_{N+1}=0} \prod_{i=1}^{N} dq_i (2\pi\varepsilon c)^{-1/2} \exp\{-\frac{1}{c}\sum_{j=1}^{N+1}(\frac{\Delta q_j^2}{2\varepsilon} - \omega^2 q_{j-1}^2)\} \tag{6.77}$$

But $J_N = \det \sin(kj\pi/N)$ can be computed directly [C4] . Put $M_{kj} = \sin(kj\pi/N)$, then $(MM^t)_{kj} = ((N+1)/2)\,\delta_{kj}$ and consequently $J_N = \det M = ((N+1)/2)^{N/2}$. Using this one easily checks that the limit in (6.76) is not one and that in fact one has $\lim F_N(t) = \lim (\pi/2)^{1/2} (e/\pi)^{N+1} \tilde{I} \to 0$, $N\to\infty$ .

The solution of this apparent paradox is just to remark that what we have done is not a change of variables. In order to do a change of variables one has to go from the variables $\{q_j\}$, $1\le j\le N$, to the variables $\{a_k\}$, $1\le k\le N$, $q_j = \sum_{k=1}^{N} a_k \sin(kj\pi/N)$, in the discretized expression (6.77). That this gives the correct result can be checked explicitly. From (6.77) one obtains now

$$I_N = J_N(\frac{4\pi c}{t})^{N/2}(2\pi\varepsilon c)^{-(N+1)/2}(\frac{\varepsilon}{2})^N \,[\prod_{k=1}^{N} \sin\frac{k\pi}{2(N+1)}]^{-1} \prod_{k=1}^{N}(1-\frac{\omega^2\varepsilon^2}{4\sin^2\frac{k\pi}{2(N+1)}})^{-1/2} \tag{6.78}$$

The last product in (6.78) equals $\sin[2(N+1)\arcsin\frac{\omega\varepsilon}{2}]/(N+1)\omega\varepsilon(1-\frac{\omega^2\varepsilon^2}{4})^{1/2}$ and the preceding product equals $2^N/(N+1)^{1/2}$ (this follows after some algebra from formula's (1.391.2) and (1.392.1) in [G12] . Then after replacing $J_N$ in (6.78) one obtains

$$I = \lim_{N\to\infty} (2\pi\varepsilon c)^{-1/2}(\frac{\varepsilon}{2t})^{N/2}(\frac{N+1}{2})^{N/2}\frac{2^N}{\sqrt{N+1}}(\frac{\sin[2(N+1)\arcsin(\omega\varepsilon/2)]}{(N+1)\omega\varepsilon(1-\frac{\omega^2\varepsilon^2}{4})^{1/2}})^{-1/2} \tag{6.79}$$

which indeed gives $(\omega/2\pi c\sin\omega t)^{1/2}$ . This calculation shows that contrarily to the claim made in [D5] discretization problems are not avoided by the use of Fourier series. In the example we have treated here the difference between the naive calculation and the correct one appears only in the normalization that can after all be fixed at the end of the computation, how general this is remains to be studied.

Let us now briefly comment on the spectral analysis of paths in Riemann normal coordinates presented by Dekker in [D6-D7] and [D30]. In these references the first of our two remarks (choice of coordinate system) is taken into account but not the second one. The result of the calculation is that the Lagrangian L in (6.70) has to be taken equal to

$$L(\dot{q},q) = \frac{1}{2}G_{\mu\nu}(q)(\dot{q}^\mu+h^\mu(q))(\dot{q}^\nu+h^\nu(q)) - \frac{c}{2}\nabla_\mu h^\mu + \frac{1}{8}c^2 R \tag{6.80}$$

in order that (6.68) is the propagator for (6.67). In [D30] it is shown that when a mixed sine cosine Fourier series is used the short time propagator obtained by integration over the Fourier components is exactly the same (without using $\doteq$) as the short time propagator obtained by Graham's scaling method.

## APPENDIX 6.1. : COVARIANCE OF FUNCTIONAL INTEGRAL REPRESENTATIONS

We treat here in some detail the change of variables in a configuration space functional integral defined trough discretization. In section 2.2.4 we found that the propagator $K(q,t;q_o,t_o)$ of (1.1) - (1.3) admits several functional integral representations of the form

$$K = \int_{Q_o}^{Q} \mathcal{D}q \left( \frac{G(q)}{c^d} \right)^{1/2} \exp \left[ - \frac{1}{c} \int_{t_o}^{t} d\tau \, L^{\gamma} (\dot{q}, G^{\mu\nu}, A^{\mu}, V) \right]. \qquad \text{(A6.1)}$$

Assuming that K is a density in Q under coordinate transformations, we should have that (A6.1) equals

$$\left| \frac{\partial q^{\alpha}}{\partial \bar{q}^{\beta}} \right|_{Q} \int_{\bar{Q}_o}^{\bar{Q}} \mathcal{D}\bar{q} \left( \frac{\bar{G}(\bar{q})}{c^d} \right)^{1/2} \exp \left[ - \frac{1}{c} \int_{t_o}^{t} d\tau \, L^{\gamma} (\dot{\bar{q}}, \bar{G}^{\mu\nu}, \bar{A}^{\mu}, \bar{V}) \right] \qquad \text{(A6.2)}$$

with $\bar{G}^{\mu\nu}$, $\bar{A}^{\mu}$, $\bar{V}$ given by (5.70) This is obvious since it is just an expression of the transformation property of K, combined with the functional integral representations. On either side it is of course possible to change to any other discretization $\gamma'$, but keeping the same discretization on both sides, we see that our definition of these functional integrals anticipated any change of variables, because $L^{\gamma}(\dot{q}, G^{\mu\nu}, A^{\mu}, V)$ was given in terms of $G^{\mu\nu}$, $A^{\mu}$ and V which are quantities in an arbitrary coordinate system.

Since $L^{\gamma}$ is not necessarily a scalar, it implies that coordinate transformations cannot be done formally. It is a mere consistency check to prove the equality of (A6.1) and (A6.2), but it is not trivial and therefore we give the simple example of a one dimensional Fokker-Planck equation (see chapter 7) in prepoint discretization.

It serves as an illustration of the rules of calculus in functional integrals, and it is also the method to be used in the cases where one has not studied the general form $L^{\gamma}$ explicitly. The example we treat is where the propagator $P(Q,t|Q_o,t_o)$ satisfies

$$\frac{\partial P}{\partial t} = \left[ \frac{\partial}{\partial Q} A(Q) + \frac{1}{2} \frac{\partial^2}{\partial Q^2} D(Q) \right] P \, , \qquad \text{(A6.3)}$$

$$P(Q,t_o|Q_o,t_o) = \delta(Q-Q_o).$$

In prepoint discretization we have the representation

$$\bar{P}(\bar{Q},t|\bar{Q}_o,t_o) = \int \prod_1^N d\bar{q}_i \prod_{j=1}^{N+1} \frac{1}{\sqrt{2\pi\varepsilon\bar{D}(\bar{q}_{j-1})}} \exp \varepsilon \sum_{j=1}^{N+1} \bar{E}_j^{\gamma_1(0)} \tag{A6.4}$$

where

$$\bar{E}_j^{\gamma_1(0)} = -\frac{\bar{\Delta}_j^2}{2\varepsilon\bar{D}(\bar{q}_{j-1})} - \frac{\bar{\Delta}_j \bar{A}(\bar{q}_{j-1})}{\bar{D}(\bar{q}_{j-1})} - \frac{\varepsilon}{2}\frac{\bar{A}^2(\bar{q}_{j-1})}{\bar{D}(\bar{q}_{j-1})}. \tag{A6.5}$$

Using the notations $\bar{q}'_j = \frac{d\bar{q}}{dq}\Big|_{q_j}$, $\bar{q}''_j = \frac{d^2\bar{q}}{dq^2}\Big|_{q_j}$, we have according to (5.70)

$$\begin{aligned}\bar{D}(\bar{q}_j) &= D(q_j)\bar{q}'^2_j \\ \bar{A}(\bar{q}_j) &= A(q_j)\bar{q}'_j - \frac{1}{2}\bar{q}''_j D(q_j).\end{aligned} \tag{A6.6}$$

We want to show the validity of $\bar{P}(\bar{Q},t|\bar{Q}_o,t_o) = |\frac{\partial q}{\partial \bar{q}}|_{\bar{Q}}\, P(Q,t|Q_o,t_o)$ transforming all the coordinates $\bar{q}_i$ in (A6.4). Since up to order $\varepsilon$

$$\frac{\bar{q}'_j}{\bar{q}'_{j-1}} = 1 + \frac{\bar{q}''_{j-1}}{\bar{q}'_{j-1}}\Delta_j + \frac{1}{2}\frac{\bar{q}'''_{j-1}}{\bar{q}'_{j-1}}\Delta_j^2$$

we obtain

$$\bar{P} = \left(\Big|\frac{d\bar{q}}{dq}\Big|_Q\right)^{-1} \int \prod_{i=1}^N dq_i\, \bar{q}'_i\, \frac{d\bar{q}}{dq}\Big|_{q_{N+1}} \prod_{j=1}^{N+1} \frac{1}{\bar{q}'_{j-1}\sqrt{2\pi\varepsilon D(q_{j-1})}} \exp \sum_{i=1}^{N+1} \bar{E}_j^{\gamma_1(0)}$$

$$= \Big|\frac{dq}{d\bar{q}}\Big|_{\bar{Q}} \int \prod_{i=1}^N dq_i \prod_{j=1}^{N+1} \frac{1}{\sqrt{2\pi\varepsilon D(q_{j-1})}} \left(1 + \frac{\bar{q}''_{j-1}}{\bar{q}'_{j-1}}\Delta_j + \frac{1}{2}\frac{\bar{q}'''_{j-1}}{\bar{q}'_{j-1}}\Delta_j^2\right) \exp \varepsilon\bar{E}_j^{\gamma_1(0)}. \tag{A6.7}$$

For the transformation of $\bar{E}_j^{\gamma_1(0)}$, we need

$$\bar{\Delta}_j = \bar{q}'_{j-1}\Delta_j + \frac{1}{2}\bar{q}''_{j-1}\Delta_j^2 + \frac{1}{6}\bar{q}'''_{j-1}\Delta_j^3 + O(\Delta_j^4) \tag{A6.8}$$

giving

$$\exp -\frac{\bar{\Delta}_j^2}{2\varepsilon\bar{D}(q_{j-1})} = \exp -\frac{\Delta_j^2}{2\varepsilon D(q_{j-1})}\Bigg(1 - \frac{1}{2}\frac{\bar{q}''_{j-1}}{\bar{q}'_{j-1}}\frac{\Delta_j^3}{\varepsilon D(q_{j-1})} - \left(\frac{1}{6}\frac{\bar{q}'''_{j-1}}{\bar{q}'_{j-1}} + \frac{1}{8}\frac{\bar{q}''^2_{j-1}}{\bar{q}'^2_{j-1}}\right)\frac{\Delta_j^4}{\varepsilon D(q_{j-1})} + \frac{1}{8}\frac{\bar{q}''^2_{j-1}}{\bar{q}'^2_{j-1}}\frac{\Delta_j^6}{\varepsilon^2 D(q_{j-1})^2} + O(\varepsilon^{3/2})\Bigg). \tag{A6.9}$$

The other parts of $\overline{E}_j^{\gamma_1(0)}$ give

$$\exp - \overline{\Delta}_j \frac{\overline{A}(\overline{q}_{j-1})}{\overline{D}(q_{j-1})} = 1 - \frac{A}{D}(q_{j-1})\Delta_j + \frac{1}{2}\frac{\overline{q}''_{j-1}}{q'_{j-1}}\Delta_j - \frac{1}{2}\frac{\overline{q}''_{j-1}}{\overline{q}'_{j-1}}\frac{A}{D}\Delta_j^2 + \frac{1}{4}\frac{\overline{q}''_{j-1}}{\overline{q}'_{j-1}}\Delta_j^2 +$$

$$+ \frac{1}{2}\left(\frac{A}{D}\right)^2(q_{j-1})\Delta_j^2 + \frac{1}{8}\frac{\overline{q}''^2_{j-1}}{\overline{q}'^2_{j-1}}\Delta_j^2 - \frac{1}{2}\frac{\overline{q}''_{j-1}}{\overline{q}'_{j-1}}\left(\frac{A}{D}\right)(q_{j-1})\Delta_j^2 + O(\varepsilon^{3/2}) \qquad (A6.10)$$

$$\exp - \frac{\varepsilon}{2}\frac{\overline{A}^2(\overline{q}_{j-1})}{\overline{D}(q_{j-1})} = 1 - \frac{\varepsilon}{2}\frac{A^2}{D}(q_{j-1}) + \frac{\varepsilon}{2}\frac{\overline{q}''_{j-1}}{\overline{q}'_{j-1}}A - \frac{\varepsilon}{8}\frac{\overline{q}''^2}{\overline{q}'^2}D(q_{j-1}) + O(\varepsilon^{3/2}). \qquad (A6.11)$$

The product of (9), (10) and (11) gives now exp $\overline{E}_j^{\gamma_1(0)}$. Replacing this in (7) and doing there the product with the preceeding term, one finds a long expression that has to be calculated using the replacements $\Delta_j^2 \triangleq \varepsilon D(q_{j-1})$, and the other replacements up to $\Delta_j^6$ (see 5.27)
Many terms cancel, and one is left with

$$\overline{P} = \left|\frac{dq}{d\overline{q}}\right|_{\overline{Q}} \int \prod_{i=1}^{N} \prod_{j=1}^{N+1} \frac{1}{\sqrt{2\pi\varepsilon D(q_{j-1})}} \exp - \frac{\Delta_j^2}{2\varepsilon D(q_{j-1})} \left(1 - \frac{A(q_{j-1})}{D(q_{j-1})}\Delta_j\right)$$

$$= \left|\frac{dq}{d\overline{q}}\right|_{\overline{Q}} \int \prod_{i=1}^{N} dq_i \prod_{j=1}^{N+1} \frac{1}{\sqrt{2\pi\varepsilon D(q_{j-1})}} \exp - \frac{\varepsilon}{2D(q_{j-1})}\left[\frac{\Delta_j}{\varepsilon} + A(q_{j-1})\right]^2$$

$$= \left|\frac{dq}{d\overline{q}}\right|_{\overline{Q}} \int_\gamma \mathcal{D}q \frac{1}{\sqrt{D}} \exp \int d\tau \; L^{\gamma_1(0)}(\dot{q},A,D) \qquad (A6.12)$$

which finishes the proof for the discretization $\gamma_1(0)$. For an arbitrary discretization $\gamma_1$ one proceeds as follows. First one goes from the discretizations $\gamma$ in the q coordinates to $\gamma_1(0)$, using consistently the relations (5.27), then one goes to the q-coordinates by relation (12) that we have show for $\gamma_1(0)$ and finally one goes back to $\gamma$ in the q-system.
It is illustrated clearly by the calculation above that it is in general not possible to do formal transformations in functional integrals.
$L^{\gamma_1(0)}$ is not a scalar under coordinate transformations, since A is not a contravariant vector.

## CHAPTER VII : FUNCTIONAL INTEGRAL METHODS IN FOKKER-PLANCK DYNAMICS

### 7.1. Introduction

The purpose of this chapter is to apply the results of the previous chapters in the case of Fokker-Planck dynamics.

We consider macroscopic physical systems, not necessarily in thermodynamic equilibrium, which can be described by a small number of variables $q^1, \ldots, q^d$. We assume that the set $q^\mu$, $\mu = 1, \ldots, d$ is complete in the sense that all other variables of the system have much shorter relaxation times. The macroscopic nature of the variables $q^\mu$ implies that they refer to a collective property of a sufficiently large number of the microscopic constituents. Therefore, they take on continuous values to a good approximation.
Furthermore, their change in time is continuous to a good approximation, since it is brought about by the local interaction of the microscopic constituents. [G18]

Often one can describe the physics of such systems by considering the deterministic time evolution of averaged quantities. Indeed, if the system is in a globally stable state, the statistical nature of the macroscopic dynamics is of minor importance and can be neglected.

In order to study instabilities however fluctuations have to be included and $q^\mu(t)$ has to be considered as a multidimensional stochastic process, characterized by a complete set of probability densities. [G13,S2,H2,N1]

If we assume that the variables $q^\mu$ satisfy the Langevin equations (summation over i, $i = 1, \ldots, m \geq d$ is performed)

$$\dot{q}^\mu + a^\mu(q(t),t) = \sigma_i{}^\mu(q(t),t) f^i(t) \ , \quad \mu=1,\ldots,d \tag{7.1}$$

where $f_i(t)$ is so-called Gaussian white noise (See Appendix 7.1) with zero mean and correlation $\delta^{ij}\delta(t-t')$ then the process $q^\mu(t)$ will be a continuous Markovian process, determined by its transition probability density $P(q,t|q_o,t_o)$ which satisfies the Fokker-Planck equation

$$\begin{aligned} &\frac{\partial}{\partial t}P(q,t|q_o,t_o) = [\partial_\mu A^\mu(q,t) + \tfrac{1}{2}\partial_\mu \partial_\nu D^{\mu\nu}(q,t)]\ P(q,t|q_o,t_o) \ , \\ &P(q,t_o|q_o,t_o) = \delta(q-q_o) \end{aligned} \tag{7.2}$$

where

$$A^\mu(q,t) \equiv a^\mu(q,t) - \tfrac{1}{2}(\partial_\nu \sigma_i^\mu)(q,t)\sigma_i{}^\nu(q,t) \tag{7.3}$$

is the drift (usually defined with the opposite sign) and

$$D^{\mu\nu}(q,t) = \sigma_i^{\ \mu}(q,t)\ \sigma_i^{\ \nu}(q,t) \tag{7.4}$$

is the diffusion matrix. This will be shown in section 7.2.

Equation (7.2) is of the form (1.1-1.3) and everything what we have said about the functional integral representations of the propagator $P(q,t|q_o,t_o)$ can be applied here. In order to understand the discussions about this functional integral representation we review part of the history.

Pioneering research on path integrals for stochastic processes has been done by Wiener for Brownian motion and by Onsager and Machlup in relation to irreversible nonequilibrium thermodynamics[O1]. The latter work is restricted to linear processes near equilibrium obeying a so-called linear Fokker-Planck equation (i.e. linear drift and constant diffusion matrix). It is reviewed nicely by Falkoff in [F2]. Onsager and Machlup showed that the conditional probability density has a path integral representation

$$P(q,t|q_o,t_o) \sim \int \mathcal{D}q \ \exp[-\int_{t_o}^{t} d\tau L(\dot{q}(\tau),q(\tau))]\ \delta(q(t)-q)\delta(q(t_o)-q_o). \tag{7.5}$$

In the case of a linear Fokker-Planck equation the ordering and discretization problem is so simple that it can only lead to an extra additive constant in L which means a constant proportionality factor in P (Cf.the calculation in section 1.3) that is of no importance now.

In words (7.5) means that the transition probability density to go from $(q_o,t_o)$ to $(q,t)$ is built up from contributions from all fluctuating paths, $\exp[-\int L(\dot{q},q)d\tau]$ being the probability density for a path $q(\tau)$. The most probable path (to be defined more precisely in section 7.5) is then obtained by minimizing $\int L(\dot{q},q)d\tau$.

In the linear theory $L(\dot{q},q)$ is quadratic and the integrand in (7.5) a Gaussian. Since for Gaussian functions integrating over its variables under given constraints gives apart from a constant factor the same result as maximizing the Gaussian under the same constraints one has that averages of paths under certain constraints are also obtained by minimizing $\int L(\dot{q},q)d\tau$.

We see then that the Onsager-Machlup function L has several meanings. Apart from being the Lagrangian in a path integral representation it acts as a Lagrangian for both most probable path and average path. In a generalization of the Onsager-Machlup theory to nonlinear Markoff processes (arbitrary drift and diffusion, as in (7.2)) each of these meanings has its own generalization. This has led to much discussion in the recent literature and it is impossible to mention here explicitly the contributions from all authors in [B1,D5,E5,E6,G11,G13-G19,H1,H7,I1,J1-J4,L32, M20,P6,S7,W4,W5,W7,Y2 and others].

The need for a generalization of the Onsager-Machlup theory was emphasized soon[T5,F2], early contributions were given by Stratonovich[S7] and Wiegel[W4].

In 1973 Graham[G13] studied the generalization of the Onsager-Machlup theory to the case of arbitrary stationary continuous Markoff processes. He derived a path integral representaion of the transition probability density directly from the Langevin equations and the path probability density for white noise. In section 7.6 we shall seethat this approach leads exactly to the same classes of functional integral representations as the ones we have derived in the previous chapters when the white noise is properly treated as the differential of the Wiener process.[L22]

The multiplicity of possible functional integral representations was studied in this context first by Haken[H1](1976) and by Leschke and Schmutz[L32](1977).

In 1977 Graham[G15] derived the new and covariant path integral representation leading to the Lagrangian $L^G$ that we have met in chapter 6 . It was proved rigorously only very recently (Fujita and Kotani[F6], Takahashi and Watanabe[T1]) that Graham's covariant Lagrangian is the Lagrangian for the most probable path also when the diffusion matrix defines a Riemannian space with nonvanishing curvature. In a flat space this result has been proved earlier.[D28,I1,L3] In section 7.5 we shall study the problem of the most probable path in more detail.

The generalization of the result of the linear theory that averages of paths under certain constraints are obtained by minimization of $\int L(\dot{q},q)d\tau$ has been treated by Graham in [G18] and we shall not consider it any further here.

In the mean time also the importance of phase space functional integrals for classical stochastic dynamics had become clear. Jansen[J1] and Phythian[P5,P6] were the first to present a functional version of the Martin-Siggia-Rose[M6] operator theory which was developed to be for classical systems as powerful and general as the methods of quantum theory. Of particular relevance is that the introduction of the p-variables conjugate to the q-variables allows for an easy and systematic treatment of response functions in addition to correlation functions. The basic quantity is the generating functional and the various approximate techniques that have been developed in quantum field theory (renormalized perturbation expansions, saddle point methods, renormalization group,...) are being applied.([A5,B2,B3,D1,L24,C3,J1-J4,Y1]).

In section 7.3 we derive the generating functional for correlation and response functions of the general Markovian process described by (7.2) and in section 7.4 we study its perturbation expansion on arbitrary initial conditions.

The topics that we treat in this chapter are chosen in accordance with our earlier work in this field. We have not considered here the WKB-type expansion for Langevin equations that we have given in [L17,L23].

## 7.2 From Langevin to Fokker-Planck equation

The Langevin equation (7.1) does not define a stochastic process before it is supplemented with an additional interpretation rule. Different rules are considered in the theory of stochastic differential equations[A 7]. Most well-known are the Ito and Stratonovich rules[M22]. In [V2] Van Kampen has reviewed the controversy about the proper treatment of equations as (7.1). Here we shall assume that the white noise in (7.1) can be considered as the limit of a stationary process with finite correlation time. This is equivalent with the Stratonovich interpretation rule and has the advantage that changes of variables can be performed in (7.1) with the usual rules of calculus. The variety of stochastic processes generated by (7.1) is studied in section 7.6.

Averages over Gaussian white noise are done by (see Appendix 7.1)

$$\{F[f]\} = \bar{N}^{-1} \int \mathcal{D}f(\tau)\ F[f] \exp[-\tfrac{1}{2}\int_{t_o}^{t} d\tau\ f^i(\tau)f^i(\tau)] \tag{7.6}$$

$$\bar{N} = \int \mathcal{D}f(\tau) \exp[-\tfrac{1}{2}\int_{t_o}^{t} d\tau\ f^i(\tau)f^i(\tau)] \quad .$$

This functional integral seems different from the ones we studied before. However replacing the white noise by the derivative of the Wiener process one obtains a configuration space path integral over the Wiener process. This is what one should do in a more refined treatment (sect.7.6). Here we shall keep on working with (7.6), where $t_o$ and $t$ can be replaced by $-\infty$ and $+\infty$ respectively.

Let $q_f(t;q_o,t_o)$ be the solution of (7.1) for a fixed function $f(\tau)$ and initial condition $q_f(t_o;q_o,t_o) = q_o$. The transition probability density for this deterministic process is $P_f(q,t|q_o,t_o) = \delta(q-q_f(t;q_o,t_o))$ and satisfies the continuity equation

$$\dot{P}_f(q,t|q_o,t_o) = \partial_\nu[a^\nu(q,t) - \sigma_i{}^\nu(q,t)f^i(t)]\ P_f(q,t|q_o,t_o). \tag{7.7}$$

The average over $f(\tau)$ of (7.7) gives the Fokker-Planck equation for $P(q,t|q_o,t_o) = \{P_f(q,t|q_o,t_o)\}$, the transition probability density of the Markovian process defined by 7.1.

Anticipating the ambiguity in the derivation we consider the white noise as a limit of a stationary process with finite correlation time $\tau_c$ : $\{f^i(\tau)f^j(\tau')\} = \delta^{ij}\Delta_{\tau_c}(\tau-\tau') \xrightarrow{\tau_c\to 0} \delta^{ij}\delta(\tau-\tau'), \Delta_{\tau_c}(\tau) = \Delta_{\tau_c}(-\tau)$. Let $\Delta_{\tau_c}^{-1}(\tau)$ be the inverse of the covariance ($\int d\tau'\ \Delta_{\tau_c}(\tau-\tau')\Delta_{\tau_c}^{-1}(\tau'-\tau'') = \delta(\tau-\tau'')$ ) then formula (7.6) is replaced by

$$\{F[f]\} = \frac{\int \mathcal{D}f(\tau)\ F[f(\tau)]\ \exp[-\frac{1}{2}\int d\tau d\tau' f(\tau)\Delta_{\tau_c}^{-1}(\tau-\tau')f(\tau')]}{\int \mathcal{D}f(\tau)\ \exp[-\frac{1}{2}\int d\tau d\tau'\ f(\tau)\Delta_{\tau_c}^{-1}(\tau-\tau')f(\tau')]} . \tag{7.8}$$

The functional integration by parts lemma ([R2] p.116)

$$\int \mathcal{D}f(\tau)\ \frac{\delta}{\delta f^i(\tau)}\ G[f(\tau)] = 0 \tag{7.9}$$

leads when applied to the integrand in (7.8) to the useful relation (Novikov's theorem)

$$\{f^i(\tau)F[f(\tau)]\} = \int_{t_o}^{t} dt'\ \Delta_{\tau_c}(t-t')\ \{\frac{\delta F[f(\tau)]}{\delta f^i(t')}\} . \tag{7.10}$$

We take now for $F[f(\tau)]$ the functional

$$\int dq\ P_f(q,t|q_o,t_o)\ \phi(q,t) = \phi(q_f(t;q_o,t_o),t) . \tag{7.11}$$

One has

$$\int_{t_o}^{t} dt'\ \Delta_{\tau_c}(t-t')\{\frac{\delta}{\delta f^i(t')}\int dq\ P_f(q,t|q_o,t_o)\ \phi(q,t)\} =$$

$$\int_{t_o}^{t} dt'\ \Delta_{\tau_c}(t-t')\{\frac{\partial\phi(q_f,t)}{\partial q_f(t)}\ \frac{\delta q_f(t;q_o,t_o)}{\delta f^i(t')}\} \tag{7.12}$$

One has also that for $t>t_o$

$$q_f(t;q_o,t_o) = q_o + \int d\tau\theta(t-\tau)\theta(\tau-t_o)(\sigma_j^{\ \nu}(q_f(\tau),\tau)f^j(\tau)-a^\nu(q_f(\tau),\tau)) \tag{7.13}$$

and consequently

$$\frac{\delta q_f(t;q_o,t_o)}{\delta f^i(t')} = \begin{cases} 0 & , \quad t'>t \\ \theta(t-t')\sigma_j^{\ \nu}(q_f(t;q_o,t_o),t)\delta^{ji} + \text{terms vanishing when } t'\to t & , \quad t'<t \end{cases} \tag{7.14}$$

Replacing (7.14) in (7.12) we obtain (As for $\tau_c\to 0$ the function $\Delta_{\tau_c}\to\delta(\tau-\tau')$ we need in (7.12) only the term written in (7.14))

$$\int_{t_o}^{t} dt'\ \Delta_{\tau_c}(t-t')\ \{\frac{\partial\phi(q_f,t)}{\partial q_f(t)}\ \sigma_j^{\ \nu}(q_f(t;q_o,t_o),t)\delta^{ji}\} . \tag{7.15}$$

In the limit $\tau_c\to 0$ one has that $\int dt'\ \Delta_{\tau_c}(t-t')\to 1/2$ and then (7.15) becomes

$$\frac{1}{2}\{\frac{\partial\phi(q_f,t)}{\partial q_f(t)}\ \sigma_j^{\ \nu}(q_f(t;q_o,t_o),t)\delta^{ji}\} =$$

$$= \frac{1}{2}\{\int dq\ \delta(q-q_f(t;q_o,t_o))\frac{\partial\phi(q,t)}{\partial q^\nu}\ \sigma_j^{\ \nu}(q,t)\delta^{ji}\}$$

$$= -\frac{1}{2}\int dq\phi(q,t)\ \frac{\partial}{\partial q^\nu}\ \sigma_j^{\ \nu}(q,t)\delta^{ji}\ \ \{P_f(q,t|q_o,t_o)\}. \tag{7.16}$$

Consequently

$$\{f^i(t)P_f(q,t|q_o,t_o)\} = -\frac{1}{2}\frac{\partial}{\partial q^\mu}\ \sigma_j^\mu(q,t)\delta^{ji}\{\ P_f(q,t|q_o,t_o)\} \tag{7.17}$$

and we obtain by averaging of (7.7) the Fokker-Planck equation

$$\dot{P}(q,t|q_o,t_o) = \partial_\nu[a^\nu(q,t) + \frac{1}{2}\sigma_i^{\ \nu}(q,t)\partial_\mu\sigma_i^{\ \mu}(q,t)]\ P(q,t|q_o,t_o) \tag{7.18}$$

which is the same as (7.2-7.4).

The Markovian property of the process $q(\tau)$ can be checked as follows : the joint probability density $(t_n>t_{n-1}>\ldots>t_o)$ is

$$W_n(q_n,t_n;\ldots;q_1,t_1) = \{\prod_j P\ (q_j,t_j|q_{j-1},t_{j-1})\}$$

and from (7.6) one sees that the average factorizes as

$$\prod_j\{P_f(q_j,t_j|q_{j-1},t_{j-1})\}= P(q_n,t_n|q_{n-1},t_{n-1})P(q_{n-1},t_{n-1}|q_{n-2},t_{n-2})\ldots$$

$$P(q_2,t_2|q_1,t_1)P(q_1,t_1|q_o,t_o)$$

which is the Markovian property.

## 7.3 Correlation and response functions

The Hamiltonian operator (2.42) is in the case of equation (7.2)

$$H(\hat{p},\hat{q},t) = -\frac{i}{2}\ \hat{p}_\mu\hat{p}_\nu D^{\mu\nu}(\hat{q},t) - \hat{p}_\mu A^\mu\hat{q},t) \tag{7.19}$$

and $P(q,t|q_o,t_o)$, which is in fact the fundamental solution or propagator of the Fokker-Planck equation, can be written as

$$P(q,t|q_o,t_o) = \langle q|U(t,t_o)|q_o\rangle\ . \tag{7.20}$$

It is advocated to use here out of the infinitely many functional integral representations the phase space functional integral in the prepoint discretization :

$$P(q,t|q_o,t_o) = \int_{\gamma_1(0)} \mathcal{D}q\mathcal{D}p \; \exp\{i\int_{t_o}^{t} d\tau[p\,\dot{q} - H^{\gamma_1(0)}(p,q,\tau)]\}\delta(q(t)-q)\delta(q(t_o)-q_o) \tag{7.21}$$

where

$$H^{\gamma_1(0)}(p,q,\tau) = -\tfrac{1}{2}ip_\mu p_\nu D^{\mu\nu}(q,\tau)-p_\mu A^\mu(q,\tau). \tag{7.22}$$

Consider now the bra-vector

$$\langle L,t| = \int dQ \; {}^{L}\langle Q,t| \tag{7.23}$$

with ${}^{L}\langle q,t|$ as defined in (3.23). Because of the conservation of probability $\langle L,t|$ is in fact independent of time and can be denoted $\langle L|$. Indeed, in the q-representation

$$\langle L,t|q\rangle = \int dQ \; \langle Q,t|q\rangle = \int dQ \; \langle Q|U(t,0)|q\rangle = \int dQ \; P(Q,t|q,0) =$$
$$= \int dQ\{\delta(Q-q_f(t;q,0))\} = 1 \tag{7.24}$$

The ket-vector $|q_o,t_o\rangle^R$ defined in (3.24) has also a simple meaning here. In the q-representation it is given by

$$\langle q|q_o,t_o\rangle^R = \langle q|U(0,t_o)|q_o\rangle = P(q,0|q_o,t_o) \tag{7.25}$$

the transition probability density at time zero.

We can show now very easily that the correlation functions of the stochastic process q(t) are given by

$$\langle L|P\hat{q}^{\mu_1}(t_1)\ldots\hat{q}^{\mu_n}(t_n)|q_o,t_o\rangle^R =$$
$$= \int_{\gamma_1(0)} \mathcal{D}q\mathcal{D}p \;\; q^{\mu_1}(t_1)\ldots q^{\mu_n}(t_n)\; \exp\{i\int_{t_o}^{t} d\tau \; p.\dot{q}-H^{\gamma_1(0)}(p,q,\tau)\}\; \delta(q(t_o)-t_o) \tag{7.26}$$

where on the left hand side appear the Heisenberg operators defined in (3.21) and where on the right hand side the integration over q has eliminated the $\delta(q(t)-q)$ appearing in (7.21). Because of the time-independence of $\langle L|$ the time t which has to be greater then all $t_i$, $i=1,\ldots,n$, can eventually be taken as $+\infty$. Suppose now $t_n \geq t_{n-1} \geq \cdots \geq t_1 \geq t_o$, one has then from (7.23) and (3.20)(3.22) and $\langle L|U(t) = \langle L|$

$$\langle L|P\hat{q}^{\mu_1}(t_1)\ldots\hat{q}^{\mu_n}(t_n)|q_o,t_o\rangle^R =$$
$$= \int dq \; \langle q|U(t_n)U^{-1}(t_n)\hat{q}^{\mu_n}U(t_n,t_{n-1})\hat{q}^{\mu_{n-1}}\ldots\hat{q}^{\mu_1}U(t_1,t_o)|q_o\rangle$$
$$= \int dq \prod_{i=1}^{n} dq_i \; \langle q|\hat{q}^{\mu_n}|q_n\rangle\langle q_n|U(t_n,t_{n-1})\hat{q}^{\mu_{n-1}}|q_{n-1}\rangle\ldots\hat{q}^{\mu_1}|q_1\rangle\langle q_1|U(t_1,t_o)|q_o\rangle$$

$$=\int dq \prod_{i=1}^{n} dq_i\ \delta(q-q_n) q_n^{\mu_n} \ldots q_1^{\mu_1}\ P(q_n,t_n|q_{n-1},t_{n-1}) \ldots P(q_1,t_1|q_o,t_o) \quad (7.26)$$

But the process is Markovian and the product of conditional probability densities is just the joint probability density $W_n(q_n,t_n;\ldots;q_1,t_1)$, hence

$$<L|P\hat{q}^{\mu_1}(t_1)\ldots\hat{q}^{\mu_n}(t_n)|q_o,t_o>^R = \int \prod_{i=1}^{n} dq_i\ q_n^{\mu_n}\ldots q_1^{\mu_1}\ W_n(q_n,t_n;\ldots;q_1,t_1) \quad (7.27)$$

which is the correlation function.

The quantities

$$<L|P\ \hat{p}_{\mu_1}(t'_1)\ldots\hat{p}_{\mu_m}(t'_m)\hat{q}^{\nu_1}(t_1)\ldots\hat{q}^{\nu_n}(t_n)|q_o,t_o>^R = \quad (7.28)$$
$$= \int_{\gamma_1(0)} \mathcal{D}q\mathcal{D}p\ p_{\mu_1}(t'_1)\ldots p_{\mu_n}(t'_m) q^{\nu_1}(t_1)\ldots q^{\nu_n}(t_n)\ \exp\{i\int_{t_o}^{t} d\tau[p\dot{q}-H^{\gamma_1(0)}(p,q,\tau)]\}\ \delta(q(t_o)-q_o)$$

are response functions. Consider for example $<L|P\hat{p}_\rho(t_1)\hat{q}^\nu(t_2)|q_o,t_o>^R$. It can be written as

$$\frac{1}{i}\frac{\delta}{\delta K^\rho(t_1)} \int_{\gamma_1(0)} \mathcal{D}q\mathcal{D}p\ q^\nu(t_2)\ \exp\{i\int_{t_o}^{t} d\tau[p\dot{q}+\tfrac{i}{2}p_\mu p_\nu D^{\mu\nu}(q,\tau)+p_\mu(A^\mu(q,\tau)+K^\mu(\tau))]\}\ \delta(q(t_o)-q_o)\Big|_{K=0} \quad (7.29)$$

but (7.29) is just $\frac{1}{i}\frac{\delta}{\delta K^\rho(t_1)} <q^\nu(t_2)>^K\Big|_{K=0}$ where $<q^\nu(t_2)>^K$ is the one point correlation function of the stochastic process defined by (7.1) with the drift $a^\mu(q,t)$ modified to $a^\mu(q,t)+K^\mu(t)$ :

$$\dot{q}^\mu(t) + a^\mu(q,t) = \sigma_i^{\ \mu}(q,t)f^i(t) - K^\mu(t), \quad (7.30)$$

in other words it is the linear response of $q^\mu(t_2)$ to the perturbation $K^\mu(t_1)$. This interpretation is easily extended to the case when one has several $p_\mu(t')$ and to q-dependent perturbations. An immediate consequence is that (7.28) must vanish if one time $t'_i$ is greater then all times $t_j$ (causality). One should remark that (7.28) is well-defined also for coinciding times $t'_i$ and $t_j$ since because of the $\gamma_1(0)$ discretization it takes the value of the limit $t'_i \to t_j+0$. Consequently the $P$-product is in fact the $P_{\gamma_1(0)}$-product as defined in chapter 3 .

We can now write the generating functional of correlation and response functions as

$$Z[J,J^*] = \int_{\gamma_1(0)} \mathcal{D}q\mathcal{D}p\ \ \exp\{i\int_{t_o}^{t} d\tau[p\dot{q}-H^{\gamma_1(0)}(p,q,t)+J.q+J^*.p\ ]\}\delta(q(t_o)-q_o). \quad (7.31)$$

One has that $Z[0,0] = \langle L|q_o,t_o\rangle^R = \int dq\, P(q,t|q_o,t_o) = 1$. In the next section we shall use the following boundary conditions on $Z[J,J^\star]$ :

$$\frac{1}{i}\frac{\delta Z[J,J^\star]}{\delta J_\mu(t')}\Big|_{t'=t_o} = \int \mathcal{D}q\mathcal{D}p\; q^\mu(t_o)\; \exp\{i\int_{t_o}^{t} d\tau[p.\dot{q}-H^{\gamma_1(0)}(p,q,\tau)+J.q+J^\star.p]\}\delta(q(t_o)-q_o)$$

$$= q_o^\mu\; Z[J,J^\star]\;, \tag{7.32}$$

$$\frac{1}{i}\frac{\delta Z[J,J^\star]}{\delta J^{\star\mu}(t')}\Big|_{t'=t} = 0\;. \tag{7.33}$$

This last formula is proved as follows : the discretized version of the functional integral representing the left hand side of (7.33) is

$$\int \prod_{i=1}^{N+1} dq_i \prod_{j=1}^{N+1}\frac{dp_j}{(2\pi)^d}\, p_{\mu\; N+1}\, \exp\{i\sum_{j=1}^{N+1}[p_{\mu j}(q_j^\mu - q_{j-1}^\mu) - H^{\gamma_1(0)}(p_j,q_{j-1},\bar{t}_j) + \varepsilon J_j q_j + \varepsilon J^\star{}_j p_j]\}. \tag{7.34}$$

Since $q^\mu_{N+1}$ only appears in $\exp[ip_{\mu\; N+1}q^\mu_{N+1}]$ the integfation over $q_{N+1}$ gives $\delta(p_{N+1})$ and as $p_{\mu\; N+1}\;\delta(p_{N+1}) = 0$ (7.34) vanishes and one has (7.33).

Suppose now that we do not only want to average over the random forces $f^i(\tau)$ but also over the initial condition $q(t_o)=q_o$. This can easily be done. Let $b(q(t_o))$ be the probability distribution at time $t_o$. Averaging over $q_o$ through $(\ldots)_b = \int dq_o\; b(q_o)\ldots$ will in the operator formalism only affect $|q_o,t_o\rangle^R$ which has to be replaced by

$$|b,t_o\rangle^R = \int dq_o\; b(q_o)\; |q_o,t_o\rangle^R \tag{7.35}$$

and in the functional integral only $\delta(q(t_o)-q_o)$ which has to be replaced by $b(q_o)$.

When drift and diffusion do not depend on time explicitly the Fokker-Planck equation has a stationary solution $p_s(q)$ satisfying

$$\partial_\mu[A^\mu(q) + \tfrac{1}{2}\partial_\nu D^{\mu\nu}(q)]\; p_s(q) = 0. \tag{7.36}$$

One has because of the stability and the uniqueness of this stationary state

$$p_s(q) = \lim_{t_o\to-\infty} P(q,t|q_o,t_o) \tag{7.37}$$

and the stationary correlation and response functions are also obtained by taking the limit $t_o\to-\infty$ in (7.25), (7.28), (7.31). Let $\hat{B}(t_i',t_j) \equiv \hat{p}_{\mu_1}(t_1')\ldots\hat{p}_{\mu_m}(t_m')\hat{q}^{\nu_1}(t_1)\ldots\hat{q}^{\nu_n}(t_n)$, then

$$\lim_{t_o \to -\infty} \langle L|P\hat{B}(t_i',t_j)|b,t_o\rangle^R = \lim_{t_o \to -\infty} \langle L|P\hat{B}(t_i',t_j)U^{-1}(t_o) \int dQ_o\, b(Q_o)\,|Q_o\rangle$$

$$= \lim_{t_o \to -\infty} \langle L|P\hat{B}(t_i',t_j) \int dq'\, |q'\rangle\langle q'|\, U(0,t_o)\,|q_o\rangle \int dq_o\, b(q_o)$$

which is using (7.37) and $\int dq_o\, b(q_o)=1$ (7.38)

$$= \langle L|P\hat{B}(t_i',t_j)|\, S\rangle$$

with

$$|S\rangle = \int dq' p_s(q')\,|q'\rangle \ . \quad (7.39)$$

Using this formalism it is easy now to derive the usual fluctuation dissipation relations between correlation and response functions but we shall not go into this topic here. (One has to use $U^{-1}(t)|S\rangle=|S\rangle$ which follows from $\hat{H}|S\rangle= 0$, see (7.36))

## 7.4 Perturbation expansion on arbitrary initial conditions

Following our work [L 7] we study now the perturbation expansion for the generating functional.

The freedom of choosing a discretization in its path integral representation allows us to obtain an expansion with vertices and Feynman rules which are as simple as possible.

In order to obtain a perturbation expansion for $Z[J,J^\star]$ we split from $H^{\gamma_1(o)}(p,q,\tau)$ a quadratic part.

Putting $D^{\mu\nu}(t,q) = C^{\mu\nu} + \bar{D}^{\mu\nu}(t,q)$, with $C^{\mu\nu}$ a constant matrix, and $A^\mu(t,q) = \lambda_{(\mu)}q^\mu + \bar{A}^\mu(t,q)$ (no summation over $\mu$ here) one writes $H^{\gamma_1(o)} = H_o + \bar{H}$ with

$$H_o = -\frac{i}{2} C^{\mu\nu} p_\mu p_\nu - \lambda_{(\mu)} p_\mu q^\mu \ .$$

Then $Z[J,J^\star]$ can be expressed as

$$Z[J,J^\star] = \exp[-i\int_{t_o}^{t} d\tau\, \bar{H}(p(\tau),q(\tau),\tau)]\Big|_{\substack{p_\mu(\tau)=\frac{1}{i}\frac{\delta}{\delta J^{\star\mu}(\tau)} \\ q^\mu(\tau)=\frac{1}{i}\frac{\delta}{\delta J_\mu(\tau)}}} . Z_o[J,J^\star] \quad (7.40)$$

with the free generating functional $Z_o$ given by

$$Z_o[J,J^\star] = \int_{\gamma_1(0)} \mathcal{D}q\mathcal{D}p \exp i\int_{t_o}^{t} d\tau[p_\mu\dot{q}^\mu - H_o(p,q,\tau)+J_\mu q^\mu + J^{\star\mu}p_\mu]\ \delta(q(t_o)-q_o). \quad (7.41)$$

We shall use here another method to compute $Z_o$ as the one used in chapter 4 .

The integration by parts lemma

$$\int \mathcal{D}q\mathcal{D}p \, \frac{\delta}{\delta p_\mu(\tau)} K[p,q] = \int \mathcal{D}q\mathcal{D}p \, \frac{\delta}{\delta q^\mu(\tau)} K[p,q] = 0$$

for any functional $K[p,q]$ , gives when applied to (7.41) the two coupled equations (no sum over $\mu$ in what follows)

$$\left(\frac{\partial}{\partial t} + \lambda_{(\mu)}\right)\frac{1}{i}\frac{\delta Z_o}{\delta J_\mu(\tau)} = - J^{\star\mu}(\tau)\, Z_o[J,J^\star] - C^{\mu\nu}\frac{\delta Z_o}{\delta J^{\star\nu}(\tau)} \quad ,\mu = 1,\dots,d \qquad (7.42)$$

$$\left(\frac{\partial}{\partial t} - \lambda_{(\mu)}\right)\frac{1}{i}\frac{\delta Z_o}{\delta J^{\star\mu}(\tau)} = J_\mu(\tau)\, Z_o[J,J^\star] \ , \ \mu = 1,\dots,d. \qquad (7.43)$$

From (7.43) one obtains

$$\frac{1}{i}\frac{\delta Z_o}{\delta J^{\star\mu}(\tau)} = [C_\mu + \int_{t_o}^{\tau} dt' \, J_\mu(t')\, Z_o[J,J^\star] \exp(-\lambda_{(\mu)}t')] \exp(\lambda_{(\mu)}\tau). \qquad (7.44)$$

The boundary condition (7.33) for $Z=Z_o$ gives $C_\mu = -\int_{t_o}^{t} dt' \, J_\mu(t')\exp(-\lambda_{(\mu)}t')Z_o$ and consequently

$$\frac{1}{i}\frac{\delta Z_o}{\delta J^{\star\mu}(\tau)} = - \int_{\tau}^{t} dt' \, \exp[-\lambda_{(\mu)}(t'-t)]\, J_\mu(t')\, Z_o$$

$$= - \int_{t_o}^{t} dt' \, D_{(\mu)}(t'-t)\, J_\mu(t')\, Z_o \qquad (7.45)$$

with $D_{(\mu)}(t) \equiv \theta(t)\exp(-\lambda_{(\mu)}t)$. From (7.42) one has

$$\frac{1}{i}\frac{\delta Z_o}{\delta J_\mu(\tau)} = (\bar{c}^\mu - \int_{t_o}^{\tau} dt' [J^{\star\mu}(t') \exp(\lambda_{(\mu)}t')Z_o + C^{\mu\nu}\frac{\delta Z_o}{\delta J^{\star\nu}(t')} \exp(\lambda_{(\mu)}t')])\exp(-\lambda_{(\mu)} \qquad (7.46)$$

and the boundary condition (7.32) for $Z=Z_o$ gives $\bar{c}^\mu = q_o^\mu \exp(\lambda_{(\mu)}t_o)Z_o$. Replacing now (7.45) in (7.46) one obtains

$$\frac{1}{i}\frac{\delta Z_o}{\delta J_\mu(\tau)} = q_o^\mu \exp[\lambda_{(\mu)}(t_o-\tau)] - \int_{t_o}^{t} dt' \, D_{(\mu)}(\tau-t')J^{\star\mu}(t')\, Z_o +$$

$$+ i \sum_{\nu=1}^{d} \int_{t_o}^{t} dt'' \, \Delta^{\mu\nu}(\tau,t'')\, J_\nu(t'')\, Z_o \qquad (7.47)$$

where $\Delta^{\mu\nu}(t',t'') = \Delta^{\nu\mu}(t'',t')$ is given by

$$\Delta^{\mu\nu}(t',t'') = \frac{C^{\mu\nu}}{\lambda_{(\mu)}+\lambda_{(\nu)}}[D_{(\mu)}(t'-t'') + D_{(\nu)}(t''-t') -$$

$$- \exp[-(\lambda_{(\mu)}t' + \lambda_{(\nu)}t'') + (\lambda_{(\mu)} + \lambda_{(\nu)})t_o]]. \qquad (7.48)$$

From (7.45) and (7.47) and using $Z_o[J,J^\star]_{J=J^\star=0}=1$ one obtains finally

$$Z_o[J,J^\star] = \exp[i \sum_{\mu=1}^{d} q_o^\mu \int_{t_o}^{t} dt' \exp[\lambda_{(\mu)} (t_o-t')]J_\mu(t') -$$

$$- i \sum_{\mu,\nu=1}^{d} \int_{t_o}^{t} dt' \int_{t_o}^{t} dt'' J_\mu(t') S^\mu{}_\nu(t'-t'') J^{\star\nu}(t'')$$

$$- \frac{1}{2} \sum_{\mu,\nu=1}^{d} \int_{t_o}^{t} dt' \int_{t_o}^{t} dt'' J_\mu(t') \Delta^{\mu\nu}(t',t'') J_\nu(t'')] \qquad (7.49)$$

with $S^\mu{}_\nu(t) = \delta^\mu{}_\nu \theta(t)\exp(-\lambda_{(\mu)}t)$. We remark that $Z_o[J,J^\star]$ given by (7.41) looks like the limit of a multidimensional integral $\int dx \exp[-\frac{1}{2}(x,Ax) + (j,x)] \sim \exp[-\frac{1}{2}(j,A^{-1}j)]$ $(x=(p,q))$ and consequently one would not expect the linear term in $J_\mu(t)$ in the exponential in (7.49). However, if one discretizes carefully (7.41) one can check that it is rather of the form $\int dx \exp[-\frac{1}{2}(x,Ax) + (j+d,x)]$ which can be calculated and in the limit gives ((7.49). (Compare with section 4.1)

We want to include now an average over the initial condition. Suppose that at time $t_o$ we are given the probability density $b(q)$, then the generating functional will be

$$Z^b[J,J^\star] = \int dq_o\, b(q_o)\, Z[J,J^\star] = \exp[-i\int_{t_o}^{t} d\tau \bar{H}] \int dq_o\, b(q_o)\, Z_o[J,J^\star].$$

Let $\tilde{b}(x)$ be the Fourier transform of $b(q)$, i.e. $\tilde{b}(x) = \int dq\, b(q) \exp[iq.x]$, then

$$Z_o^b[J,J^\star] = \int dq_o\, b(q_o)\, Z_o[J,J^\star] = \tilde{b}(\int_{t_o}^{t} dt' \exp[\lambda_{(\mu)}(t_o-t')]\, J_\mu(t')). \qquad (7.50)$$

$$.\exp(-\sum_{\mu,\nu} \int_{t_o}^{t} dt'dt''\, J_\mu(t')\,[iS^\mu{}_\nu(t'-t'')J^{\star\nu}(t'') + \frac{1}{2}\Delta^{\mu\nu}(t',t'')J_\nu(t'')])$$

$$Z^b[J,J^\star] = \exp[-i\int_{t_o}^{t} d\tau\, \bar{H}(\frac{1}{i}\frac{\delta}{\delta J^\star(\tau)}, \frac{1}{i}\frac{\delta}{\delta J(\tau)}, \tau)]\, Z_o^b[J,J^\star]. \qquad (7.51)$$

By developing the exponential in (7.51) one obtains the perturbation expansion of $Z^b[J,J^\star]$, and consequently through the relation (3.17) also of the correlation and response functions. When the functional derivatives act on the exponential in $Z_o^b[J,J^\star]$ one generates as usual the propagators $i\,S^\mu{}_\nu(t'-t'')$ and $\Delta^{\mu\nu}(t',t'')$ shown in figure 7.1. The undulated line is associated to $p_\nu(t)$ and the straight one to $q^\mu(t)$ and the arrow in fig.7.1 indicates that time increases in that direction due to the factor $\theta(t'-t'')$.

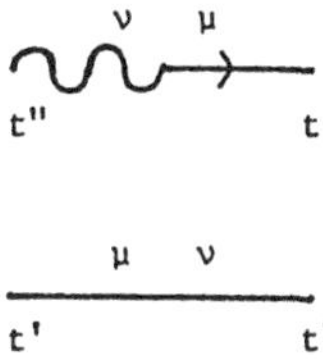


Figure 7.1

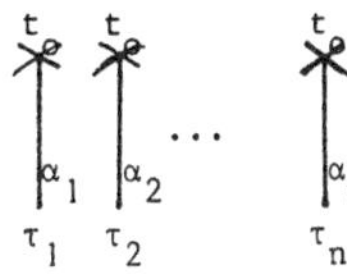


Figure 7.2

In order to see what happens when the functional derivatives act on $\tilde{b}$ in $Z_o^b$ we compute

$$\frac{1}{i^n}\frac{\delta^n}{\delta J_{\alpha_1}(\tau_1)\delta J_{\alpha_2}(\tau_2)\ldots\delta J_{\alpha_n}(\tau_n)}\tilde{b}\Big(\int_{t_o}^{t} dt' \exp[\lambda_{(\mu)}(t_o-t')]J_\mu(t')\Big)\Big|_{J=0}$$

$$= \frac{1}{i^n}\frac{\partial^n \tilde{b}(x)}{\partial x_{\alpha_1}\ldots\partial x_{\alpha_n}}\Big|_{x=0}\prod_{k=1}^{n}\exp[\lambda_{(\alpha_k)}(t_o-\tau_k)] \quad . \tag{7.52}$$

Since $\frac{1}{i^n}\frac{\partial b(x)}{\partial x_{\alpha_1}\ldots\partial x_{\alpha_n}}\Big|_{x=0} = \int dq\, q^{\alpha_1}\ldots q^{\alpha_n}\, b(q) \equiv \langle q^{\alpha_1}(t_o)\ldots q^{\alpha_n}(t_o)\rangle_b$

the averages at time $t_o$, (7.52) is $\langle q^{\alpha_1}(t_o)\ldots q^{\alpha_n}(t_o)\rangle_b \prod_{k=1}^{n}\exp[\lambda_{(\alpha_k)}(t_o-\tau_k)]$. The corresponding diagram is shown in figure 7.2.

One should remark that due to the factor $\theta(t'-t")$ all closed loops constructed only with $iS^\mu{}_\nu$ vanish, and this also applies to tadpoles, i.e. graphs proportional to $S^\mu{}_\nu(0)\sim\theta(0)$ since the $\theta(0)$ appearing in this context is to be interpreted as zero as a consequence of the $\gamma_1(0)$ discretization which tells us that $S^\mu{}_\nu(t'-t')$ has to be interpreted as $S^\mu{}_\nu(t'-(t'+\delta)) = S^\mu{}_\nu(-\delta) = 0$. We also note that at zero order in $\bar{H}$ one has (the [0] indicates this)

$$\langle p_\nu(t)q^\mu(t')\rangle_b^{[0]} = i\, S^\mu{}_\nu(t'-t),$$

but

$$\langle q^\mu(t)q^\nu(t')\rangle_b^{[0]} = \Delta^{\mu\nu}(t,t') + \langle q^\mu(t_o)q^\nu(t_o)\rangle_b \exp[\lambda_{(\mu)}(t_o-t)+\lambda_{(\nu)}(t_o-t')]$$

which is shown graphically in figure 7.3.a and b respectively.

In [L7] we have illustrated the expansion by an example that we shall not repeat here.

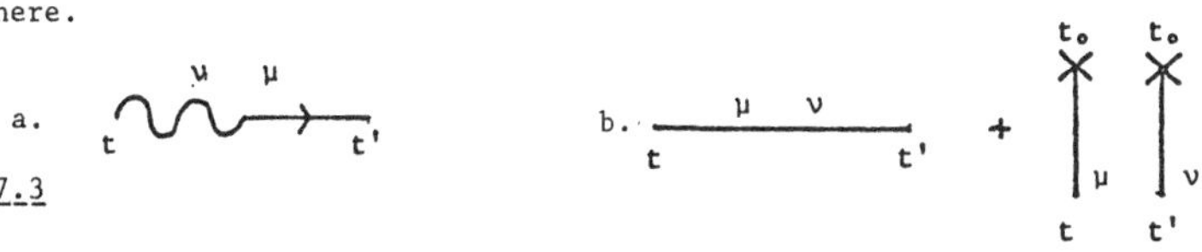


Figure 7.3

The stationary correlation and response functions are obtained taking the limit $t_o \to -\infty$. ($a^\mu$ and $D^{\mu\nu}$ must be independent of t now) One sees immediately that $S^\mu{}_\nu(\tau)$ remains unchanged, while $\Delta^{\mu\nu}$ tends to (all $\lambda_{(\mu)} > 0$)

$$\Delta^{\mu\nu}_{\text{stat.}}(t',t") = \frac{C^{\mu\nu}}{\lambda_{(\mu)}+\lambda_{(\nu)}} [D_{(\mu)}(t'-t")+D_{(\nu)}(t"-t')], \qquad (7.53)$$

and one obtains a result for $Z_{o\ \text{stat}}$ which is independent of the initial condition. We remark that for the limit to exist the splitting of the quadratic part $H_o$ must be done in such a way that all $\lambda_{(\mu)} > 0$.

In [L4] we have studied in the case of a one dimensional process the perturbation expansion in $\gamma_1(\alpha)$ and $\gamma_2(\alpha)$ discretization and shown the cancellation between $\alpha$-dependent tadpoles and contributions from $\alpha$-dependent vertices, a phenomenon that we have treated systematically in chapter 4 of this volume. The choice of $\gamma_1(0)$ that we have made here is clearly the best one since it gives the simplest Hamiltonian and Feynman rules.

The perturbation expansion that we have obtained here is the same as the one obtained by Garrido and San Miguel in [G3] but the use of functional integral techniques has avoided some complications of the series generated through a Wick expansion in an operator formalism.

The deterministic situation (absence of the stochastic force $f^i(t)$) corresponds to putting $D^{\mu\nu}(t,q)=0$ and consequently ( In (7.48) $C^{\mu\nu}=0$) $\Delta^{\mu\nu}(t',t")=0$. The perturbation expansion reduces then to the tree graphs (graphs without closed loops) only.

## 7.5 The Onsager-Machlup function as Lagrangian for the most probable path

The interpretation of the configuration path integral representation of the transition probability density as a sum over all fluctuating paths such that $q(t_o)=q_o, q(t)=q$ strongly suggest the possibility of giving a meaning to the most probable path. This trajectory would then be given by the Euler-Lagrange equations derived from an Onsager-Machlup-Lagrangian $L_{OM}(\dot{q},q)$. The nondifferentiability of the paths of diffusion processes however has led us to many different Lagrangians in the functional integrals defined through discretization. Nevertheless it is possible to obtain a unique $L_{OM}(\dot{q},q)$ by simple and intuitive techniques based on the use of these integrals, at least when the diffusion matrix, considered as the metric of a Riemannian manifold (See section 5.3), has vanishing curvature. This we have shown in [L3] which we shall follow here.

It should be clear from the beginning that it only makes sense to speak of the probability of a small region around a path in q-space since the probability of a single path is zero. Furthermore the most probable path should be twice differentiable being the solution of Euler-Lagrange equations and our work must consist therefore in comparing the probabilities of small regions around differentiable paths. We also make the natural assumption that the notion of most probable path is covariant and that the Lagrangian $L_{OM}$ is a scalar.

If the manifold described by the diffusion matrix has zero curvature we can work in Cartesian coordinates (i.e.constant diffusion). Let us suppose then that we are in an orthogonal rectilinear coordinate system where $D^{\mu\nu}\equiv\delta^{\mu\nu}$ , $D = 1$ and let us start from the configuration space path integral for $P(q,t|q_o,t_o)$ in $\gamma(\alpha)$ discretization

$$P(q,t|q_o,t_o) = \int_{\gamma(\alpha)} \mathcal{D}q\ \exp\{-\int_{t_o}^{t} d\tau[\sum_{\mu=1}^{d} \frac{1}{2}(\dot{q}^{\mu}+A^{\mu}(q))^2-\alpha\partial_{\mu}A^{\mu}(q)]\ \} \tag{7.54}$$

where $\gamma(\alpha)$ is either $\gamma_1(\alpha)$ or $\gamma_2(\alpha)$ which are equivalent here (See section 2.3 ). We make now a distinction between the case when $A^{\mu}(q)$ is a gradient $A^{\mu}(q)=\partial_{\mu}\phi(q)$ and the case when $A^{\mu}(q)$ is not a gradient. In the former case we integrate by parts the crossterms $\dot{q}^{\mu}A^{\mu}(q)$ using the generalization to more dimensions of the formula (A4.19) i.e.

$$\sum_{\mu=1}^{d} \int_{t_o}^{t} d\tau\ \dot{q}^{\mu}A^{\mu}(q(\tau)) = \phi(q(t))-\phi(q(t_o))-(\frac{1}{2}-\alpha) \sum_{\mu=1}^{d} \int_{t_o}^{t} d\tau\partial_{\mu}A^{\mu}(q(\tau)). \tag{7.55}$$

Doing this we obtain from (7.54) that

$$P(q,t|q_o,t_o) = \exp[\phi(q_o)-\phi(q)]\int\mathcal{D}q\ \exp[-\int_{t_o}^{t} d\tau\ \frac{1}{2}\sum_{\mu=1}^{d} (\dot{q}^{\mu})^2+V(q)]\ \delta(q(t)-q)\delta(q(t_o)-q_o) \tag{7.56}$$

where

$$V(q) = \frac{1}{2} \sum_{\mu=1}^{d} [A^{\mu}(q)^2-\partial_{\mu}A^{\mu}(q)\ ]. \tag{7.57}$$

The path integral (7.56) is independent of the discretization. (The corresponding operator in the operator formalism has no ordering ambiguities). We now claim that the most probable path is the one that makes the action in (7.56) stationary and that $L_{OM}$ can be taken, when we add to the Lagrangian in (7.56) the total derivative $\dot{\phi}(q(\tau)) =\Sigma\ \dot{q}^{\mu}A^{\mu}(q(\tau))$, as the Lagrangian $L^{\gamma(1/2)}$ in(7.54). The proof goes as follows : the probability of the paths in a small region of volume characterised by a small parameter $\delta$ around the differentiable curve $y = y(\tau)$ will be ($y(t_o)=y_o$ $=q_o$ fixed)

$$P[y(\tau),\delta] = \lim_{N\to\infty} \int_{\Gamma_\delta} \prod_{i=1}^{N} \prod_{\mu=1}^{d} dq_i^{\mu} \prod_{j=1}^{N+1} P(q_j,t_j|q_{j-1},t_{j-1}) \tag{7.58}$$

where the integration is over the region $\delta$ such that ($\dot{y}^{\mu}(t_i)\equiv y_i^{\mu}$) $y_i^{\mu}-\delta\leq q_i^{\mu}\leq y_i^{\mu}+\delta$. From (7.56) one has

$$P(q_j,t_j|q_{j-1},t_{j-1}) = \exp[\phi(q_{j-1})-\phi(q_j)]\,[2\pi\varepsilon]^{d/2}\exp\{-\sum_{\mu=1}^{d}\frac{(\Delta q_j)^2}{2\varepsilon} - \varepsilon V(q_{j-1})\}. \tag{7.59}$$

We now perform the change of variables $q_j = y_j + x_j$ and substitute in (7.58) :

$$P[y(\tau),\delta] = \lim_{N\to\infty} \exp[-\varepsilon \sum_{j=1}^{N+1} ( \sum_{\mu=1}^{d} \frac{(\dot{y}_{j-1})^2}{2} + V(y_{j-1}))] .$$

$$. \int_{\Gamma'_\delta} \prod_{i=1}^{N+1} \prod_{\mu=1}^{d} (dx_i^\mu (2\pi\varepsilon)^{-1/2}) \exp[\,\phi(y_o)-\phi(y_{N+1}+x_{N+1})] .$$

$$.\exp\{ \sum_{j=1}^{N+1} [- \sum_{\mu=1}^{d} (\frac{(\Delta x_j)^2}{2\varepsilon} + \dot{y}_{j-1}\Delta x_j^\mu) - \varepsilon(V(y_{j-1}+x_{j-1})-V(y_{j-1}))]\} \tag{7.60}$$

with $\Gamma'_\delta$ such that $-\delta \le x_i \le \delta$ and where we have used $\Delta q_j^\mu = \varepsilon\dot{y}_{j-1} + \Delta x_j^\mu + O(\varepsilon^2)$ since the curve $y(\tau)$ is differentiable.

Recall now that we are only interested in (7.60) when $\delta\to 0$ since the quantity of interest is the quotient $P[y(\tau),\delta]/P[z(\tau),\delta]$ around two differentiable curves $y(\tau)$ and $z(\tau)$ when $\delta\to 0$. This implies that $\phi(y_{N+1}+x_{N+1})$ in (7.60) can be replaced by $\phi(y(t))$ since $|x_{N+1}| = O(\delta)$.

We consider now for any function $f^\mu(\tau)$ the sum $\sum_{j,\mu} f^\mu(t_{j-1})\Delta x_j^\mu$. One has (note that $x_o = 0$)

$$\sum_{j=1}^{N+1} \sum_{\mu=1}^{d} f_{j-1}^\mu (x_j^\mu - x_{j-1}^\mu) = \sum_{\mu=1}^{d} f_{N+1}^\mu\, x_{N+1}^\mu + \sum_{j=1}^{N+1} \sum_{\mu=1}^{d} x_j^\mu (f_j^\mu - f_{j-1}^\mu) . \tag{7.61}$$

The first term on the right hand side of (7.61) is of $O(\delta)$ on so is the second since it is

$$\sum_{j=1}^{N+1} \sum_{\mu=1}^{d} x_j^\mu\, (\varepsilon \frac{\partial f^\mu(t_{j-1})}{\partial t_{j-1}} + O(\varepsilon^2)) = \sum_{j-1}^{N+1} \varepsilon O(\delta) = (t-t_o)O(\delta) = O(\delta). \tag{7.62}$$

The exponential of a sum like (7.61) will then only contribute a term $\exp(O(\delta))$ and need not to be considered. This allows us to get rid of the term $\dot{y}_{j-1}^\mu \Delta x_j^\mu$ in (7.60). The last term there is

$$\sum_{j=1}^{N+1} \varepsilon (V(y_{j-1}+x_{j-1}) - V(y_{j-1})) = \sum_{j=1}^{N+1} \varepsilon O(\delta) = O(\delta) \tag{7.63}$$

and does not contribute either when $\delta\to 0$, Since $A^\mu = \partial_\mu \phi$ we have

$\phi(y(t))-\phi(y_o) = \int d\tau\, \dot{y}^\mu(\tau) A^\mu(y(\tau))$ and using also (7.57)

$$P[y(\tau),\delta]\Big|_{\delta\to 0} = \exp[- \int_{t_o}^{t} d\tau \sum_{\mu=1}^{d} \frac{1}{2}(\dot{y}^\mu + A^\mu(y(\tau))^2 - \frac{1}{2}\partial_\mu A^\mu(y(\tau))]\, P_W(\Gamma'_\delta) \tag{7.64}$$

with $P_W(\Gamma'_\delta)$ the Wiener measure of the paths $x(\tau)$, $|x(\tau)|\leq\delta$ in the region $\Gamma'_\delta$ around the path $x(\tau)\equiv 0$ :

$$P_W(\Gamma'_\delta) = \lim_{N\to\infty} \int_{\Gamma'_\delta} \prod_{i=1}^{N+1} \prod_{\mu=1}^{d} (dx_i^\mu (2\pi\varepsilon)^{-1/2}) \exp[-\sum_{j=1}^{N+1} \sum_{\mu=1}^{d} \frac{(x_j^\mu - x_{j-1}^\mu)^2}{2\varepsilon}]. \tag{7.65}$$

It is independent of the curve $y(\tau)$ and consequently this term cancels out in the quotient of the probabilities around two differentiable curves thus showing that

$$Q = \lim_{\delta\to 0} \frac{P[y(\tau),\delta]}{P[z(\tau),\delta]} = \exp[-\int_{t_0}^{t} d\tau (L_{OM}(\dot{y},y) - L_{OM}(\dot{z},z))] \tag{7.66}$$

where

$$L_{OM}(\dot{q},q) = \sum_{\mu=1}^{d} [\frac{1}{2}(\dot{q}^\mu + A^\mu(q))^2 - \frac{1}{2}\partial_\mu A^\mu(q)] \tag{7.67}$$

The most probable path is by definition the curve $y(\tau)$ such that $Q>1$ for any other curve $z(\tau)$ and (7.66) shows that (7.67) is the Lagrangian for the most probable path. In arbitrary coordinates $L_{OM}(\dot{q},q)$ is the given by (Cf.chapters 5 and 6 )

$$L_{OM}(\dot{q},q) = \frac{1}{2}D_{\mu\nu}(q)(\dot{q}^\mu + h^\mu(q))(\dot{q}^\nu + h^\nu(q)) - \frac{1}{2}\nabla_\mu h^\mu(q). \tag{7.68}$$

It follows from the transformation properties of the Fokker-Planck equation (7.2) which are described in detail in section 5.4 that (7.68) is a scalar which reduces to (7.67) in Cartesian coordinates. We remark that for one degree of freedom (d=1) the problem is completely solved by the previuous considerations since there is no curvature and the drift is always a gradient.

We note that (7.66) is not the classical Lagrangian in the sense of a WKB-expansion associated to the problem (section (5.3)) but differs from it by the term in the divergence of the drift $\frac{1}{2}\nabla_\mu h^\mu(q)$. It is in fact the flat space version of both Feynman Lagrangian and Graham's Lagrangian.

Let us treat now the case when $A^\mu$ is not a gradient. The probability $P[y(\tau),\delta]$ is given by (7.58) but we use now instead of (7.59) the short time propagator obtained by Graham's scaling method

$$\begin{aligned} P(q_j,t_j|q_{j-1},t_{j-1}) = (2\pi\varepsilon)^{-d/2} \exp[-\sum_{\mu=1}^{d} \frac{(\Delta q_j)^2}{2\varepsilon}][1 - \Delta q_j^\mu A^\mu(q_{j-1}) + \\ + \frac{1}{2}\varepsilon(\partial_\mu A^\mu(q_{j-1}) - A^\mu(q_{j-1})A^\mu(q_{j-1})) - \frac{1}{2}\Delta q_j^\mu \Delta q_j^\nu (\partial_\mu A^\nu(q_{j-1}) - A^\mu A^\nu(q_{j-1})) \\ + O(\varepsilon^{3/2})] \end{aligned} \tag{7.69}$$

We remark however immediately that without using any replacements $\doteq$ (7.67) can be rewritten as the short time propagator for mid-point discretization $(\bar{q}_{j-1} = \frac{1}{2}(q_j + q_{j-1}))$

$$P(q_j,t_j|q_{j-1},t_{j-1}) = (2\pi\varepsilon)^{-d/2} \exp[\sum_{\mu=1}^{d} (-\frac{(\Delta q_j^\mu)^2}{2\varepsilon} - \Delta q_j^\mu A^\mu(\bar{q}_{j-1}) - \varepsilon V(q_{j-1})] \quad (7.70)$$

with V(q) given by (7.57). Replacing (7.70) in (7.58) and performing as before the change of variables $q_i = y_i + x_i$ one obtains (after dropping terms $O(\varepsilon\delta)$ or $O(\varepsilon^{3/2})$)

$$P[y(\tau),\delta] = \lim_{N\to\infty} \exp \sum_{j=1}^{N+1} [\sum_{\mu=1}^{d} (-\frac{1}{2}\varepsilon(\dot{y}_{j-1}^\mu)^2 - \dot{y}_{j-1}^\mu A^\mu(y_{j-1}) - \varepsilon V(y_{j-1})]. \quad (7.71)$$

$$.\int \prod_{i=1}^{N+1} \prod_{\mu=1}^{d} dx_i^\mu (2\pi\varepsilon)^{1/2} \exp\{\sum_{j=1}^{N+1} [\sum_{\mu=1}^{d} (-\frac{(\Delta x_j^\mu)^2}{2\varepsilon} - \dot{y}_{j-1}^\mu \Delta x_j^\mu - \Delta x_j^\mu A^\mu(\bar{y}_{j-1} + \bar{x}_{j-1}))$$

$$- \varepsilon(V(y_{j-1} + x_{j-1}) - V(y_{j-1}))]\} .$$

We repeat now the arguments after (7.60) (Formulae (7.61)-(7.62)) to eliminate the same terms as there. We note that $\Delta x_j^\mu A^\mu(\bar{y}_{j-1} + \bar{x}_{j-1}) = \Delta x_j^\mu A^\mu(y_{j-1} + \bar{x}_{j-1}) + O(\varepsilon^{3/2})$ $= \Delta x_j^\mu f^\mu(t_{j-1}, \bar{x}_{j-1})$ and we obtain then from (7.71)

$$P[y(\tau),\delta]\Big|_{\delta\to 0} = \exp\{-\int_{t_0}^{t} d\tau \sum_{\mu=1}^{d} [\frac{1}{2}(\dot{y}^\mu + A^\mu(y(\tau)))^2 - \frac{1}{2}\partial_\mu A^\mu(y(\tau))]\}. \quad (7.72)$$

$$.\lim_{N\to\infty} \int \prod_{i=1}^{N+1} \prod_{\mu=1}^{d} (dx_i (2\pi\varepsilon)^{-1/2}) \exp\{\sum_{j=1}^{N+1} \sum_{\mu=1}^{d} -[\frac{(x_j^\mu - x_{j-1}^\mu)^2}{2\varepsilon} - (x_j^\mu - x_{j-1}^\mu) f^\mu(t_{j-1}, \bar{x}_{j-1})]\}$$

The term $\sum_{j,\mu} \Delta_j f^\mu(t_{j-1}, \bar{x}_{j-1})$ being discretized in the mid-point is a stochastic integral in the Stratonovich sense, that we had eliminated by partial integration in the case that $A^\mu$ was a gradient. Here we have to use a result of Ito ([I2] formula (A.3)) who proved by probabilistic methods that the conditional expectation value in $\Gamma'_\delta$ of the exponential of a Stratonovich integral i.e. the quantity in (7.72) divided by the Wiener measure of $\Gamma'_\delta$ has the limit 1 when $\delta\to 0$. We have then

$$P[y(\tau),\delta]\Big|_{\delta\to 0} = \exp\{-\int_{t_0}^{t} d\tau \sum_{\mu=1}^{d} [\frac{1}{2}(\dot{y}^\mu + A^\mu(y(\tau)))^2 - \frac{1}{2}\partial_\mu A^\mu(y(\tau))]\} P_W(\Gamma'_\delta) \quad (7.73)$$

which is again (7.64) thus showing that the Lagrangian (7.67) (inarbitrary curvilinear coordinates (7.68)) is the Onsager-Machlup function in all cases where $D^{\mu\nu}$ has vanishing curvature.

This result has also been proved using probabilistic methods by Dürr and Bach [D28] for one degree of freedom and by Ito [I1] in the general case that we have considered here. Bach and Dürr have made applications in [D29] and [B1].

As already mentionned in the introduction it was shown only recently ([F6,T1 ]) that Graham's Lagrangian $L^G(\dot{q},q)$ is the Lagrangian for the most probable path in curved spaces also as we had conjectured in [L13].

## 7.6 Derivation of functional integral representations from stochastic differential equations

If one is given a Langevin equation in the form ($\nu,\kappa=1,\dots,d$)

$$dq^{\nu} + a^{\nu}(q(t))dt = \sqrt{\eta}\, \sigma^{\nu}{}_{\kappa}(q(t))\, dw^{\kappa} \tag{7.74}$$

where $dw^{\kappa}$ is the differential of the Wiener process, formally $dw^{\kappa} = f^{\kappa}\, dt$ and $f^{\kappa}(t)$ Gaussian white noise, with a $\delta$-correlation $\{f^{\kappa}(t)f^{\lambda}(t')\} = \delta(t-t')\delta^{\kappa\lambda}$, one can consider the transition probability P, of the stochastic process q(t) and construct its functional integral representation from the associated Fokker-Planck equation. The procedure to do this has been discussed before. Here the question arises to construct the representation immediately from (1). It is well known however that this equation, as given here, is not sufficiently specified [V2] , and a precise definition is given in the theory of stochastic processes [A7,D20 ]. Up to our level of description, it suffices to present a lattice version of (74) in the form

$$q^{\nu}_{j} - q^{\nu}_{j-1} + \varepsilon a^{\nu}(q^{r}_{j-1}) = \sqrt{\eta}\, \sigma^{\nu}{}_{\kappa}(q^{s}_{j-1})(w^{\kappa}_{j} - w^{\kappa}_{j-1}) \tag{7.75}$$

in order to introduce the additional structure. We discretized the time $t_j = t_\circ + j\varepsilon$ , and used the notation $q(t_j) = q_j, w(t_j)=w_j . q^{r}_{j-1}=q_{j-1}+r\, \Delta q_j$ , where $\Delta q_j = q_j - q_{j-1}$ ,and r,s are real parameters. As initial condition for (75) we take $q_\circ = \Omega_\circ$. The stochastic process defined by (75) in the limit $\varepsilon \to 0$ depends on s, but not on r. For s=0, the limiting equation is an Ito stochastic differential equations, while for s=1/2 , it is named after Stratonovich.$^{(*)}$ Exploiting the discretized form (75) as a transformation rule to the density of the Wiener process, or considering $P(\Omega,t|\Omega_\circ,t_\circ)$ as obtained from the probability of the deterministic equations by averaging over the Wiener process, we derive in two different ways functional integral representations, and compare these to the ones obtained from the associated Fokker-Planck equation

$$\frac{\partial P}{\partial T} = \frac{\partial}{\partial \Omega^{\mu}} [(a^{\mu}(\Omega) - \eta s \sigma^{\rho}{}_{\kappa}(\Omega)\, \partial_{\rho}\sigma^{\mu}{}_{\kappa}(\Omega))P] + \frac{\eta}{2}\frac{\partial}{\partial \Omega^{\mu}}\frac{\partial}{\partial \Omega^{\nu}}(\sigma^{\mu}{}_{\kappa}(\Omega)\sigma^{\nu}{}_{\kappa}(\Omega)P) \tag{7.76}$$

using certain correspondence rules.

---

(*) The Stratonovich interpretation is also obtained in the limit from (1), if one takes for $\dot{w}(t)$ a stationary stochastic process with zero mean and short autocorrelation time $\tau \to 0$. [V2,W8 ] For a heuristic proof of this, see section 7.2.

Considering now the interval $[t_\circ,T]$ and taking $\varepsilon$ such that $t_{N+1}=T$, the probability $P_1$ that

$$\{w(t_\circ)=0, w_1\leq w(t_1)\leq w_1+dw_1, w_2\leq w(t_2)\leq w_2+dw_2,\dots,w_{N+1}\leq w(T)\leq w_{N+1}+dw_{N+1}\}$$

is given by

$$P_1(\{w_i\})=\prod_{j=1}^{N+1}(2\pi\varepsilon)^{-d/2}\prod_{\kappa=1}^{d}dw_j^{\kappa}\exp-\frac{1}{2\varepsilon}\sum_{j=1}^{N+1}\sum_{\kappa=1}^{d}(w_j^{\kappa}-w_{j-1}^{\kappa})^2\quad(w_o=0)\quad(7.77)$$

Now, through the stochastic differential equation, one can transform $P_1$ in a probability for paths in q-space. Making first the change of variables $z_j^{\kappa}=w_j^{\kappa}-w_{j-1}^{\kappa}$, $j=1,2,\dots,N+1$, one has $\mathrm{Det}(\partial z_j^{\kappa}/\partial w_i^{\ell})=1$, and consequently we write

$$P_1(\{w_i\})=\prod_{j=1}^{N+1}(2\pi\varepsilon)^{-d/2}\prod_{\kappa=1}^{d}dz_j^{\kappa}\exp-\frac{1}{2}\sum_{j=1}^{N+1}\sum_{\kappa=1}^{d}(z_j^{\kappa})^2.\qquad(7.78)$$

Defining $\sigma_{\mu\kappa}$ by $\sigma^{\mu}{}_{\ell}\,\sigma_{\mu\kappa}=\delta_{\kappa\ell}$ and introducing $g_{\mu\nu}=\sigma_{\mu\kappa}\sigma_{\nu\kappa}$, one has $\sigma_{\mu\kappa}=g_{\mu\nu}\sigma^{\nu}{}_{\kappa}$.

(75) is then

$$z_j^{\kappa}=\eta^{-1/2}\sigma_{\nu\kappa}(q_{j-1}^{s})\,[\Delta q_j^{\nu}+\varepsilon a^{\nu}(q_{j-1}^{r})]\qquad(7.79)$$

and we compute the Jacobian $J=\mathrm{Det}\left(\frac{\partial z_j^{\kappa}}{\partial q_i^{\alpha}}\right)$. Since $\frac{\partial z_j^{\kappa}}{\partial q_k^{\alpha}}=0$ for $k>j$, one has

$J=\prod_{j=1}^{N+1}\det\left(\frac{\partial z_j^{\kappa}}{\partial q_j^{\alpha}}\right)$, the last determinant being of the dxd matrix of indices $(\kappa,\alpha)$. Explicitly

$$\frac{\partial z_j^{\kappa}}{\partial q_j^{\alpha}}=\frac{1}{\sqrt{\eta}}\,\sigma_{\nu\kappa}(q_{j-1}^{s})[\delta^{\nu}{}_{\alpha}+\varepsilon r\partial_{\alpha}a^{\nu}(q_{j-1}^{r})+s\,\sigma^{\nu}{}_{\ell}(q_{j-1}^{s})\partial_{\alpha}\sigma_{\gamma\ell}(q_{j-1}^{s})(\Delta^{\gamma}q_j+\varepsilon a^{\gamma}(q_{j-1}^{r})]$$
$$(7.80)$$

Consequently, $\det\left(\frac{\partial z_j^{\kappa}}{\partial q_j^{\alpha}}\right)=\eta^{-d/2}[\det\sigma_{\nu\kappa}(q_{j-1}^{s})]\det(1+X)$ and X can be read from (80).

Now $\det(1+X)=\exp[\mathrm{tr}\,\ell n(1+X)]=\exp[\mathrm{tr}(X-X^2/2)+O(\varepsilon^{3/2})]$, where we used $\Delta q_j=O(\varepsilon^{1/2})$ since the dominant term in the argument of the exponential in (78) will be $-\frac{1}{\varepsilon}g_{\mu\nu}(q_{j-1}^{s})\Delta q_j^{\mu}\Delta q_j^{\nu}$ [see(79)] and we shall integrate over the $q_j$ afterwards. Therefore keeping terms up to $O(\varepsilon)$

$$\det(1+X)=\exp\{s\,\sigma^{\nu}{}_{\ell}(q_{j-1}^{s})\partial_{\nu}\sigma_{\gamma\ell}(q_{j-1}^{s})[\Delta q_j^{\gamma}+\varepsilon a^{\gamma}(q_{j-1}^{r})]-\frac{\varepsilon}{2}s^2\eta\sigma^{\nu}{}_{\ell}(q_{j-1})\sigma^{\beta}{}_{\kappa}(q_{j-1})$$

$$g^{\gamma\gamma'}(q_{j-1})\partial_{\beta}\sigma_{\gamma\ell}(q_{j-1})\partial_{\nu}\sigma_{\gamma'\kappa}(q_{j-1})+\varepsilon r\,\partial_{\alpha}a^{\alpha}(q_{j-1})\}\qquad(7.81)$$

Here we used the replacements $\Delta q_j^{\gamma}\Delta q_j^{\gamma'}\doteq\varepsilon\eta g^{\gamma\gamma'}(q_{j-1}^{s})$ (5.35) and we have put $q_{j-1}$ instead of $q_{j-1}^{s}$ or $q_{j-1}^{r}$ in terms of order $\varepsilon$. One has $\det\sigma_{\rho\kappa}=(\det g_{\rho\lambda})^{1/2}$ and the Jacobian is finally

$$J = \prod_{j=1}^{N+1} [\eta^{-d} \det(g_{\beta\lambda}(q^s_{j-1}))]^{1/2} \exp \sum_1^{N+1} s\, \sigma^\nu{}_\kappa(q^s_{j-1}) \partial_\nu \sigma_{\gamma\kappa}(q^s_{j-1}) [\Delta q^\gamma_j + a^\gamma(q^r_{j-1})] -$$

$$- \frac{\varepsilon}{2} s^2 \eta \partial_\beta \sigma^\nu{}_\kappa(q_{j-1}) \partial_\nu \sigma^\beta{}_\kappa(q_{j-1})$$

$$\exp \sum_1^{N+1} \varepsilon\, r \partial_\alpha a^\alpha (q_{j-1}) \tag{7.82}$$

since $\sigma^\nu_\ell \, \sigma^\beta{}_\kappa \partial_\beta(\sigma_{\gamma\ell}) \partial_\nu(\sigma_{\gamma'\kappa}) g^{\gamma\gamma'} = \partial_\beta \sigma^\nu{}_\kappa \partial_\nu \sigma^\beta{}_\kappa$. Replacing now (79) in (78), and with the expression for the Jacobian, $P_1$ is obtained as ($q_\circ = Q_\circ$)

$$P_1 = \prod_{j=1}^{N+1} [(2\pi\varepsilon\eta)^{-d} \det g_{\mu\nu}(q^s_{j-1})]^{1/2} \prod_{\mu=1}^{d} dq^\mu_j \exp -\varepsilon/2 \sum_{j=1}^{N+1} g_{\mu\nu}(q^s_{j-1})$$

$$[\frac{\Delta q^\mu_j}{\varepsilon} + a^\mu(q^r_{j-1})][\frac{\Delta q^\nu_j}{\varepsilon} + a^\nu(q^r_{j-1})]$$

$$\exp \sum_{j=1}^{N+1} s\sigma^\nu{}_\kappa(q^s_{j-1}) \partial_\nu \sigma_{\gamma\kappa}(q^s_{j-1}) [\Delta q^\gamma_j + \varepsilon a^\gamma(q^r_{j-1})] + \varepsilon\, r\, \partial_\alpha a^\alpha(q_{j-1}) -$$

$$- \frac{\varepsilon}{2} s^2 \eta \partial_\beta \sigma^\nu{}_\kappa(q_{j-1}) \partial_\nu \sigma^\beta{}_\kappa(q_{j-1}) \,. \tag{7.83}$$

Introducing integrations $\int \prod_{j=1}^{N+1} dp_j$ it's possible to write this in the form of a phase space functional integral

$$P_1 = \prod_{j=1}^{N+1} \int \prod_{\mu=1}^{d} \frac{dq^\mu_j dp_{j\mu}}{(2\pi)^d} \exp\{i\varepsilon \sum_{j=1}^{N+1} p_{j\mu} \frac{\Delta q^\mu_j}{\varepsilon} + \frac{i\eta}{2} g^{\alpha\beta}(q^s_{j-1}) p_{j\alpha} p_{j\beta} + p_{j\mu}[a^\mu(q^r_{j-1}) +$$

$$+ \eta s \partial_\alpha g^{\alpha\mu}(q^s_{j-1}) - \eta s \sigma^\lambda{}_\kappa(q^s_{j-1}) \partial_\lambda \sigma^\mu{}_\kappa(q^s_{j-1}) ] \}$$

$$\exp\{i\varepsilon \sum_{j=1}^{N+1} - ir\partial_\alpha\, a^\alpha(q_{j-1}) + \frac{i\eta}{2} s^2 [\partial_\beta \sigma^\nu{}_\kappa(q_{j-1}) \partial_\nu \sigma^\beta{}_\kappa(q_{j-1}) -$$

$$- \partial_\nu \sigma^\nu{}_\kappa(q_{j-1}) \partial_\lambda \sigma^\lambda{}_\kappa(q_{j-1})] \,. \tag{7.84}$$

Fixing here $q_{N+1} = Q$, i.e. integrating $\delta(q_{N+1} - Q) P_1$ over $dq_i$ $i = 1, \dots, N+1$, one obtains the transition probability density $P(Q,T|Q_\circ,t_\circ)$ of the process $q(t)$ as a functional integral over paths with fixed endpoints $q(t_o) = Q_\circ, q(T) = Q$.

$$P(Q,T|Q_\circ,t_\circ) = \lim_{N\to\infty} \int_{q_\circ = Q_\circ}^{q_{N+1} = Q} \prod_{i=1}^{N} dq_i \prod_{j=1}^{N+1} \frac{dp_j}{(2\pi)^d} \exp\{i\varepsilon \sum_{j=1}^{N+1} p_{j\mu} \frac{\Delta q^\mu_j}{\varepsilon} - h^{[r,s]}(p_j, q_j, q_{j-1})\} \tag{7.85}$$

$$h^{[r,s]}(p,q',q) = - i\eta g^{\mu\nu}(q^s)p_\mu p_\nu + p_\mu[\eta s\sigma^\lambda{}_\kappa(q^s)\partial_\lambda\sigma^\mu{}_\kappa(q^s) - \eta s\partial_\alpha g^{\alpha\mu}(q^s) - a^\mu(q^r)] + i\, r\partial_\alpha a^\alpha(q) - \frac{i\eta s^2}{2}[\partial_\beta\sigma^\nu{}_\kappa(q)\,\partial_\nu\sigma^\beta{}_\kappa(q) - \partial_\nu\sigma^\nu{}_\kappa(q)\partial_\lambda\sigma^\lambda{}_\kappa(q)] \tag{7.86}$$

with $q^s = q + s(q'-q)$

We have obtained (85) from the probability density (77) of the Wiener process, using (75) as a transformation, keeping terms up to order $\varepsilon$ in the functional integral, and applying there the replacements $\doteq$. We want to show now that this expression satisfies the Fokker-Planck equation

$$\frac{\partial P}{\partial T} = \frac{\partial}{\partial Q^\mu}(A^\mu(Q) + \eta/2\ \frac{\partial}{\partial Q^\nu}\, g^{\mu\nu}(Q))P(Q,T;Q_0,t_0) \tag{7.87}$$

$$A^\mu(Q) = a^\mu(Q) - \eta s\sigma^\rho{}_\kappa(Q)\partial_\rho\sigma^\mu{}_\kappa(Q)$$

This could be done by developing in (85), or in its configuration space version, all quantities around the prepoint, and using the replacements (5.35) for the powers of $\Delta q_j$. It can then be compared with the functional integral representation in the prepoint which can be derived from (87). We want to make the comparison here by introducing a discretization, depending on two parameters r and s, with the same s as in (87). To that end we rewrite the "Hamiltonian" corresponding to (87)

$$\hat{H} = -\frac{i\eta}{2}\hat{p}_\mu\hat{p}_\nu g^{\mu\nu}(\hat{q}) - \hat{p}_\mu A^\mu(\hat{q}) \tag{7.88}$$

as $\hat{H} = \hat{H}_1 + \hat{H}_2$, where

$$\hat{H}_1 = -\frac{i\eta}{2}\hat{p}_\mu\hat{p}_\nu g^{\mu\nu}(\hat{q}) + \eta s\,\hat{p}_\mu\sigma^\nu{}_\kappa(\hat{q})\partial_\nu\sigma^\mu{}_\kappa(\hat{q})$$

$$\hat{H}_2 = -\hat{p}_\mu a^\mu(\hat{q})\,. \tag{7.89}$$

Then using the results, obtained in chapter 2, we can associate to $\hat{H}_1, \hat{H}_2$ the phase space functions $H_1^{\Omega s}(p,q)$, $H_2^{\Omega r}(p,q)$ and the functions $h_1^{\Omega s}(p,q',q)$, $h_2^{\Omega r}(p,q',q)$ where the correspondence rule $\Omega_\alpha(u,v) = \exp i(\frac{1}{2} - \alpha)\, u^\mu v_\mu$ is used, with $\alpha = s$ for $\hat{H}_1$ and $\alpha = r$ for $\hat{H}_2$

All this yields a functional integral representation in the form

$$P(Q,T;Q_0,t_0) = \int_{\gamma_{12}} \mathcal{D}q\mathcal{D}p\ \exp i\int_{t_0}^{T} d\tau[p_\mu\dot{q}^\mu - H_{12}(p,q)]\ \delta(q(T)-Q)\delta(q(t_0)-Q_0) \tag{7.90}$$

where $\gamma_{12}$ discretizes $H_{12} = H_1^{\Omega s} + H_2^{\Omega r}$ as $h_1^{\Omega s} + h_2^{\Omega r}$. In fact we have $h_1^{\Omega s} = H_1^{\Omega s}(p,q^s)$

and $h_2^{\Omega r} = H_2^{\Omega r}(p,q^r)$ (i.e. $\gamma_1(s)$ and $\gamma_1(r)$ discretization respectively) and an explicit calculation gives

$$h_1^{\Omega s}(p,q',q) + h_2^{\Omega r}(p,q',q) = -\frac{i\eta}{2} g^{\mu\nu}(q^s)p_\mu p_\nu + \eta s p_\mu \sigma^\nu{}_\kappa(q^s)\partial_\nu\sigma^\mu{}_\kappa(q^s) - \eta s p_\nu \partial_\mu g^{\mu\nu}(q^s)$$

$$- i\eta s^2 \partial_\mu[\sigma^\nu{}_\kappa \partial_\nu \sigma^\mu{}_\kappa](q^s) + \frac{i\eta s^2}{2}\partial_\mu\partial_\nu g^{\mu\nu}(q^s)$$

$$- p_\mu a^\mu(q^r) + ir\partial_\alpha a^\alpha(q^r) \,. \tag{7.91}$$

Taking into account that we can replace $q^r, q^s$ by a $q$ in terms of order $\varepsilon$, and using $2\partial_\mu[\sigma^\nu{}_\kappa\partial_\nu\sigma^\mu{}_\kappa] - \partial_\mu\partial_\nu g^{\mu\nu} = (\partial_\nu\sigma^\mu{}_\kappa)(\partial_\mu\sigma^\nu{}_\kappa) - (\partial_\mu\sigma^\mu{}_\alpha)(\partial_\nu\sigma^\nu{}_\alpha)$ we see that (90) is the same as (74).

The same result can also be obtained more easily using the so called Ito rules in the stochastic differential equation. These say that terms of order higher than $O(dt) = O(\varepsilon)$ should be dropped while $dw = O(dt^{1/2})$ with $dw^k dw^\ell = \delta^{k\ell}\, dt$. Expanding (75) around the prepoint we have up to order $\varepsilon$

$$\Delta q_j^\nu + \varepsilon a^\nu(q_{j-1}) = \sqrt{\eta}\,[\sigma^\nu{}_\kappa(q_{j-1}) + s\Delta q_j^\mu \partial_\mu\sigma^\nu{}_\kappa(q_{j-1})]\,\Delta w_j^\kappa$$

$$= \sqrt{\eta}\,[\sigma^\nu{}_\kappa(q_{j-1}) + s\eta\partial_\mu\sigma^\nu{}_\kappa(q_{j-1})\sigma^\mu{}_\ell(q_{j-1})]\Delta w_j^\ell\,\Delta w_j^\kappa$$

leading to

$$\Delta q_j{}^\nu + \varepsilon[a^\nu(q_{j-1}) - s\eta\partial_\mu\sigma^\nu{}_\kappa(q_{j-1})\sigma^\mu{}_\kappa(q_{j-1})] = \sqrt{\eta}\sigma^\nu{}_\kappa(q_{j-1})\,\Delta w_j{}^\kappa . \tag{7.92}$$

This corresponds to the Ito form of the original equation (75). The Stratonovich form is obtained if we expand around $q_{j-1}^{1/2}$ in the coefficient of $\Delta w$.

$$\Delta q_j{}^\nu + \varepsilon[a^\nu(q_{j-1}) - \eta(s-\tfrac{1}{2})\partial_\mu\sigma^\nu{}_\kappa(q_{j-1})\sigma^\mu{}_\kappa(q_{j-1})] = \sqrt{\eta}\sigma^\nu{}_\kappa(q_{j-1}^{1/2})\,\Delta w_j{}^\kappa$$

but will not be of further interest to us here. The Ito form (91) is particularly convenient because we can compare it immediately to the functional integral in prepoint discretization, by the same procedure that we used to go from (79) to (84). Indeed, the Jacobian is now very simple,

$$J = \mathrm{Det}\left(\frac{\partial w_j{}^\beta}{\partial q_i{}^\mu}\right) = \prod_{j=1}^{N+1} \eta^{-d/2}\,[\det g_{\mu\nu}(q_{j-1})]^{1/2} \tag{7.93}$$

and replacing also the argument of the exponential, we obtain a functional integral representation in the prepoint discretization

$$P(Q,T|Q_\circ,t_\circ) = \lim_{N\to\infty} \int_{q_\circ=Q_\circ}^{q_{N+1}=Q_{N+1}} \prod_1^{N+1} dq_i \prod_{j=1}^{N+1} [\,2\pi\varepsilon\eta\,]^{-d/2}(\det g_{\mu\nu}(q_{j-1}))^{1/2}$$

$$\exp - \frac{\varepsilon}{2\eta}\sum_{j=1}^{N+1} g_{\mu\nu}(q_{j-1})\left(\frac{\Delta q_j^\mu}{\varepsilon} + A^\mu(q_{j-1})\right)\left(\frac{\Delta q_j^\nu}{\varepsilon} + A^\nu(q_{j-1})\right) \tag{7.94}$$

with $A^\mu = a^\mu - \eta s\, \sigma^\rho{}_\kappa \partial_\rho \sigma^\mu{}_\kappa$

just like in (87), and this is the representation that follows at once from (88) where $\hat{H}$ is written in antistandard order(p.19) . The replacement used in(92) to obtain the different forms of the (discretized) stochastic differential equations and the ones used in the functional integral have the same origin, and it's worth remarking that while the simple rule $\Delta w_j^\kappa\, \Delta w_j^\ell = \varepsilon\delta^{\kappa\ell}$ is enough to compare the different forms of (92),one needs the replacements up to $(\Delta q)^6$ to compare the related functional integral representations (83) and (94).

We want to give now still another method to obtain the same functional integral representation. It has a nice formal analogue [J3,L4], but can be made precise when we use the discretized form (92), that we write as

$$\Delta q_j^\nu + A^\nu(q_{j-1})\varepsilon = \sigma^\nu{}_i(q_{j-1})\, \Delta w^i . \tag{7.95}$$

Formally the method consists in solving

$$\dot{q}^\nu + A^\nu(q) = \sigma^\nu{}_i(q) f^i(t) \tag{7.96}$$

for fixed f and doing the average over $f^\mu(t)$ afterwards, using the formula for the average of a functional $F(f)$ (7.6)

$$\{F(f)\} = \bar{N}^{-1} \int \mathcal{D}f\, F(f) \exp - \frac{1}{2}\int d\tau \sum_{\mu=1}^{d} (f^\mu(\tau))^2 . \tag{7.97}$$

Let $q_f(t,Q_\circ,t_\circ)$ be the solution of (96) for fixed $f(t)$ and initial condition $Q_\circ$ at $t_\circ$ . Then the probability density $P(Q,T|Q_\circ,t_\circ)$ is $\{P_f(Q,T|Q_\circ,t_\circ)\}$ , where $P_f = \delta(Q-q_f(T,Q_\circ,t_\circ))$ is the probability density of the deterministic process with $f(t)$ fixed. For any functional $F(q_f(\tau))$ depending on $q_f(\tau)$ in the interval $[t_\circ,T]$ , one has

$$F(q_f(\tau)) = \int \mathcal{D}q(\tau)\, F(q(\tau))\ \ \delta(q(\tau) - q_f(\tau))\Big|_{t_\circ}^{T} \tag{7.98}$$

Here $\delta(q(\tau)-q_f(\tau)) = \prod_i \delta(q(t_i)-q_f(t_i))$ just expresses that the original differential equation (96) is satisfied in the time interval $[t_\circ,T]$ , and consequently we can write

$$\delta(q(\tau)-q_f(\tau))\Big|_{t_o}^{T} = \delta[\dot{q}^\nu+A^\nu(q(\tau))-\sqrt{\eta}\sigma^\nu{}_\mu(q(\tau))f^\mu(\tau)]\Big|_{t_o}^{T}\ \delta(q^\nu(t_o)-Q_o^\nu)\,J \tag{7.99}$$

with J the Jacobian due to the change of argument of the $\delta$-function . Using the representation $\delta[K(\tau)]\Big|_{t_o}^{T} = \int \mathcal{D}p\ \exp i\int_{t_o}^{T} d\tau\ p(\tau)K(\tau)$ for this $\delta$-function one has in the case that $F(q_f) = \delta(Q-q_f(T))$

$$\delta(Q-q_f(T)) = \int \mathcal{D}q\mathcal{D}p\ \ \exp\{i\int d\tau\ p_\mu(\tau)[\dot{q}^\mu+ A^\mu(q) - \sqrt{\eta}\sigma^\mu{}_\nu(q)f^\nu(\tau)]\}J\ \delta(q(T)-Q)\delta(q(t_o)-Q_o) \tag{7.100}$$

We obtain the representation of $P(Q,T\ Q_o,t_o)$ by doing the average $\{\delta(Q-q_f(T))\}$ . From (97) we see that this is just a Gaussian integration which can be done exactly.

This scheme will be made more rigorous now, by using discretized forms for all the formal functional integrations. We replace (98) , for $F(q_f(\tau)) = \delta(Q-q_f(T))$ by

$$\delta(Q - q_f(T)) = \lim_{N\to\infty} \int \prod_{i=1}^{N+1} dq_i\ \delta(q_{N+1}-Q) \prod_{j=1}^{N+1} \delta(q_j - q_f(t_j)) \tag{7.101}$$

with the usual time division of $[t_o,T]$, $t_j= t_o + j\varepsilon$ ,$t_{N+1}=T$. The discretized form (95) of the stochastic differential equation is used to transform the product of $\delta$-functions into ($q_o=Q_o$)

$$\begin{aligned}\prod_{j=1}^{N+1} \delta(q_j-q_f(t_j)) &= \mathrm{Det}\{ \frac{\partial}{\partial q^\mu{}_\kappa}[ \frac{q_j^\nu- q_{j-1}^\nu}{\varepsilon} + A^\nu(q_{j-1})-\sqrt{\eta}\ \sigma^\nu{}_i(q_{j-1})\frac{\Delta w_i}{\varepsilon} ] \}\\ &\times \prod_{j=1}^{N+1} \prod_{\nu=1}^{d} \delta( \frac{q_j^\nu-q_{j-1}^\nu}{\varepsilon} + A^\nu(q_{j-1}) -\sqrt{\eta}\sigma^\nu{}_i(q_{j-1})\frac{\Delta w_i}{\varepsilon} )\end{aligned} \tag{7.102}$$

which is the discretized version of (99) including the Jacobian. Since everything is discretized in the prepoint, this Jacobian is

$$J = \prod_{j=1}^{N+1} \frac{1}{\varepsilon^d}$$

and writing the N+1 $\delta$-functions $\delta(K_j)$ in (102) as $\int dp_j(\frac{\varepsilon}{2\pi})^d \exp i\ \varepsilon p_j K_j$ ;

$$\delta(Q-q_f(T)) = \lim_{N\to\infty} \int \prod_{i=1}^{N+1} \frac{dq_i dp_i}{(2\pi)^d}\ \exp i\varepsilon \sum_{j=1}^{N+1} p_{j\mu}( \frac{\Delta q_j^\mu}{\varepsilon} + A^\mu(q_{j-1})-\sqrt{\eta}\sigma^\mu{}_i(q_{j-1})\frac{\Delta w_i}{\varepsilon})\ \delta(q_{N+1}-Q) \tag{7.103}$$

The average over Brownian motion is done by $\int \prod_{j=1}^{N+1} (2\pi\varepsilon)^{-d/2} \prod_{\kappa=1}^{d} dw_j^\kappa \exp\{-\frac{1}{2\varepsilon}\sum_{1}^{N+1}\sum_{\kappa=1}^{d} (\Delta w_j^\kappa)^2\}$ ($w_o=0$) , corresponding to (97) .

Changing variables $z_j^k = \Delta w_j^k$, the Gaussian integration in (103) gives the phase space integral version of (94)

$$P(Q,T|Q_o,t_o) = \int_{q_o=Q_o}^{q_{N+1}=Q} \prod_{i=1}^{N} dq_i \prod_{j=1}^{N+1} dp_j (2\pi)^{-d} \exp\{i\varepsilon \sum_{j=1}^{N+1} p_{j\mu}\frac{\Delta q_j^{\mu}}{\varepsilon} + \frac{i\eta}{2} g^{\mu\nu}(q_{j-1}) p_{j\mu}p_{j\nu} + p_{j\mu}A^{\mu}(q_{j-1})\} . \qquad (7.104)$$

The same calculation can also be done starting from equation (75). The essential difference is that the expression for J is now more complicated. In fact we have in that case

$$J = \prod_{j=1}^{N+1} \frac{1}{\varepsilon^d} \det\{\delta^{\nu}{}_{\mu} + \varepsilon r \partial_{\mu} a^{\nu}(q_{j-1}^r) - \sqrt{\eta}\, s \partial_{\mu}\sigma^{\nu}{}_{\kappa}(q_{j-1}^s)\Delta w_j^{\kappa}\} \qquad (7.105)$$

which can be evaluated as in (81), using the replacements for $\Delta w_j^{\kappa}\,\Delta w_j^{\rho}$ (these are allowed since one must average afterwards). The remaining Gaussian integral leads to representation (85) of the transition probability P.[L21]

## Appendix 7.1 : Gaussian white noise and the Wiener process

White noise is a stochastic process $f(\tau) = (f^1(\tau),\dots,f^m(\tau))$ for which ($\{\dots\}$ denotes average)

$$\{f^i(\tau)\} = 0$$
$$\{f^i(\tau)\, f^j(\tau')\} = \delta^{ij}\delta(\tau-\tau') \quad , \quad i,j=1,\dots,m. \tag{A7.1}$$

In the case of Gaussian white noise higher moments are given by the rules of Gaussian processes : odd moments vanish and even moments are the sum of terms one obtains by breaking them up in all possible ways into a product of pairs and applying (A7.1) to each pair.

In a strict sense however because of the $\delta$-function in (A7.1) this process is badly defined. Its integral however $w(t) = \int_0^t f(\tau)d\tau$ is the well-defined Wiener process.

We shall derive now by a limiting procedure the probability density function of $f(\tau)$, $t_o \le \tau \le t$. We divide the interval $[t_o, t]$ into N+1 intervals of length $\varepsilon$. The discrete version of (A7.1) is

$$\{f^i_k\, f^j_l\} = \{f^i(t_k)\, f^j(t_l)\} = \delta_{ij}\, \frac{1}{\varepsilon}\delta_{\ell k}. \tag{A7.2}$$

Because of the Gaussian property one has that the joint probability function $W_{N+1}(f_1,t_1,\dots,f_{N+1},t_{N+1})$ is given by

$$W_{N+1}(f_1,t_1;\dots;f_{N+1},t_{N+1})df_1df_2\dots df_{N+1} = \tag{A7.3}$$
$$= \frac{1}{\overline{N}} \exp[-\frac{1}{2}\sum_{j,k=1}^{N+1} f_j \varepsilon\delta_{jk} f_k]\, df_1df_2\dots df_{N+1}$$

where $\overline{N}$ is a normalization constant such that $\int \prod_{j=1}^{N+1} df_j\, W_{N+1} = 1$ hence

$$\overline{N} = \int \prod_{i=1}^{N+1} df_i\, \exp[-\frac{1}{2}\varepsilon \sum_{j,k=1}^{N+1} f_j\delta_{jk}f_k] . \tag{A7.4}$$

Consider now a function $F(f_1,f_2,\dots,f_{N+1})$. Its expectation value is given by

$$\{F(f_1,f_2,\dots,f_{N+1})\} = \frac{\int \prod_{i=1}^{N+1} df_i\, F(f_1,\dots,f_{N+1})\exp[-\frac{\varepsilon}{2}\sum_{j=1}^{N+1} f_j^2]}{\int \prod_{i=1}^{N+1} df_i\, \exp[-\frac{\varepsilon}{2}\sum_{j=1}^{N+1} f_j^2]}. \tag{A7.5}$$

When we take formally the limit $N\to\infty$ ($\varepsilon\to 0$ in (A7.5) then $F(f_1,\dots,f_{N+1})$ becomes a functional $F[f(\tau)]$ . When we set formally $\mathcal{D}f(\tau) = \lim_{N\to\infty} \prod_{i=1}^{N+1} df_i$ and $\int_{t_o}^{t} d\tau\, f(\tau)^2 = \lim_{N\to\infty} \varepsilon\sum_{j=1}^{N+1} f_j^2$ we obtain (7.6).

In a second part of this appendix we treat as an example of (7.1) the stochastic differential equation

$$\frac{dw^i(t)}{dt} = f^i(t), \qquad w(t_o)=0 \tag{A7.6}$$

and show that its solution $w^i(t)$ is indeed the Wiener process defined through its joint probability densities

$$W_n(w_1,t_1;w_2,t_2;\ldots;w_n,t_n)dw_1\ldots dw_n =$$

$$=[\prod_{j=1}^{n}(2\pi(t_j-t_{j-1}))^{-m/2}\exp[-\sum_{j=1}^{n}\sum_{\mu=1}^{m}\frac{(w_j^\mu-w_{j-1}^\mu)^2}{2(t_j-t_{j-1})}]dw_1\ldots dw_n \tag{A7.7}$$

Let $\tilde{W}_n(w_1,t_1;\ldots;w_n,t_n)$ be the joint probability density of the solution of (A7.6). Because of the Markov property one has

$$\tilde{W}_n(w_1,t_1;\ldots;w_n,t_n) = P(w_1,t_1|w_{n-1},t_{n-1})P(w_{n-1},t_{n-1}|w_{n-2},t_{n-2})\ldots$$
$$P(w_1,t_1|0,t_o) \tag{A7.8}$$

with $P(w',t'|w,t)$ the transition probability density which must have the value

$$(2\pi(t'-t))^{-m/2}\exp[-\frac{1}{2(t'-t)}\sum_{\mu=1}^{m}(w'^\mu-w^\mu)^2] \tag{A7.9}$$

in order that (A7.8) coincides with (A7.7). That this is so follows by explicit calculation of the path integral

$$P(w,'t'|w,t) = \int \mathcal{D}w\,\exp[-\int_t^{t'}d\tau\,\frac{1}{2}\sum_{\mu=1}^{m}(\frac{dw^\mu}{d\tau})^2]\,\delta(w(t')-w')\,\delta(w(t)-w) =$$

$$= \lim_{N\to\infty}\int\prod_{i=1}^{N}dq_i\prod_{j=1}^{N+1}(2\pi\varepsilon)^{-m/2}\exp[-\sum_{j=1}^{N+1}\sum_{\mu=1}^{m}\frac{(q_j^\mu-q_{j-1}^\mu)^2}{2\varepsilon}] \tag{A7.10}$$

where $\varepsilon=(t'-t)/(N+1)$ and $q_o=w, q_{N+1}=w'$ (See [F5]p42) or directly from the Fokker-Planck equation

$$\dot{P}(w',t'|w,t) = \frac{1}{2}\sum_{\mu=1}^{m}\frac{\partial}{\partial w^\mu}\frac{\partial}{\partial w^\mu}P(w',t'|w,t)\ . \tag{A7.11}$$

## CHAPTER VIII: PRODUCT INTEGRALS

### 8.1. Introduction and relation with the usual functional integrals

Product integrals are a generalization of functional integrals that arise for example when one computes the propagator associated to a evolution equation

$$i \frac{\partial U(t,t')}{\partial t} = \hat{H}(t)\, U(t,t')\ ,\ U(t,t) = 1, \tag{8.1}$$

where $\hat{H}(t)$ is an operator acting on a Hilbert space $\mathcal{H} = \mathcal{H}_1 \otimes E$; where $\mathcal{H}_1$ is the familiar space of functions $f(Q) = (Q^1,Q^2,\dots,Q^d)$ and E a vector space of finite dimension n. In $\mathcal{H}_1$ act the usual operators $\hat{Q}^\alpha, \hat{P}_\beta = -i\ \partial/\partial Q^\beta$, with commutation relations $[\hat{Q}^\alpha,\hat{P}_\beta] = i\ \delta^\alpha{}_\beta$. Let $|\nu\rangle$, $\nu = 1,2,\dots n$, be an orthonormal basis of the space E. We use letters of the beginning of the Greek alphabet for components in $\mathcal{H}_1$ and of the end for E. In $\mathcal{H}_1$ we have the basis $|q\rangle$, $\hat{Q}^\alpha|q\rangle = q^\alpha|q\rangle$, and the basis $|p\rangle$, $\hat{P}_\beta\ |p\rangle = p_\beta|p\rangle$. In $\mathcal{H}$ we have the basis $|q,\nu\rangle = |q\rangle \otimes|\nu\rangle$ and $|p,\nu\rangle = |p\rangle \otimes |\nu\rangle$ normalized as

$$\langle q',\nu'\ |q\ ,\nu\ \rangle = \delta_{\nu\nu'}\ \delta(q'-q),\quad \langle p',\nu'|\ p,\nu\rangle = \delta_{\nu\nu'}\ \delta(p'-p). \tag{8.2}$$

We define in $\mathcal{H}$ the operators $\hat{q}^\alpha = \hat{Q}^\alpha \otimes I_E$, $\hat{p}_\beta = \hat{P}_\beta \otimes I_E$, where $I_E$ is the identity on E, and one has then

$$\hat{q}^\alpha|q,\nu\rangle = q^\alpha|q,\nu\rangle\ ,\ \hat{p}_\beta|p,\nu\rangle = p_\beta\ |p,\nu\rangle\quad . \tag{8.3}$$

One also has

$$\langle q,\nu|p,\nu'\rangle = \delta_{\nu\nu'}\ (2\pi)^{-d/2}\quad \exp[i\ p.q]. \tag{8.4}$$

One can then represent the operators in $\mathcal{H}$ by nxn matrices whose elements will be functions of $\hat{q}^\alpha$ and $\hat{p}_\beta$. If these functions are in antistandard order (operators $\hat{q}^\alpha$ to the right of the operators $\hat{p}_\beta$) then one has

$$\langle q',\nu'|\ \hat{H}(t)\ |q,\nu\ \rangle = h_{\nu'\nu}\ (p,q,t) \tag{8.5}$$

where $h_{\nu'\nu}$ is an nxn matrix.

Let us compute now $\langle Q,\nu\ |U(t,t_\circ)|\ Q_\circ,\nu_\circ\ \rangle$. Using the completeness relation

$$\sum_{\nu} \int |q,\nu\rangle \, dq \, \langle q,\nu| = 1 \tag{8.6}$$

one has ($t_j = t_\circ + j\varepsilon$, $t_{N+1} = t$)

$$I_{\nu\nu_\circ} \equiv \langle Q,\nu|U(t,t_\circ)|Q_\circ,\nu_\circ\rangle = \sum_{\mu_1\ldots\mu_N} \int_{\substack{q_\circ=Q_\circ\\ \mu_\circ=\nu_\circ}}^{\substack{q_{N+1}=Q\\ \mu_{N+1}=\nu}} \prod_{i=1}^{N} dq_i$$

$$\cdot \prod_{j=1}^{N+1} \langle q_j,\mu_j |U(t_j,t_{j-1})|q_{j-1}, \mu_{j-1}\rangle . \tag{8.7}$$

From $U(t_j,t_{j-1}) = 1 - i\varepsilon\hat{H}(t_{j-1})$ (we only need to keep terms up to $O(\varepsilon)$ since we shall consider the limit $N \to \infty$ ) one has ($\Delta q_j \equiv q_j - q_{j-1}$)

$$\langle q_j,\mu_j |U(t_j,t_{j-1})| q_{j-1},\mu_{j-1}\rangle = \int \frac{dp}{(2\pi)^d} \, [1_E \exp (ip.\Delta q_j).$$

$$\cdot \exp (-i\varepsilon\tilde{h} (p_j,q_{j-1},t_{j-1}))]_{\mu_j\mu_{j-1}}, \tag{8.8}$$

where $\hat{h}$ is the nxn matrix $h_{\mu\nu}$ defined by formula (5). Let us call $(I_N)_{\nu\nu_\circ}$ the result obtained replacing (8) in the right hand side of (7). One has

$$(I_N)_{\nu\nu_\circ} = \int_{\mu_\circ=\nu_\circ}^{\mu_{N+1}=\nu} \prod_{i=1}^{N} dq_i \prod_{j=1}^{N+1} \frac{dp_j}{(2\pi)^d} [1_E \exp i\varepsilon \sum_{j=1}^{N+1} p_j . \frac{\Delta q_j}{\varepsilon} .$$

$$\cdot P \prod_{j=1}^{N+1} \exp[-i\varepsilon\tilde{h}(p_j,q_{j-1},t_{j-1})]_{\mu_j\mu_{j-1}} ] , \tag{8.9}$$

where the symbol P means that the product of matrices is to be done chronologically. The limit $N \to \infty$ of (9), which is $I_{\nu\nu_\circ}$, is written formally as

$$I_{\nu\nu_\circ} = \int_{\nu_\circ}^{\nu} \mathcal{D}q \, \mathcal{D}p \quad P \exp i \int_{t_\circ}^{t} d\tau \, [p.\dot{q} - \tilde{h}(p(\tau),q(\tau),\tau))] \quad .$$

$$\cdot \delta(q(t) - Q).\delta(q(t_\circ)-Q_\circ) \tag{8.10}$$

and is called a product integral which by definition is the limit of $(I_N)_{\nu\nu_\circ}$ when $N \to \infty$ .We remark that we should indicate in (10) explicitly the prepoint discretization $\gamma_1(0)$ (see formula (9)) corresponding to the antistandard order we are using, but as in this chapter we shall always work with this discretization we omit it. The reason of the symbol P in (9) is of course that $\tilde{h}$ is now a matrix and consequently one cannot write $\prod_j \exp - i\varepsilon\tilde{h}_j$ as $\exp - i\varepsilon \sum_j \tilde{h}_j$, since the $\tilde{h}$ do not necessarily commute at unequal times.

We shall see now that it is possible to reduce the product integral (10) to an infinite series of standard functional integrals. In order to do this we decompose the matrix $\tilde{h}$ as

$$\tilde{h}(p,q,\tau) = \tilde{A}(p,q,\tau) + \tilde{B}(p,q,\tau) + \tilde{C}(p,q,\tau) \tag{8.11}$$

where

$$\tilde{A}_{\mu\nu} = A(p,q,\tau)\delta_{\mu\nu} \ , \ \tilde{B}_{\mu\nu} = B_\nu(p,q,\tau)\delta_{\mu\nu} \ , \tag{8.12}$$

$$\tilde{C}_{\mu\nu} = C_{\mu\nu}(p,q,\tau) \ , \ C_{\mu\nu} \neq 0 \ , \ \mu\neq\nu \ , \ C_{\mu\mu} = 0. \tag{8.13}$$

Then (10) can be written as (we put $\tilde{A}(\tau) = \tilde{A}(p(\tau),q(\tau),\tau)$, etc.)

$$I_{\nu\nu_0} = \int_{\nu_0}^{\nu} \mathcal{D}q\mathcal{D}p \exp i \int_{t_0}^{t} d\tau [\, p.\dot{q} - \tilde{A}(\tau)\,] \, . \, P \exp -i \int_{t_0}^{t} d\tau [\tilde{B}(\tau) + \tilde{C}(\tau)\,] \, .$$

$$. \ \delta(q(t)-Q) \ . \ \delta(q(t_0)-Q_0), \tag{8.14}$$

since $\tilde{A}(\tau)$ is diagonal and is not affected by the P-symbol. We call

$$\tilde{V}(t,t_0) \equiv P \exp - i \int_{t_0}^{t} d\tau [\, \tilde{B}(\tau) + \tilde{C}(\tau) \,] \ . \tag{8.15}$$

Going now to an interaction picture representation with $\tilde{B}(\tau)$ as the free operator one easily proves the standard formula

$$\tilde{V}(t,t_0) = \tilde{V}_0(t,t_0) \, P \exp[\, - i \int_{t_0}^{t} d\tau \ \tilde{C}^{(ip)}(\tau) \,] \tag{8.16}$$

where

$$\tilde{C}^{(ip)}(\tau) \equiv \tilde{V}_0(\tau,t_0)^{-1} \, \tilde{C}(\tau)\tilde{V}_0(\tau,t_0) \tag{8.17}$$

with

$$\tilde{V}_0(t'',t') \equiv P \exp [-i \int_{t'}^{t''} d\tau \ \tilde{B}(\tau)] \ . \tag{8.18}$$

Here we have

$$[\tilde{V}_0(t'',t')]_{\mu\nu} = \delta_{\mu\nu}\Delta_\nu(t'',t') \ , \tag{8.19}$$

$$\Delta_\nu(t'',t') = \exp\left[-i\int_{t'}^{t''} d\tau\, B_\nu(\tau)\right] . \tag{8.20}$$

Thenfrom (14) one obtains

$$I_{\nu\nu_\circ} = \int \mathcal{D}q\mathcal{D}p \exp i\int_{t_\circ}^{t} d\tau[p.\dot{q}-A(\tau)\,].\delta(q(t)-Q)\delta(q(t_\circ)-Q_\circ) .$$

$$. \left[\delta_{\nu\nu}\,\Delta_\nu\ (t,t_\circ) + \sum_{k=1}^{\infty} (-i)^k \int \prod_{i=1}^{k} d\tau_i \prod_{j=1}^{k-1} \theta(\tau_j - \tau_{j+1}) \right.\, .$$

$$. \exp\left[-i\int_{\tau_1}^{t} d\tau B_\nu(\tau)\right] . C_{\nu\mu}\ (\tau_1)\ . \exp\left[-i\int_{\tau_2}^{\tau_1} d\tau B_{\mu_1}\ (\tau)\right]\ . \ \ldots$$

$$. \left. C_{\mu_{k-1}\ \nu_\circ}(\tau_k).\exp\left[-i\int_{\tau_\circ}^{\tau_k} d\tau\, B_{\nu_\circ}(\tau)\right]\right], \tag{8.21}$$

which expresses the product integral in terms of standard functional integrals.

We shall do now two applications of product integrals.

## 8.2. First example: a modified Langevin equation

Consider the modified Langevin equation [B10] ($\sigma^\alpha_{\ i}$ is a constant matrix)

$$\dot{q}^\alpha(t) + A^\alpha(q(t)\ ,m(t)) = \sigma^\alpha_{\ i}\ \xi^i(t)\ , \tag{8.22}$$

where $q=(q^1,\ldots,q^d)$ , $\xi^i(t)$ is a Gaussian white noise with correlation $\{\xi^i(t)\xi^j(t')\} = \delta^{ij}\ \delta(t-t')$, and $m(t)$ is an independently defined Markovian process taking values $f(\nu),\nu = 1,2,\ldots n$, characterized by the transition matrix $M_{\mu\nu}$ in the sense that the probability that $m(t+s) = f(\mu)$ if $m(t)=f(\nu)$ is $[\exp(-s\tilde{M})]_{\mu\nu}$ where $\tilde{M}$ is the nxn matrix of elements $M_{\mu\nu}$ . In Ref [B10] it is proved that the generating functional $Z_{\nu_\circ}[J,J^\star]$ of the non-Markovian process $q(t)$ for initial conditions $m(t_\circ)=\nu_\circ$, $q(t_\circ) = Q_\circ$ , is given by the product functional integral

$$Z_{\nu_\circ}[J,J^\star] = \int_{\mu_\circ = \nu_\circ} \mathcal{D}q\mathcal{D}p\ P \exp i\int_{t_\circ}^{t} d\tau[p.\dot{q} - \tilde{h}(p,q)+J.q + J^\star.p]\ .\ \delta(q(t_\circ)-Q_\circ), \tag{8.23}$$

where the matrix $\tilde{h}$ has elements

$$h_{\mu\nu}(p,q) = \delta_{\mu\nu}[-\tfrac{i}{2}\, D^{\alpha\beta} p_\alpha p_\beta - p_\alpha A^\alpha(q,f(\nu))] - i\, M_{\mu\nu}\ , \tag{8.24}$$

with $D^{\alpha\beta} = \sum_i \sigma^\alpha_i \sigma^\beta_i$. We remark that in the generating functional $Z_{\nu_\circ}$ only the initial conditions are fixed $(q(t_\circ)=Q_\circ, \mu_\circ = \nu_\circ)$ (cf. section 7.3). Putting $Z_{\nu_\circ} = \sum_\nu Z_{\nu\nu_\circ}$ one has that $Z_{\nu\nu_\circ}$ is defined as $\lim I_N^{(J)}$, $N \to \infty$, with $(J_j=J(t_j), J^\star_j = J^\star(t_j))$

$$(I_N^{(J)})_{\nu\nu_\circ} = \sum_{\mu_1\ldots\mu_N} \int_{\substack{q_\circ=Q_\circ\\ \mu_\circ=\nu_\circ}}^{\mu_{N+1}=\nu} \prod_{i=1}^{N+1} dq_i \prod_{j=1}^{N+1} \frac{dp_j}{(2\pi)^d} \, P \prod_{j=1}^{N+1} \exp i\varepsilon[p_j \frac{\Delta q_j}{\varepsilon} -$$

$$- \tilde{h}(p_j,q_{j-1}) + J_j.q_j + J^\star_j.p_j]\Big|_{\mu_j\mu_{j-1}}. \tag{8.25}$$

We apply now formula (21). In this example one has

$$A(\tau) = -\frac{1}{2} D^{\alpha\beta} p_\alpha(\tau) p_\beta(\tau) - J(\tau).q(\tau) - J^\star(\tau).p(\tau) \tag{8.26}$$

$$B_\nu(\tau) = - p_\alpha(\tau) A^\alpha(q(\tau), f(\nu)) - i\, a_\nu , \tag{8.27}$$

where $a_\nu = M_{\nu\nu} > 0$, and $C_{\mu\nu}(\tau) = - i\, M_{\mu\nu}$, $\mu \neq \nu$.
Using (21) we have

$$Z_{\nu\nu_\circ}[J,J^\star] = \int \mathcal{D}q\mathcal{D}p \;\exp i \int_{t_\circ}^{t} d\tau \;[p.\dot{q} + \frac{i}{2} D^{\alpha\beta} p_\alpha p_\beta + J.q + J^\star.p\;].$$

$$. \;\delta(q(t_\circ)-Q_\circ) \;.\,[\;\delta_{\nu\nu_\circ} \exp[-a_\nu(t-t_\circ)] \exp i \int_{t_\circ}^{t} d\tau [p_\alpha A^\alpha(q,f(\nu)) +$$

$$+ \sum_{k=1}^{\infty} \int \prod_{i=1}^{k} d\tau_i \prod_{j=1}^{k-1} \theta(\tau_j - \tau_{j+1}) .\, \exp i \int_{\tau_1}^{t} d\tau [p_\alpha A^\alpha(q,f(\nu)) + i\, a_\nu] \;.$$

$$. \;P \prod_{j=1}^{k} (- M_{\mu_{j-1}\mu_j}) \;.\, \exp i \int_{\tau_{j+1}}^{\tau_j} d\tau \; p_\alpha A^\alpha(q,f(\mu_j)) + i a_{\mu_j} \;] , \tag{8.28}$$

where $\mu_\circ = \nu$, $\mu_k = \nu_\circ$, $\tau_{k+1} = t_\circ$. In (28) the symbol P just indicates that the product that follows is to be done as in (21). Formula (28) has a simple probabilistic interpretation. Consider the transition probability density $W(Q,\nu, t \mid Q_\circ,\nu_\circ,t_\circ)$ of the Markovian process $(q(t), m(t))$ which is given by (28) with $J=J^\star = 0$ and when we fix $q(t) = Q$ (this means that in (25) we put $q_{N+1} = Q$ and we do not integrate over $q_{N+1}$). One has then from (28) that

$$W(Q,\nu,t;Q_o;\nu_o,t_o) = \delta_{\nu\nu_o} w_\nu(Q,t;Q_o,t_o) + \sum_{k=1}^{\infty} \sum_{\mu_1 \ldots \mu_{k-1}}$$

$$\int \prod_{i=1}^{k} d\tau_i \prod_{j=1}^{k-1} \theta(\tau_j - \tau_{j+1}) .$$

$$. \prod_{i=1}^{k} dq_i \; w_\nu(Q,t;q_1,\tau_1).(-M_{\nu\mu_1}) \; . \; w_{\mu_1}(q_1,\tau_1;q_2,\tau_2) \ldots$$

$$\ldots (- M_{\mu_{k-1}\nu_o}) \; w_{\nu_o}(q_k,\tau_k;Q_o,t_o), \tag{8.29}$$

where $w_\mu(q',t';q,t)$ is the probability density for going from $q(t)=q$, to $q(t')=q'$ with $m(\tau) = \mu$ for $t < \tau < t'$ which is given by

$$w_\mu(q',t';q,t) = \int_{q(t)=q}^{q(t')=q'} \mathcal{D}q\mathcal{D}p \; \exp i \int_t^{t'} d\tau \; [p.\dot{q} + \frac{i}{2} D^{\alpha\beta} p_\alpha p_\beta + p_\alpha A^\alpha(q,f(\mu)) + $$
$$+ \, ia_\mu \; ] \; . \tag{8.30}$$

Considering that $(-M_{\mu\nu} d\tau)$ is the probability of the transition $\nu \to \mu$ in time $d\tau$ we see that the first term in the right hand side of (29) is the probability of going from $(Q_o,\nu_o,t_o)$ to $(Q,\nu,t)$ with no transition of $m(\tau)$, while the term with k integrations $d\tau_i$ is the probability of going from $(Q_o,\nu_o,t_o)$ to $(Q,\nu,t)$ with k transitions of $m(\tau)$ at times $t > \tau_1 > \tau_2 > \ldots > \tau_k > t_o$. The non-Markovian process we have considered here is a special case of the composite processes studied by Van Kampen in[V1].

## 8.3. Second example: partition function of a spin 1/2 system

### 8.3.1. In general

The second example we treat is the computation of the partition function of a system with spin 1/2.[B9]
Let $\hat{H}(\hat{p},\hat{q})$ be the quantum mechanical Hamiltonian of such a system. It is a 2x2 matrix which is in general no diagonal if a magnetic field is present. The partition function of such a system is given by [F4] $(\beta = (kT)^{-1})$

$$Z = \mathrm{Tr} \exp - \beta \, \hat{H} \; . \tag{8.31}$$

One has $\mathrm{Tr}\,\hat{A} = \int dq <q\,|\mathrm{tr}\,\hat{A}|q>$, where $\mathrm{tr}\,\hat{A} = \sum_\alpha \hat{A}_{\alpha\alpha}$, the trace in the twodimensional space and consequently

$$Z = \mathrm{tr} \int dq < q \mid \exp - \beta\hat{H}(p,q) \mid q >. \tag{8.32}$$

We can immediately construct a functional integral representation of this quantity in the same way as we have done for evolution operators before. We obtain the following product functional integral ($\tilde{H}$ is a 2 dimensional matrix)

$$Z = \mathrm{tr} \int dq \int \mathcal{D}q\mathcal{D}p \;\; P \exp \int_0^\beta du[ip.\dot{q} - \tilde{H}(p,q)] \;.$$

$$.\; \delta(q(\beta)-q).\delta(q(0)-q). \tag{8.33}$$

where the symbol P acts now on the "time" u . This formula is obtained from the one for the evolution operator by the substitutions $it \to \beta$ , $i\int d\tau \to \int du$ , and the point on q in (33) indicates derivative with respect to u. In order to compute (33) we could directly apply the general formula (21), however it is instructive to treat this simpler two dimensional case directly. We go back to the notation for evolution operators to compare with our previous calculation ($\beta \to it$, $\int du \to i\int d\tau$ ). Let $\tilde{H}(p(\tau),q(\tau))$ in (33) be of the form ($d(\tau) = d(p(\tau),q(\tau))$, etc.)

$$\tilde{H} = - d(\tau)\; 1 - a(\tau)\;\; \hat{a}^+a - b(\tau)\hat{a}^+ - c(\tau)\hat{a} \tag{8.34}$$

where 1 is the unit 2x2 matrix and

$$\hat{a} = \begin{bmatrix} 0 & 0 \\ 1 & 0 \end{bmatrix} \;,\; \hat{a}^+ = \begin{bmatrix} 0 & 1 \\ 0 & 0 \end{bmatrix} \;. \tag{8.35}$$

Clearly, any general H can be written as in (34). We use the following notations

$$A = \begin{bmatrix} A_{11} & A_{10} \\ A_{01} & A_{00} \end{bmatrix} \;,\; |0\rangle = \begin{bmatrix} 0 \\ 1 \end{bmatrix} \;,\; |1\rangle = \begin{bmatrix} 1 \\ 0 \end{bmatrix} \;. \tag{8.36}$$

Formula (33) is now ($t_0$ will be zero in this example)

$$Z = \mathrm{tr} \int dq \int \mathcal{D}q\mathcal{D}p \exp i \int_{t_0}^{t} d\tau[p.\dot{q} + d(\tau)] \;.\; \delta(q(t) - q).$$

$$.\; \delta(q(t_0)-q).P \exp [- i \int_{t_0}^{t} d\tau [\tilde{h}_0(\tau) + \tilde{h}_I(\tau)]], \tag{8.37}$$

where $\tilde{h}_0(\tau) = - a(\tau)\hat{a}^+a$, $\tilde{h}_I(\tau) = - b(\tau)\hat{a}^+ - c(\tau)\hat{a}$.

We want to compute $\tilde{A} = P \exp[-i\int_{t_0}^{t} d\tau[\tilde{h}_0(\tau) + \tilde{h}_I(\tau)]]$. Using formula (16) one has

$$\tilde{A} = P\exp[-i\int_{t_0}^{t} d\tau\,\tilde{h}_0(\tau)].\ P\exp[-i\int_{t_0}^{t} d\tau\,\tilde{h}_I^{(ip)}(\tau)], \tag{8.38}$$

with

$$\tilde{h}_I^{(ip)}(\tau) = [P\exp[-i\int_{t_0}^{\tau} ds\,\tilde{h}_0(s)]]^{-1}\,\tilde{h}_I(\tau)\,[P\exp[-i\int_{t_0}^{\tau} ds\,\tilde{h}_0(s)]]. \tag{8.39}$$

Using $\hat{a}^+\hat{a}^n = \hat{a}^+\hat{a}$ one easily shows that

$$P\exp i\int_{t_0}^{\tau} ds\,a(s)\,\hat{a}^+a = \begin{bmatrix} \exp i\int_{t_0}^{\tau} ds\,a(s) & 0 \\ 0 & 1 \end{bmatrix}. \tag{8.40}$$

Putting $\bar{\Delta}(\tau,t_0) = \exp i\int_{t_0}^{\tau} ds\,a(s)$ one sees that $(\bar{\Delta}(\tau,t_0))^{-1} = \bar{\Delta}(t_0,\tau)$ and from (38) one obtains

$$\tilde{A} = \begin{bmatrix} \bar{\Delta}(t,t_0) & 0 \\ 0 & 1 \end{bmatrix}.\ P\exp\int_{t_0}^{t} d\tau \begin{bmatrix} \bar{\Delta}(t_0,\tau) & 0 \\ 0 & 1 \end{bmatrix}. \\ .\begin{bmatrix} 0 & ib(\tau) \\ ic(\tau) & 0 \end{bmatrix}.\begin{bmatrix} \bar{\Delta}(\tau,t_0) & 0 \\ 0 & 1 \end{bmatrix}, \tag{8.41}$$

and consequently

$$\tilde{A} = \begin{bmatrix} \bar{\Delta}(t,t_0) & 0 \\ 0 & 1 \end{bmatrix}.\ P\exp\int_{t_0}^{t} d\tau \begin{bmatrix} 0 & i\bar{\Delta}(t_0,\tau)b(\tau) \\ ic(\tau)\bar{\Delta}(\tau,t_0) & 0 \end{bmatrix}. \tag{8.42}$$

We develop now the P-exponential in a Dyson series

$$\tilde{A} = \begin{bmatrix} \bar{\Delta}(t,t_0) & 0 \\ 0 & 1 \end{bmatrix} \sum_{n=0}^{\infty} \int \prod_{i=1}^{n} d\tau_i \prod_{j=1}^{n-1} \theta(\tau_j - \tau_{j+1})\ . \\ .\ P\prod_{i=1}^{n} \begin{bmatrix} 0 & i\bar{\Delta}(t_0,\tau_i)b(\tau_i) \\ ic(\tau_i)\bar{\Delta}(\tau_i,t_0) & 0 \end{bmatrix}. \tag{8.43}$$

The term $\tilde{A}^{(2p)}$ of the series (43) is given by

$$\tilde{A}^{(2p)} = (-1)^p \int \prod_{i=1}^{2p} d\tau_i \prod_{j=1}^{2p-1} \theta(\tau_j - \tau_{j+1})\ .$$

$$\cdot \begin{bmatrix} \vec{\bar{\Delta}}(t,t_\circ) \prod_{j=1}^{p} c(\tau_{2j})\bar{\Delta}(\tau_{2j},\tau_{2j-1})b(\tau_{2j-1}) & 0 \\ 0 & \prod_{j=1}^{p} c(\tau_{2j-1})\bar{\Delta}(\tau_{2j-1},\tau_{2j})b(\tau_{2j}) \end{bmatrix}, \tag{8.44}$$

and the term $\tilde{A}^{(2p+1)}$ is

$$\tilde{A}^{(2p+1)} = (-1)^p \int \prod_{i=1}^{2p} d\tau_i \, d\tau \prod_{j=1}^{2p-1} \theta(\tau_j - \tau_{j+1})\theta(\tau_{2p} - \tau) \,.$$

$$\cdot \begin{bmatrix} 0 & i\bar{\Delta}(t,\tau)b(\tau) \prod_{j=1}^{p} c(\tau_{2j})\bar{\Delta}(\tau_{2j},\tau_{2j-1})b(\tau_{2j-1}) \\ ic(\tau)\bar{\Delta}(\tau,t_\circ) \prod_{j=1}^{p} c(\tau_{2j-1})\bar{\Delta}(\tau_{2j-1},\tau_{2j})b(\tau_{2j}) & 0 \end{bmatrix}. \tag{8.45}$$

These last two formulas (44) and (45) could of course have been obtained directly from (21). The diagonal terms in (45) vanish and we see then that only (44) will contribute to the trace in (37). Going back to the previous notation ($it \to \beta$, $i \int d\tau \to \int du$) one obtains from (37) for Z that

$$Z = \sum_{p=o}^{\infty} \int dq \int_{q(0)=q}^{q(\beta)=q} \mathcal{D}q\mathcal{D}p \, \exp \int_0^\beta du(ip.\dot{q} + d(u)) \,.$$

$$\cdot \int \prod_{i=1}^{2p} du_i \prod_{j=1}^{2p-1} \theta(u_j - u_{j+1}).[\, \bar{\Delta}(\beta,0) \prod_{j=1}^{p} c(u_{2j}) \,.$$

$$\cdot \, \bar{\Delta}(u_{2j},u_{2j-1})b(u_{2j-1}) + \prod_{j=1}^{p} c(u_{2j-1})\bar{\Delta}(u_{2j-1},u_{2j})b(u_{2j}) \,], \tag{8.46}$$

where $\bar{\Delta}(u_1,u_2) = \exp \int_{u_2}^{u_1} du \, a(u)$. Formula (46) gives us then the partition function of the system in terms of standard functional integrals.

8.3.2. Application to the linear Jahn-Teller effect

We shall apply now this formalism to the linear Jahn-Teller effect with Hamiltonian (see Ref. [H3])

$$\hat{H}_1 = \left(\frac{1}{2}(p^2 + \omega^2 q^2) + E_\circ^{(0)}\right) 1 + \lambda(-q_1\sigma_z + q_2\sigma_x) \ . \tag{8.47}$$

The factor $E_\circ^{(0)}$ only changes the partition function by exp $(-\beta E_\circ^{(0)})$ and consequently we shall work with the Hamiltonian $\hat{H} = \hat{H}_\circ + \hat{H}_I$, with

$$\hat{H}_\circ = \frac{1}{2}(p^2 + \omega^2 q^2) 1 \tag{8.48}$$

$$\hat{H}_I = \lambda(-q_1\sigma_z + q_2\sigma_x) \ , \tag{8.49}$$

where $p^2 = p_1^2 + p_2^2$, $q^2 = q_1^2 + q_2^2$, and $\sigma_x, \sigma_y, \sigma_z$ are the Pauli matrices. One has $[\hat{q}_\ell, \hat{p}_j] = i\,\delta_{\ell j}$. We first remark that one can, without changing the partition function, replace $\hat{H}$ by $U^+\hat{H}U$, where U is a (2x2) unitary matrix. This is because tr exp $[-\beta U^+\hat{H}U]$ = tr exp $[-\beta\hat{H}]$, $U^+U=1$. Consequently we can replace in (39) $\hat{H}_I$ by

$$\hat{H}_I = \lambda[q_1\sigma_y + q_2\sigma_x] \tag{8.50}$$

since the transformation $\sigma_x \to \sigma_x, \sigma_y \to \sigma_z, \sigma_z \to -\sigma_y$, is a rotation around the x axis and is represented by a unitary matrix U such that

$$U^+\sigma_x U = \sigma_x, U^+\sigma_y U = \sigma_z, U^+\sigma_z U = -\sigma_y. \tag{8.51}$$

Our partition function is then that of the system $\hat{H}_2 = \hat{H}_\circ + \hat{\hat{H}}_I$, with

$$\hat{\hat{H}}_I = \lambda[q_1\sigma_y + q_2\sigma_x] \ . \tag{8.52}$$

We can now apply formula (46), but previously we have to identify d,a,b, and c from $\hat{H}_2$. One has

$$a=0 \ , \ b = -\lambda(q_2 - i\,q_1), \tag{8.53}$$

$$d = -h_\circ = -\frac{1}{2}(p^2 + \omega^2 q^2) \ , \ c = -\lambda(q_2 + i\,q_1). \tag{8.54}$$

Using this in (46) with $\bar{\Delta}(u_1,u_2) = 1$, one obtains

$$Z = \sum_{p=0}^{\infty} \int dq \int_{q(0)=q}^{q(\beta)=q} \mathcal{D}q\mathcal{D}p \ \exp \int_0^\beta du \ [ip.\dot{q} - h_\circ(p,q)] \ .$$

$$. \int \prod_{i=1}^{2p} du_i \prod_{j=1}^{2p-1} \theta(u_j - u_{j+1}) \ . [ \prod_{j=1}^{p} c(u_{2j})b(u_{2j-1}) +$$

$$+ \prod_{j=1}^{p} c(u_{2j-1})b(u_{2j}) \ ] \ , \tag{8.55}$$

with c and b given by (53) and (54). We evaluate this integral by the usual method

$$Z = \sum_{p=0}^{\infty} \int_0^\beta \prod_{i=1}^{2p} du_i \prod_{j=1}^{2p-1} \theta(u_j - u_{j+1}) \ .$$

$$.[ \prod_{j=1}^{p} c(u_{2j})b(u_{2j-1}) + \prod_{j=1}^{p} c(u_{2j-1})b(u_{2j})] \Big|_{q=\frac{1}{i}\frac{\delta}{\delta j}} \ . \ Z_\circ[j] \Big|_{j=0} \ , \tag{8.56}$$

where

$$Z_\circ[j] = \int dq \int_{q(0)=q}^{q(\beta)=q} \mathcal{D}q\mathcal{D}p \ \exp \int_0^\beta du \ [ip.\dot{q} - h_\circ(p,q) + j.q] \ . \tag{8.57}$$

One obviously has $Z_\circ[j] = Z_\circ^{(1)} [j_1] . Z_\circ^{(2)}[j_2]$ with

$$Z_\circ^{(i)}[j_i] = \int dq \int_{q(0)=q}^{q(\beta)=q} \mathcal{D}p\mathcal{D}q \ \exp \int_0^\beta du \ [ip\dot{q} - \tfrac{1}{2}(p^2 + \omega^2 q^2) + j_i q] . \tag{8.58}$$

Doing the Gaussian integral over p one has

$$Z_\circ^{(i)}[j_i] = \int dq \int_{q(0)=q}^{q(\beta)=q} \mathcal{D}q \ \exp \int_0^\beta du \ [- \tfrac{1}{2}(\dot{q}^2 + \omega^2 q^2) + j_i q] \ . \tag{8.59}$$

This integral is calculated in Ref. F 4] (page 83) and has the value

$$Z_\circ^{(i)} [j_i] = N^{(i)} \ \exp \frac{1}{2} \int_0^\beta du \ dv \ j_i \ \underline{\Delta} \ (u,v) j_i \ , \tag{8.60}$$

where $N^{(i)}$ is the partition function of the harmonic oscillator and

$$\underline{\Delta}(u,v) = \frac{1}{2\omega} \frac{\mathrm{ch}\ \omega(|u-v| - \frac{1}{2}\beta)}{\mathrm{sh}\ \frac{1}{2}\omega\beta} . \tag{8.61}$$

Using now in (56) the usual identity $F\ [\frac{\delta}{\delta j}]\ G[j]|_{j=o} = G\ [\frac{\delta}{\delta q}]\ F[q]\Big|_{q=o}$ one obtains

$$Z = \exp \frac{1}{2} \int du\, dv\, [\frac{\delta}{\delta q_1(u)} \underline{\Delta}(u,v) \frac{\delta}{\delta q_1(v)} + \frac{\delta}{\delta q_2(u)} \underline{\Delta}(u,v) \frac{\delta}{\delta q_2(v)}] \cdot \sum_{p=0}^{\infty} \int_0^{\beta} \prod_{i=1}^{2p} du_i \prod_{j=1}^{2p-1} \theta(u_j - u_{j-1}) \cdot$$

$$\cdot [\prod_{j=1}^{p} \lambda^2 (q_2(u_{2j}) + iq_1(u_{2j}))(q_2(u_{2j-1}) - iq_1(u_{2j-1})) +$$

$$+ \prod_{j=1}^{p} \lambda^2 (q_2(u_{2j-1}) + iq_1(\dot{u}_{2j-1}))(q_2(u_{2j}) - iq_1(u_{2j}))]\Big|_{\substack{q_1=0\\ q_2=0}} \cdot N , \tag{8.62}$$

where $N = N^{(1)}N^{(2)}$ is the partition function of the harmonic oscillator with two degrees of freedom. We can simplify (62) doing the change of variables $q_2 + iq_1 = L$, $q_2 - iq_1 = L^\star$. Using $\underline{\Delta}(u,v) = \underline{\Delta}(v,u)$ and putting $\Delta(u,v) = 2\underline{\Delta}(u,v)$ one can write (62) as

$$Z = \exp \int du\, dv\, [\frac{\delta}{\delta L^\star(u)} \Delta(u,v) \frac{\delta}{\delta L(v)}] \cdot$$

$$\cdot \sum_{p=0}^{\infty} \int_0^{\beta} \prod_{i=1}^{2p} du_i \prod_{j=1}^{2p-1} \theta(u_j - u_{j+1})\ [\prod_{j=1}^{p} \lambda^2 L^\star(u_{2j}) . L(u_{2j-1})$$

$$+ \prod_{j=1}^{p} \lambda^2 L^\star(u_{2j-1})\ L(u_{2j})]\Big|_{\substack{L=0\\ L^\star=0}} \cdot N . \tag{8.63}$$

Making the change if variables $u_{2\ell-1} \to u_\ell$ , $u_{2\ell} \to v_\ell$ one can write (63) in the form

$$Z = \exp \int du\, dv \left[ \frac{\delta}{\delta L^{\star}(u)} \Delta(u,v) \frac{\delta}{\delta L(v)} \right] .$$

$$\cdot \sum_{p=o}^{\infty} \int_0^{\beta} \prod_{i=1}^{p} du_i\, dv_i \cdot \prod_{i=1}^{p} \theta(u_i - v_i) \prod_{i=1}^{p-1} \theta(v_i - u_{i+1}) \; .$$

$$\cdot \left[ \prod_{\ell=1}^{p} \lambda^2 L^{\star}(v_\ell) L(u_\ell) + \prod_{\ell=1}^{p} \lambda^2 L^{\star}(u_\ell) L(v_\ell) \right] \Bigg|_{\substack{L=0 \\ L^{\star}=0}} \cdot N \; . \tag{8.64}$$

The first exponential in (64) generates all the contractions $\overline{L^{\star}(u)L(v)} = \Delta(u,v)$ and as $\Delta(u,v) = \Delta(v,u)$ one has $\overline{L^{\star}(u)L(v)} = \overline{L(u)L^{\star}(v)}$. Consequently we can write (64) as ($Z = 2N\tilde{Z}$)

$$\tilde{Z} = \exp \int du\, dv \left[ \frac{\delta}{\delta L^{\star}(u)} \Delta(u,v) \frac{\delta}{\delta L(v)} \right] .$$

$$\cdot \sum_{p=o}^{\infty} \lambda^{2p} \int_0^{\beta} \prod_{i=1}^{p} du_i\, dv_i \prod_{i=1}^{p} \theta(u_i - v_i) \prod_{i=1}^{p-1} \theta(v_i - u_{i+1}) \; .$$

$$\cdot \prod_{\ell=1}^{p} L^{\star}(u_\ell) L(v_\ell) \Bigg|_{\substack{L=0 \\ L^{\star}=0}} \cdot \tag{8.65}$$

One can give a graphical representation of (65) . Putting

u   v $= \Delta\,(u,v)$

x  y $= \theta\,(x-y)$,

one draws for the term $Z^{(2p)}$ the line

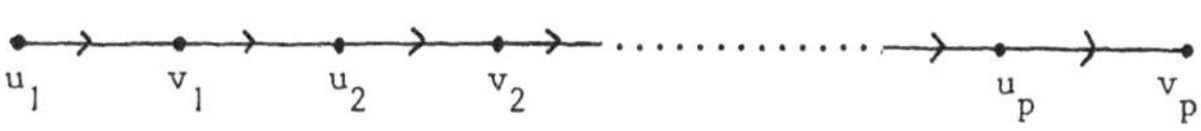


and makes p contractions 

in all possible ways on this line. Finally one must multiply by $\lambda^{2p} \int_0^{\beta} \prod_{i=1}^{p} du_i\, dv_i$ .

We draw now the graphs up to $\tilde{Z}^{(4)}$ .

$\tilde{Z}^{(2)}$ =

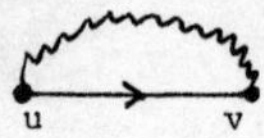

$\tilde{Z}^{(4)}$ =

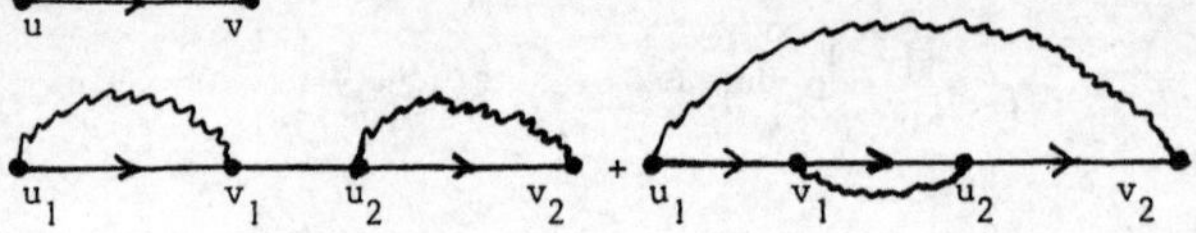

We give now the result of the calculation of these two contributions.

One obtains

$$\tilde{Z}^{(2)} = \frac{\lambda^2}{\omega^2}\beta \tag{8.66}$$

$$\tilde{Z}^{(4)} = \frac{\lambda^4}{\omega^4}\left[\frac{\beta^2}{2}\,\frac{1+\exp(-2\omega\beta)}{(1-\exp(-\omega\beta))^2} - \frac{\beta}{2\omega}\coth\frac{\omega\beta}{2}\right] \tag{8.67}$$

Recalling that the energy of the ground state is $\beta^{-1}\,\ell n\, Z,\ \beta\to\infty$ , we can obtain from (66) and (68) the first terms. One has

$$E_\circ = \lim_{\beta\to\infty}\left(-\frac{1}{\beta}\,\ell n\, Z\right) = E_\circ^{(0)} + \Delta E_\circ \ , \tag{8.68}$$

$$\Delta E_\circ = \lim_{\beta\to\infty}\left(-\frac{1}{\beta}\,\ell n\,\tilde{Z}\right). \tag{8.69}$$

From (68) and (69) one obtains

$$\Delta E_\circ = -\frac{\lambda^2}{\omega^2} + \frac{\lambda^4}{2\omega^5} + O(\lambda^6) \ , \tag{8.70}$$

in agreement with known computations.

APPENDIX 8.1.

In this chapter we studied product functional integrals and we showed how to reduce them to standard functional integrals. Here we shall explain a different technique based on functional integration over anticommuting variables[B9] following Berezin[B5].

Suppose we have the product integral

$$\tilde{Z} = \int_{\gamma_1(1)} \mathcal{D}q\mathcal{D}p \;\; P \exp i \int_{t_0}^{t} d\tau \left(p.\dot{q} - \tilde{H}(p(\tau),q(\tau),\tau\,)\right) \;. \;\; \delta(q(t)-Q)\;\delta(q(t_0)-Q_0) \tag{A8.1}$$

where the finite dimensional space E is of dimension 2 and consequently $\tilde{H}$ and $\tilde{Z}$ are 2 x 2 matrices. We write $\tilde{H}$ in terms of the usual fermion creation and annihilation operators satisfying the anticommutation relations $[\hat{a},\hat{a}^+]_+ = 1$ that we wrote explicitly in (8.35). We have

$$\tilde{H}(p,q,t) = -\, d(p,q,t)\; 1 - a(p,q,t)\hat{a}^+\hat{a} - b(p,q,t)\hat{a}^+ - c(p,q,t)\hat{a}. \tag{A8.2}$$

Then from (A8.1) we have that $\tilde{Z}$ is $\lim \tilde{Z}_N$, $N \to \infty$ with (recall we are in the $\gamma_1(0)$ prepoint discretization)

$$\tilde{Z}_N = \int_{q_0=Q_0}^{q_{N+1}=Q} \prod_{i=1}^{N} dq_i \prod_{j=1}^{N+1} \frac{dp_j}{(2\pi)^d} \exp i\varepsilon \sum_{j=1}^{N+1} \left[p_j . \frac{\Delta q_j}{\varepsilon} + d_j\right] . \; 1$$

$$. \; P \prod_{j=1}^{N+1} \exp i\varepsilon\left[\, a_j\hat{a}^+\hat{a} + b_j\hat{a}^+ + c_j\hat{a}\,\right] \tag{A8.3}$$

where $d_j \equiv d(p_j,q_{j-1},t_{j-1})$, etc., and $t_j = t_0 + j\varepsilon$, $t_{N+1}=t$.

Let $\tilde{B}$ be any 2 x 2 matrix, then $\tilde{B} = \sum_{\mu,\nu=0}^{1} |\mu\rangle\, B_{\mu\nu} \langle\nu|$ and the kernel of $\tilde{B}$ is defined by $B(\bar{\eta},\eta) = \sum_{\mu,\nu=0}^{1} B_{\mu\nu}\, \bar{\eta}^{\mu}\eta^{\nu}$, where $\bar{\eta}$ and $\eta$ are independent anticommuting variables,

$\bar{\eta}\eta = -\eta\bar{\eta}$, $\eta^2 = \bar{\eta}^2 = 0^{[\quad]}$. Let $\tilde{B} = \tilde{B}_1\tilde{B}_2$ be a product of two (2x2) matrices, then if $B_i(\bar{\eta},\eta)$ is the kernel of $\tilde{B}_i$, the kernel $B(\bar{\eta},\eta)$ of $\tilde{B}$ is

$$B(\bar{\eta},\eta) = \int d\bar{\alpha}\, d\alpha \exp\left[-\bar{\alpha}\alpha\,\right] . \;\; B_1(\bar{\eta},\alpha)\, B_2(\bar{\alpha},\eta) \tag{A8.4}$$

where $\bar{\alpha},\alpha$ are new Grassman (anticommuting) variables and integration is defined as being equivalent to left derivation, i.e.

$$\int d\alpha 1 = \int d\bar{\alpha} 1 = 0, \quad \int d\alpha\alpha = \int d\bar{\alpha}\bar{\alpha} = 1 \tag{A8.5}$$

and multiple integrals are defined as iterated integrals. Putting $\widetilde{B}_j =$ $\exp i\varepsilon\,[a_j\hat{a}^+\hat{a} + b_j\hat{a}^+ + c_j\hat{a}]$ one can write the last term in the integrand of (A8.3) as

$$B(\bar{\eta},\eta) = \int_{\alpha_o=\eta}^{\bar{\alpha}_{N+1}=\bar{\eta}} \prod_{i=1}^{N+1} d\bar{\alpha}_i\, d\alpha_i \exp\,[-\sum_{i=1}^{N} \bar{\alpha}_i\alpha_i + \sum_{j=1}^{N+1} \bar{\alpha}_j\, \alpha_{j-1}(1 + \varepsilon a_j)] \;.$$

$$.\ P \prod_{j=1}^{N+1} \exp\,[\ i\varepsilon\,(b_j\bar{\alpha}_j + c_j\,\alpha_{j-1})\ ] \tag{A8.6}$$

We see then that we can formally write now the kernel $Z(\bar{\eta},\eta)$ of $\widetilde{Z}$ in formula (A1) as (we omit the $\gamma_1(0)$ symbol)

$$Z(\eta,\eta) = \int_{\eta}^{\bar{\eta}} \mathcal{D}q\mathcal{D}p\mathcal{D}\bar{\alpha}\mathcal{D}\alpha \exp i\int_{t_o}^{t} d\tau\ [\ p.\dot{q} - H(p,q,\bar{\alpha},\alpha,\tau)]\ .$$

$$.\delta(q(t)-Q)\ .\ \delta(q(t_o)-Q_o) \tag{A8.7}$$

which expresses $Z(\bar{\eta},\eta)$ as a functional integral over anticommuting Grassman variables. Coming back to (A8.6) we see that if we compute $B(\bar{\eta},\eta) = B_{oo} + B_{10}\bar{\eta} + B_{01}\eta + B_{11}\,\bar{\eta}\eta$ we know all the matrix elements $B_{\mu\nu}$ of $\widetilde{B}$ and consequently from (A8.1) all the matrix elements $Z_{\mu\nu}$ of $\widetilde{Z}$, but as $B_{01}$ and $B_{11}$ are simply related to $B_{10}$ and $B_{00}$, respectively, we need only to compute in (A8.6) the constant term $B_{00}$ and the linear term in $\bar{\eta}$, i.e. $B_{10}$. We put $s_j = 1 + i\varepsilon a_j$, $\bar{K}_\ell = \bar{j}_\ell$, $\ell \neq N$, $\bar{K}_N = \bar{j}_N + \frac{\bar{\eta}}{i\varepsilon}\ s_{N+1}$, $K_\ell = j_\ell$, $\ell \neq 1$, $K_1 = j_1 + \frac{\eta}{i\varepsilon}\ s_1$, where $\bar{j}_\ell, j_\ell$ are anticommuting sources, and we replace in (A8.6)

$$P \prod_{j=1}^{N+1} \exp\,[i\varepsilon\,(b_j\bar{\alpha}_j + c_j\alpha_{j-1})] \longrightarrow P \prod_{j=1}^{N} \exp[i\varepsilon\,(b_j\bar{\alpha}_j + c_j\alpha_j)] \tag{A8.8}$$

(this is allowed up to terms that do not contribute in the limit $\varepsilon \to 0$). Then

$$B(\bar{\eta},\eta) = P \prod_{j=1}^{N} \exp\,[c_j\frac{\partial}{\partial\bar{K}_j} - b_j\frac{\partial}{\partial K_j}\,]\ .\ Z_o\,[K,\bar{K}]\ \Big|_{j=\bar{j}=0} \tag{A8.9}$$

where the derivatives are left derivatives, and

$$Z_o[K,\bar{K}] = \int \prod_{i=1}^{N} d\bar{\alpha}_i d\alpha_i \exp\,[-\sum_{i,j=1}^{N} \bar{\alpha}_i\,\Delta^{-1}_{ij}\,\alpha_j + i\varepsilon\sum_{\ell=1}^{N}(\bar{K}_\ell\alpha_\ell + \bar{\alpha}_\ell K_\ell)] \tag{A8.10}$$

with $\Delta^{-1}_{ij} = \delta_{ij} - s_i\,\delta_{i,j+1}$. The value of the Gaussian (A8.10) is[B5,B6]

$$Z_o[K,\bar{K}] = (\det \Delta^{-1}) \exp - \varepsilon^2 \sum_{i,j} \bar{K}_i\Delta_{ij}K_j \tag{A8.11}$$

where $\Delta_{jk} = \delta_{jk} + \theta(j-k)s_{k+1}\, s_{k+2} \, \ldots\ldots \, s_j$, with $\theta(i) = 0$, $i \leqslant 0$, $\theta(i) = 1$, $i > 0$. But one can show the general formula[B9]

$$P \prod_{j=1}^{N} \exp\left[c_j \frac{\partial}{\partial \bar{K}_j} - b_j \frac{\partial}{\partial K_j}\right]. F\left[\bar{K}_i \Delta_{ij} K_j\right] =$$

$$= P_1 F\left[(\bar{K}_i + c_i)\, \Delta_{ij}\, (K_j + b_j)\right], \tag{A8.12}$$

where the symbol $P_1$ acts on the indices (the times) of $c_i$ and $b_j$ only in the Taylor series around $(\bar{K}_i, K_j)$ of the right hand side and is defined by

$$P_1 \left[A_1(t_1)\, A_2(t_2) \ldots . A_p(t_p)\right] =$$

$$= \varepsilon_{1,2 \,\ldots\ldots\ldots\, p}^{\pi(1),\pi(2),..\pi(p)} \, A_1(t_1) A_2(t_2) \ldots . \, A_p(t_p) \tag{A8.13}$$

with $\varepsilon_{1 \ldots\ldots\ldots p}^{\pi(1) \ldots . \pi(p)}$ the completely antisymmetric tensor and $\pi$ a permutation such that $t_{\pi(1)} > t_{\pi(2)} > \ldots\ldots > t_{\pi(p)}$. The $P_1$ product is not defined for equal time but this will play no role in the final result we are looking for. Note also that $P_1$ has appeared because $b_j$ and $c_j$ are normal commuting numbers and consequently $(c_j \frac{\partial}{\partial \bar{K}_j} - b_j \frac{\partial}{\partial K_j})$ is an anticommuting quantity. Applying (A8.12) in (A8.9) we obtain immediately that ($\det \Delta^{-1} = 1$)

$$B_{oo} = P_1 \exp\left[- \varepsilon^2 \sum_{k,\ell=1}^{N} c_k\, \Delta_{k\ell} b_\ell\right] \tag{A8.14}$$

$$B_{10} = i\, P_1 \exp\left[- \varepsilon^2 \sum_{k,\ell}^{N} c_k\, \Delta_{k\ell}\, b_\ell\right] . \, \varepsilon \sum_{j=1}^{N} \Delta_{nj} b_j \, . \tag{A8.15}$$

Noting that in the limit $\varepsilon \to 0$, one has

$$\Delta_{k\ell} \longrightarrow \Delta(t_k, t_\ell) = \theta(t_k - t_\ell) \exp i \int_{t_\ell}^{t_k} d\tau\, a(\tau) \tag{A8.16}$$

and replacing (A8.14) and (A8.15) in (A8.7) one obtains

$$Z_{oo} = P_1 \int \mathcal{D}q\mathcal{D}p \exp i \int_{t_o}^{t} dt' \left[p.\dot{q} + d(t') + ic(t') \int_{t_o}^{t} dt'' b(t'')\, \Delta(t',t'')\right] \tag{A8.17}$$

$$Z_{10} = iP_1 \int \mathcal{D}q\mathcal{D}p \int_{t_o}^{t} d\tau \;\; \Delta(t,\tau)\, b(\tau) .$$

$$. \exp i \int_{t_o}^{t} dt' [p.\dot{q} + d(t') + ic(t') \int_{t_o}^{t} dt''\, b(t'') \;\; \Delta(t',t'')] \tag{A8.18}$$

In the present case, and this due to the form of the function $\Delta(t',t'')$, formules (A17) and (A18) can be further simplified replacing the $P_1$ product by a $P_2$ product acting also only on the times of $b(t')$, $c(t'')$, and which is defined in the case of the term of order n in (A18) as

$$P_2 \int d\tau\, \Delta(t,\tau)\, b(\tau)\, \exp - \int dt'\, dt''\, c(t')\, b(t'')\, \Delta(t',t'')\Big|_{\text{order } n} =$$

$$= (-1)^n \int_{t_o}^{t} d\tau \prod_{i=1}^{n} dt_i'\, dt_i''\, \Delta(t,\tau)\, b(\tau) \prod_{i=1}^{n} c(t_i')b(t_i'') .$$
$$. \Delta(t_i',t_i'') .\ \theta(\tau - t_1') \prod_{i=1}^{n-1} \theta(t_i'' - t_{i+1}') . \qquad \text{(A8.19)}$$

Replacing (A8.20) in (A8.17) and (A8.18) one obtains the final formulas which are of course the same as formula (8.21). The motivation for the appendix is that integration over anticommuting variables is a method currently used to treat fermion fields in Quantum Field Theory, and approximation methods have been developed for this kind of integrals.[T2]
An interesting work on the use of anticommuting variables integrals in statistical mechanics is [S1]

## CHAPTER IX : THE SEMICLASSICAL EXPANSION IN PHASE SPACE

### A. Introduction

A general way to state the problem, we want to treat , is to start from a second-order differential operator of the form

$$\mathcal{L}(q, \frac{\partial}{\partial q}, \eta) = \frac{\eta c}{2} g^{\mu\nu}(q)\partial_\mu\partial_\nu + \alpha(\eta) A^\mu(q)\partial_\mu + \beta(\eta)V(q) \tag{1}$$

where $q = (q^1, \ldots, q^d)$

$$\mu,\nu = 1,\ldots,d$$
$$\alpha(\eta) = \alpha_\circ + a_1\eta + a_2\eta^2 + \ldots.$$
$$\beta(\eta) = \frac{1}{c}\left(\frac{\bar{\beta}}{\eta} + b_1 + b_2\eta + b_3\eta^2 + \ldots\right) \tag{2}$$

We are interested in the fundamental solution of the equation

$$\frac{\partial}{\partial T} I(X,T;X_\circ,t_\circ) = \mathcal{L}(X, \frac{\partial}{\partial X}, \eta)\, I(X,T;X_\circ,t_\circ) \tag{3}$$

with initial condition $I(X,t_\circ;X_\circ,t_\circ) = \delta(X-X_\circ)$ , and we want to construct this solution as an asymptotic series in $\eta$ , by means of a perturbation expansion, using phase space functional integrals. It is well known that there is an associated classical problem, giving the dominant behaviour for small $\eta$ . More precisely, there exists a "classical action" $A_{cl}(X,T;X_\circ,t_\circ)$ such that I has the form

$$I = (\eta)^{-d/2} \exp -\frac{1}{\eta c} A_{cl}(X,T;X_\circ,t_\circ)\, \tilde{I}$$
$$\tilde{I} = \bar{N}\left(1 + \sum_1^\infty \eta^n K_n\right) \tag{4}$$

We show how to obtain this classical action,and how one can compute $\tilde{I}$, using the Feynman-graph expansion method

$$\tilde{I} = \exp \int_{t_\circ}^{T} H_{int}\left(\frac{1}{i}\frac{\delta}{\delta J_\lambda}, \frac{1}{i}\frac{\delta}{\delta J^{\star\mu}}\right) \bar{Z}_\circ(J,J^\star)\Big|_{J=J^\star=0}$$

with $$\bar{Z}_\circ(J,J^\star) = \bar{N}\exp -\int_{t_\circ}^{T} d\tau \int_{t_\circ}^{T} d\tau'\Big[\frac{1}{2} J_\mu(\tau)\Delta^{\mu\nu}(\tau,\tau')J_\nu(\tau') + J^{\star\mu}(\tau)S_\mu{}^\nu(\tau,\tau')J_\nu(\tau') + \frac{1}{2} J^{\star\mu}(\tau)D_{\mu\nu}(\tau,\tau')J^{\star\nu}(\tau')\Big] \tag{5}$$

That this cannot be done naively is due to discretization problems in the functional integrals (or operator ordering problems in an operator scheme). In Ch.4 we have seen thatone is able to define the series unambiguously if one keeps track of the discretization. We shall not pursue the generality that has been explained there, but make an obvious and simple choice, which is enough to illustrate the procedure. We also present the computation of the propagators and the normalization factor in (5), which can be determined through the solution of the classical problem.

We first deal with the expansion in a straightforward manner. Afterwards we restrict ourselves to equation (9.C(1)) which is also of the form (1), but use our knowledge of transformation properties and some methods of Riemannian geometry to obtain the series in a covariant and gauge invariant way. Formulas are given for the first order correction, which will be used in two applications: computation of the short time propagator from (4), and the first order correction for a free particle on the sphere $S_3$ . In order to avoid repetition, we are brief in the treatment of the general method, and give some essential material on the Jacobi equation, and the computation of general Gaussian functional integrals in appendix 9.1 and 9.2.

B. The general method

1.

Introducing operators $q^\mu \to \hat{q}^\mu$ , $-i\eta \frac{\partial}{\partial q^\mu} \to \hat{p}_\mu$ so that $[\hat{q}^\mu,\hat{p}_\nu]= i\eta\, \delta^\mu{}_\nu$, we write $I(X,T,X_\circ,t_\circ)$ as a transition matrix element of the evolution operator $\hat{U}$

$$I(X,T;X_\circ,t_\circ) = \langle X|\, \hat{U}(T,t_\circ)\,|X_\circ\rangle \tag{1}$$

where $\hat{U}$ satisfies the evolution equation

$$i \frac{\partial}{\partial t}\ \hat{U}(t,t'') = \hat{H}\, \hat{U}(t,t') \tag{2}$$

$$\hat{U}(t,t) = 1 \tag{3}$$

and $\hat{H}= i\mathcal{L}\ (q, \frac{\partial}{\partial q}\ ,\eta\ )$ can be read from (A.1)

$$\hat{H}= \frac{1}{\eta}[\ - \frac{ic}{2}\, g^{\mu\nu}(q)\hat{p}_\mu\hat{p}_\nu - \alpha_\circ A^\mu(\hat{q})\hat{p}_\mu + \frac{i\bar{\beta}}{c}\ V(\hat{q})\ -a(\eta)A^\mu(\hat{q})\hat{p}_\mu + \frac{i}{c}\, b(\eta)V(\hat{q})\,] \tag{4}$$

where $\quad a(\eta) = a_1\eta + a_2\eta^2 + \ldots$

$$b(\eta) = b_1\eta + b_2\eta^2 + \ldots \tag{5}$$

We have obtained (4) immediately in standard order, since in (A.1) we had written all the derivatives at the right. This is of course no restriction, since a second order differential operator can always be brought to this form by making the suitable commutators. To the standard ordered Hamiltonian we can associate immediately a functional integral representation in postpoint discretization (Ch.2)

$$I(X,T,X_\circ,t_\circ) = \int_{\gamma_1(1)} \mathcal{D}Q\mathcal{D}_\eta P\ \exp \frac{i}{\eta} \int_{t_\circ}^{T} d\tau[P_\mu\dot{Q}^\mu - \bar{H}^{\gamma_1(1)}(P,Q)]\delta(Q(T)-X)\delta(Q(t_\circ)-X_\circ) \tag{6}$$

where $\bar{H}^{\gamma_1(1)}(P,Q)$ is obtained from $\eta\hat{H}$ in (4) by $\hat{q} \to Q$ ,$\hat{p} \to P$. We recall that (6) has to be interpreted as the limit of a multidimensional integral$(\Delta Q_j=Q_j-Q_{j-1})$

$$I = \lim_{N\to\infty} I_N$$

$$I_N = \int_{Q_\circ=X_\circ}^{Q_{N+1}=X} \prod_{i=1}^{N} dQ_i \prod_{j=1}^{N+1} \frac{dP_j}{(2\pi\eta)^d}\ \exp \frac{i\varepsilon}{\eta} \sum_{j=1}^{N+1} P_{j_\mu} \frac{\Delta Q_j^\mu}{\varepsilon} - \bar{H}^{\gamma_1(1)}(P_j,Q_j) \tag{7}$$

and $\bar{H}^{\gamma_1(1)}(P,Q) =[- \frac{ic}{2}\, g^{\mu\nu}(Q)P_\mu P_\nu - \alpha_\circ A^\mu(Q)P_\mu + \frac{i\bar{\beta}}{c}V(Q)-a(\eta)A^\mu(Q)P_\mu + \frac{i}{c}\, b(\eta)V(Q)]$

(8)

The classical Hamiltonian associated to this problem is obtained by taking $\eta = 0$ in (8), i.e.

$$\bar{H}_{cl}(P,Q) = -\frac{ic}{2} g^{\mu\nu}(Q)P_\mu P_\nu - \alpha_\circ A^\mu(Q)P_\mu + \frac{i\bar{\beta}}{c} V(q) \tag{9}$$

We indicate by $(x^\mu(t), y_\rho(t))$ the solutions of Hamiltons equations, corresponding to (9) and satisfying the boundary conditions $x^\mu(t_\circ) = X_\circ{}^\mu$, $x^\mu(T) = X^\mu$

$$\dot{x}^\mu = \frac{\partial\bar{H}_{cl}(y,x)}{\partial y_\mu} = -ic\, g^{\mu\nu}(x)y_\nu - \alpha_\circ A^\mu(x) \tag{10}$$

$$\dot{y}_\mu = -\frac{\partial\bar{H}_{cl}(y,x)}{\partial x^\mu} = \frac{ic}{2}\partial_\mu g^{\lambda\rho}(x)y_\lambda y_\rho + \alpha_\circ\partial_\mu A^\lambda(x)y_\lambda - \frac{i\bar{\beta}}{c}\partial_\mu V \tag{11}$$

Since $P_\mu = \frac{i}{c} g_{\mu\rho}(Q)[\dot{Q}^\rho + \alpha_\circ A^\rho(Q)]$ the associated Lagrangian has the form

$$\mathcal{L}_{cl} = P_\mu\dot{Q}^\mu - \bar{H}_{cl}(P,Q) = \frac{i}{c}\{\frac{1}{2} g_{\mu\nu}(Q)[\dot{Q}^\mu + \alpha_\circ A^\mu(Q)][\dot{Q}^\nu + \alpha_\circ A^\nu(Q)] - \bar{\beta}V(Q)\}$$

$$= \frac{i}{c} L_{cl}(Q,\dot{Q}) \tag{12}$$

We separated here i/c to indicate the real Lagrangian L (the typical cases are c=i,c=1, corresponding respectively to Schrodinger and Fokker-Planck equations) and we can also introduce a real Hamiltonian

$$H_{cl} = \frac{c}{i}\,\bar{H}_{cl}(\frac{i}{c}P,Q).$$

In the functional integral representation (6), or more precisely (7), we make now the change of variables

$$\begin{cases} P_\mu(\tau) = y_\mu(\tau) + \sqrt{\eta}\, p_\mu(\tau) \\ Q^\mu(\tau) = x^\mu(\tau) + \sqrt{\eta}\, q^\mu(\tau) \end{cases} \tag{13}$$

Under this transformation, the volume element transforms as

$$\mathcal{D}Q\mathcal{D}_\eta P = \prod_{i=1}^{N} dQ_i \prod_{j=1}^{N+1} \frac{dP_j}{(2\pi\eta)^d} = \eta^{-d/2} \prod_{i=1}^{N} dq_i \prod_{j=1}^{N+1} \frac{dp_j}{(2\pi)^d} = \eta^{-d/2}\mathcal{D}q\mathcal{D}p \tag{14}$$

Writing $P_\mu\dot{Q}^\mu - \bar{H}^{\gamma_1(1)} = E_1 + E_2$

where $E_1(\tau) = P_\mu\dot{Q}^\mu - \bar{H}_{cl}(P,Q)$ , we find in the new coordinates

$$E_1(\tau)=y_\mu\dot{x}^\mu-\bar{H}_{cl}(y,x)+\eta[p_\mu\dot{q}^\mu-[\frac{1}{2}\frac{\partial^2\bar{H}_{cl}(y,x)}{\partial y_\mu\,\partial y_\nu}p_\mu p_\nu+\frac{\partial^2\bar{H}_{cl}(y,x)}{\partial y_\mu\,\partial x^\nu}p_\mu q^\nu+\frac{1}{2}\frac{\partial^2\bar{H}_{cl}(y,x)}{\partial x^\mu\partial x^\nu}q^\mu q^\nu]]$$

$$-\eta\sum_{n=1}^{\infty}(\sqrt{\eta})^n\,\tilde{W}_n(x,y,p,q) \tag{15}$$

The term of order $\sqrt{\eta}$ in (15) vanishes, owing to the equations of motion (10,11) On the other hand, the development of $E_2(\tau)$ is of the form

$$E_2(\tau)=a(\eta)A^\mu(x+\sqrt{\eta}q)\,(y_\mu+\sqrt{\eta}p_\mu)-\frac{i}{c}\,b(\eta)V(x+\sqrt{\eta}q)$$

$$=a_1\eta\,A^\mu(x)y_\mu-\frac{i}{c}\,b_1\eta\,V(x)-\eta\sum_{1}^{\infty}(\sqrt{\eta})^n\bar{W}_n(x,y,p,q) \tag{16}$$

Taking $W_n=\bar{W}_n+\tilde{W}_n$ and $\int_{t_\circ}^{T}d\tau\;L_{cl}(x,\dot{x})=A_{cl}$, we then find the following representation for I

$$I(X,T;X_\circ,t_\circ)=\eta^{-d/2}\exp-\frac{1}{\eta c}A_{cl}$$

$$\int_{\gamma_1(1)}\mathcal{D}q\mathcal{D}p\,\exp i\int_{t_\circ}^{T}d\tau[p_\mu\dot{q}^\mu-(\frac{1}{2}\frac{\partial^2\bar{H}_{cl}}{\partial y_\mu\partial y_\nu}p_\mu p_\nu+\frac{\partial^2\bar{H}_{cl}}{\partial y_\mu\partial x^\nu}p_\mu q^\nu+\frac{1}{2}\frac{\partial^2\bar{H}_{cl}}{\partial x^\mu\partial x^\nu}q^\mu q^\nu)$$

$$+a_1A^\mu(x)y_\mu-\frac{i}{c}\,b_1V(x)-\sum_{1}^{\infty}(\sqrt{\eta})^n\,W_n(x,y,p,q)]\;\;\delta(q(T))\;\delta(q(t_\circ))$$

$$\tag{17}$$

This is the appropriate form to do the perturbation expansion, since we have quadratic terms of order $\eta^\circ$ coming from the classical Hamiltonian which define a free generating functional,terms of order $\eta^\circ$ depending only on the classical solution, which come out of the functional integral, and a whole series of $\eta$ - dependent terms which define the interaction Hamiltonian.

Explicitly, we have

$$I(X,T,X_\circ,t_\circ)=\eta^{-d/2}\exp-\frac{1}{\eta c}A_{cl}\;\exp i\int_{t_\circ}^{T}d\tau\;a_1A^\mu(x)y_\mu-\frac{i}{c}\,b_1V(x)$$

$$\exp-i\sum_{1}^{\infty}(\sqrt{\eta})^n\int_{t_\circ}^{T}d\tau W_n(x(\tau),y(\tau),p(\tau),q(\tau+\delta))\Big|_{\delta\to 0^+}\quad\mathcal{Z}_\circ(J,J^\star)\Big|_{J=J^\star=0} \tag{18}$$

$$p_\mu=\frac{1}{i}\,\frac{\delta}{\delta J^{\star\mu}}$$

$$q^\mu=\frac{1}{i}\,\frac{\delta}{\delta J_\mu}$$

$$\tilde{Z}_\circ(J,J^\star) = \int_{\gamma_1(1)} \mathcal{D}q\mathcal{D}p \; \exp i \int_{t_\circ}^{T} d\tau[p_\mu \dot{q}^\mu - (\frac{1}{2} \frac{\partial^2 \bar{H}_{cl}(y,x)}{\partial y_\mu \partial y_\nu} p_\mu p_\nu + \frac{\partial^2 \bar{H}_{cl}(y,x)}{\partial y_\mu \partial x^\nu} p_\mu q^\nu +$$

$$+ \frac{1}{2} \frac{\partial^2 \bar{H}_{cl}(y,x)}{\partial x^\mu \partial x^\nu} q^\mu q^\nu) + J_\mu(\tau) q^\mu(\tau) + J^{\star\mu}(\tau) p_\mu(\tau)]$$

$$\delta(q(T))\; \delta(q(t_\circ)) \tag{19}$$

We indicated the fact that the interaction is discretized in the postpoint, by $q(\tau+\delta)|_{\delta\to 0^+}$ , and this will define the perturbation series unambiguously. Indeed, $\tilde{Z}_\circ(J,J^\star)$ is a Gaussian functional integral, which has a solution of the form

$$\tilde{Z}_\circ(J,J^\star) = \tilde{N} \exp - \int_{t_\circ}^{T} dt'dt''[\frac{1}{2} J_\mu(t')\Delta^{\mu\nu}(t',t'')J_\nu(t'') + J^{\star\mu}(t')S_\mu{}^\nu(t',t'')J_\nu(t'') +$$

$$+ \frac{1}{2} J^{\star\mu}(t')D_{\mu\nu}(t',t'')J^{\star\nu}(t'')\,] \tag{20}$$

where $\tilde{N} = \tilde{Z}_\circ(0,0)$ is a normalization factor independent of $J,J^\star$.

The ambiguity in the perturbation series comes as usual from the $\overline{p\, q}$ contractions at equal time. Whereas $\Delta^{\mu\nu}(t',t'')$ and $D_{\mu\nu}(t',t'')$ are continuous at $t' = t''$, $\lim_{\delta\to 0} S_\mu{}^\nu(t,t+\delta) \neq \lim_{\delta\to 0} S_\mu{}^\nu(t,t-\delta)$.

This problem is solved, keeping in mind the discretization in (18). The postpoint discretization used there, and indicated by $q(\tau+\delta)|_{\delta\to 0^+}$, tells us that equal time contractions $\overline{p_\mu(\tau)q^\nu}(\tau)$ have to be interpreted as $\lim_{\delta\to 0^+} S_\mu{}^\nu(\tau,\tau+\delta)$. Use of a different discretization would have led to a different interaction Hamiltonian and a different interpretation of the contraction, but finally giving the same perturbation expansion, as is clear from what has been said in (Ch.4)

To summarize, what we have done: $I(X,T;X_\circ,t_\circ)$ can be written as

$$I(X,T;X_\circ,t_\circ) = \eta^{-d/2}\, \bar{N} \exp - \frac{1}{\eta c} A_{cl}\, (1 + \sum_1^\infty \eta^n K_n) \tag{21}$$

$$\bar{N} = \tilde{Z}_\circ(0,0) \exp i \int_{t_\circ}^{T} d\tau[a_1 A^\mu(x(\tau))\, y_\mu(\tau) - \frac{i}{c} b_1 V(x)] \tag{22}$$

and where $K_n$ can be computed from (18) as a series in $\eta$ . It's also easy to see that there will be no odd powers of $\sqrt{\eta}$ in the resulting series, since these terms contain an odd number of p's and q's. This result can be completed by calculating the propagators $\Delta^{\mu\nu}$, $S_\mu{}^\nu$ and $D_{\mu\nu}$ , more precisely the equations they satisfy and their expression in terms of the classical solution, and by evaluating the normalization factor $\tilde{N}$ which is given by the functional integral (18) for $J=J^\star = 0$. This will be done in the subsequent sections.

2. The equations for the propagators

Introducing $B^{\mu}{}_{\nu}(t) = \dfrac{\partial^2 \bar{H}_{cl}(y,x)}{\partial y_{\mu}\,\partial x^{\nu}}$, $C_{\mu\nu}(t) = \dfrac{\partial^2 \bar{H}_{cl}(y,x)}{\partial x^{\mu}\,\partial x^{\nu}}$, we write the free generating functional (19) as

$$\tilde{Z}_o(J,J^{\star}) = \int_{\gamma_1(1)} \mathcal{D}q\mathcal{D}p \exp i \int_{t_o}^{T} d\tau \,(p_{\mu}\dot{q}^{\mu} - [-\tfrac{ic}{2} g^{\mu\nu}(t)p_{\mu}p_{\nu} + B^{\mu}{}_{\nu}(t)p_{\mu}q^{\nu} + \tfrac{1}{2} C_{\mu\nu}(t)q^{\mu}q^{\nu}])$$

$$+ i \int_{t_o}^{T} d\tau \; J_{\mu}q^{\mu} + J^{\star\mu}p_{\mu} \tag{23}$$

or , using the explicit form (9)

$$\tilde{Z}_o(J,J^{\star}) = \int_{\gamma_1(1)} \mathcal{D}q\mathcal{D}p \;\; \exp i \int_{t_o}^{T} d\tau \; p_{\mu}\dot{q}^{\mu} - [-\tfrac{ic}{2} g^{\mu\nu}p_{\mu}p_{\nu} - (\alpha_o\partial_{\nu}A^{\mu} + ic\, y_{\sigma}\partial_{\nu} g^{\sigma\mu})p_{\mu}q^{\nu}$$

$$-\tfrac{1}{2}\,(\tfrac{ic}{2} y_{\lambda} y_{\rho}\partial_{\mu}\partial_{\nu}g^{\lambda\rho} + \alpha_o y_{\lambda}\partial_{\mu}\partial_{\nu}A^{\lambda} - \tfrac{i\bar{\beta}}{c}\partial_{\mu}\partial_{\nu}V)q^{\mu}q^{\nu}]$$

$$\exp i \int_{t_o}^{T} d\tau (J_{\mu}q^{\mu} + J^{\star\mu}p_{\mu}) \quad \delta[q(T)]\delta[q(t_o)] \tag{24}$$

The three propagators $\Delta^{\mu\nu}$, $S_{\mu}{}^{\nu}$, $D_{\mu\nu}$ are defined by (20). The fact that $\tilde{Z}_o(J,J^{\star})$ is quadratic in J and $J^{\star}$ is a direct consequence of the boundary conditions $\delta(q(T))$ $\delta(q(t_o))$. This is immediately seen in the discretized version of (23). If nonvanishing boundary conditions are used, there will appear linear terms in J and $J^{\star}$[L7,L8]

An easy way to compute the propagators, is to go back to the operator formalism. Fromthe form of the exponential in (23) ,and the fact that it is discretized in the postpoint, we know that the associated Hamiltonian operator $\hat{H}_o$ is, (now $[\hat{q}^{\mu},\hat{p}_{\lambda}] = i\delta^{\mu}{}_{\lambda}$ )

$$\hat{H}_o = -\frac{ic}{2} g^{\mu\nu}(t)\hat{p}_{\mu}\hat{p}_{\nu} + B^{\mu}{}_{\nu}(t)\hat{q}^{\nu}\hat{p}_{\mu} + \frac{1}{2} C_{\mu\nu}(t)\hat{q}^{\mu}\hat{q}^{\nu} \tag{25}$$

Defining then the evolution operator $U_o(t,t')$ by

$$i \frac{\partial U_o(t,t')}{\partial t} = \hat{H}_o(t)U_o(t,t') \qquad U_o(t,t')\big|_{t=t'} = 1 \tag{26}$$

and $U_o(t) = U(t,0)$ (27)

we can introduce new operators and new states by taking

$$\hat{q}^{\mu}(t) = U_o^{-1}(t)\, \hat{q}^{\mu}U_o(t) \tag{28}$$

$$\hat{p}_{\mu}(t) = U_o^{-1}(t)\, \hat{p}_{\mu}U_o(t)$$

$$^{L}\langle q,t| = \langle q|U_\circ(t) \tag{29}$$

$$|q,t\rangle^{R} = U_\circ^{-1}(t)|q\rangle$$

The time ordered products of several operators $\hat{q}^{\mu}(t),\hat{p}_{\lambda}(t')$ between these new states, have a functional integral representation of the form (P stands for chronological product).

$$^{L}\langle Q,T|\ P\ \hat{p}_{\mu_1}(t'_1)\ldots\ \hat{p}_{\mu_n}(t'_n)\hat{q}^{\nu_1}(t_1)\ldots\ \hat{q}^{\nu_m}(t_m)\ |Q_\circ,t_\circ\rangle^{R}$$

$$= \int_{\gamma_1(1)} \mathcal{D}q\mathcal{D}p\ p_{\mu_1}(t'_1)\ldots\ q^{\nu_m}(t_m)\ \exp\ i\int dt\ W(t)\quad \delta(q(T)-Q)\ \delta(q(t_\circ)-Q_\circ) \tag{30}$$

where $i\int dt\ W(t)$ is the argument of the exponential in (23) without the source terms. For completeness, we must add here that the right hand side of (30) is in fact better defined than the left hand side, unless we introduce an explicit ordering for the product of operators $\hat{p}_{\mu}(t),\hat{q}^{\mu}(t)$ at equal time [L31,L32]

Since $\tilde{N} = \tilde{Z}_\circ(0,0)$ , we find by comparing to (23) (31)

$$\tilde{N} = {}^{L}\langle 0,T|\ 0,t_\circ\rangle^{R}$$

$$\tilde{N}\Delta^{\alpha\beta}(t,t') = {}^{L}\langle 0,T|P\ \hat{q}^{\alpha}(t)\hat{q}^{\beta}(t')\ |0,t_\circ\rangle^{R} \qquad \text{(a)} \tag{32}$$

$$\tilde{N}\ S_{\mu}{}^{\nu}(t,t') = {}^{L}\langle 0,T|P\quad \hat{p}_{\mu}(t)\hat{q}^{\nu}(t')|\ 0,t_\circ\rangle^{R} \qquad \text{(b)}$$

$$\tilde{N}\ D_{\alpha\beta}(t,t') = {}^{L}\langle 0,T|P\quad \hat{p}_{\alpha}(t)\hat{p}_{\beta}(t')\ |0,t_\circ\rangle^{R} \qquad \text{(c)}$$

The equations for the propagators are now found by taking the derivatives of these expressions, and follow from the equations of motion of $\hat{q}^{\lambda}(t),\hat{p}_{\mu}(t)$.

$$\dot{\hat{q}}^{\lambda}(t) = i\ U_\circ^{-1}(t)\ [\hat{H}_\circ,\hat{q}^{\lambda}]U_\circ(t)$$
$$= -\ ic\ g^{\lambda\alpha}(t)\ \hat{p}_{\alpha}(t) + B^{\lambda}{}_{\nu}(t)\ \hat{q}^{\nu}(t) \tag{33}$$

$$\dot{\hat{p}}_{\lambda}(t) = i\ U_\circ^{-1}(t)\ [\hat{H}_\circ,\hat{p}_{\lambda}]U_\circ(t)$$
$$= -\ B^{\mu}{}_{\lambda}(t)\hat{p}_{\mu}(t) - C_{\lambda\nu}(t)\hat{q}^{\nu}(t) \tag{34}$$

Using then $\frac{d}{dt_1}\ P[\hat{A}(t_1)\hat{B}(t_2)] = \delta(t_1-t_2)\ [\hat{A}(t_1),\hat{B}(t_1)] + P\ \dot{\hat{A}}(t_1)\hat{B}(t_2)$ (35)

we have $\tilde{N}\ \frac{d}{dt}\ \Delta^{\lambda\beta}(t,t') = {}^{L}\ \langle 0,T|P\ \dot{\hat{q}}^{\lambda}(t)\hat{q}^{\beta}(t')\ |0,t_\circ\rangle^{R}$ (a) (36)

$$\tilde{N}\frac{d}{dt}S_\alpha{}^\beta(t,t') = \delta(t-t')\ {}^L\langle 0,T|[\hat{p}_\alpha(t),\hat{q}^\beta(t)]|0,t_\circ\rangle^R + {}^L\langle 0,T|P\ \dot{\hat{p}}_\alpha(t)\hat{q}^\beta(t')|0,t_\circ\rangle^R \quad \text{(b)}$$

$$\tilde{N}\frac{d}{dt'}S_\alpha{}^\beta(t,t') = \delta(t-t')\ {}^L\langle 0,T|[\hat{q}^\beta(t)\hat{p}_\alpha(t)]|0,t_\circ\rangle^R + {}^L\langle 0,T|P\ \dot{\hat{q}}^\beta(t')\hat{p}_\alpha(t)|0,t_\circ\rangle^R \quad \text{(c)}$$

which implies (32,33,34)

$$\frac{d}{dt}\Delta^{\lambda\beta}(t,t') = -ic\ g^{\alpha\lambda}(t)\ S_\alpha{}^\beta(t,t') + B^\lambda{}_\nu(t)\Delta^{\nu\beta}(t,t') = 0 \quad \text{(a)}$$

$$\frac{d}{dt}S_\alpha{}^\beta(t,t') = \delta(t-t')\ (-i\delta_\alpha{}^\beta) - B^\mu{}_\alpha(t)S_\mu{}^\beta(t,t') - C_{\alpha\nu}(t)\Delta^{\nu\beta}(t,t') \quad \text{(b)} \quad (37)$$

$$\frac{d}{dt'}S_\alpha{}^\beta(t,t') = \delta(t-t')\ (i\delta_\alpha\beta) - ic\ g^{\beta\lambda}(t')D_{\alpha\lambda}(t,t') + B^\beta{}_\nu(t')S_\alpha{}^\nu(t,t') \quad \text{(c)}$$

(a) determines $S_\alpha{}^\beta(t,t')$ as a function of $\Delta^{\mu\nu}(t,t')$

$$S_\alpha{}^\beta(t,t') = \frac{i}{c}\,[g_{\alpha\nu}(t)\ \frac{d}{dt} - g_{\alpha\lambda}(t)B^\lambda{}_\nu(t)]\Delta^{\nu\beta}(t,t') \quad (38)$$

Combination of (a) and (b) leads to the equation for $\Delta^{\mu\nu}(t,t')$.

$$[\frac{d}{dt}\ \delta_\rho{}^\alpha + B^\alpha{}_\rho(t)][g_{\alpha\nu}(t)\ \frac{d}{dt} - g_{\alpha\lambda}(t)B^\lambda{}_\nu(t)]\ \Delta^{\nu\beta}(t,t') - ic\ C_{\rho\nu}\Delta^{\nu\beta}(t,t') = -\ c\ \delta_\rho{}^\beta\ \delta(t-t') \quad (39)$$

This determines $\Delta^{\nu\beta}(t,t')$ as the Green's function of the Jacobi equation, obtained fromthe Lagrangian $L_{cl}$ (app.9.1.) The relation between $\Delta^{\mu\nu}(t,t')$ and $D_{\mu\nu}(t,t')$ is found by combining (a) and (c).

$$D_{\alpha\lambda}(t,t') = \frac{i}{c}(g_{\lambda\beta}(t')\ \frac{d}{dt'} - g_{\lambda\rho}(t')B^\rho{}_\beta(t'))\ \frac{i}{c}(g_{\alpha\nu}(t)\frac{d}{dt} - g_{\alpha\sigma}(t)B^\sigma{}_\nu(t))\ \Delta^{\nu\beta}(t,t') + \frac{1}{c}\ g_{\alpha\lambda}(t)\ \delta(t-t'). \quad (40)$$

The representation of the propagators in the operator formalism gives us moreover the symmetry relations

$$\Delta^{\mu\nu}(t,t') = \Delta^{\nu\mu}(t',t) \quad (41)$$

$$D_{\mu\nu}(t,t') = D_{\nu\mu}(t',t) \quad (42)$$

and the boundary conditions for equation (39).

$$\Delta^{\mu\nu}(T,t') = \frac{1}{\tilde{N}} \; {}^{L}\langle 0,T|P\, \hat{q}^{\mu}(T)\hat{q}^{\nu}(t')|0,t_\circ\rangle^{R} = 0 \tag{43}$$

$$\Delta^{\mu\nu}(t,t_\circ) = \frac{1}{\tilde{N}} \; {}^{L}\langle 0,T|P\, \hat{q}^{\mu}(t)\, \hat{q}^{\nu}(t_\circ)|0,t_\circ\rangle^{R} = 0$$

The representation of $\Delta^{\mu\nu}(t,t')$ in terms of the classical solution $x(t)$ (10) can be given through the Jacobi fields $\bar{J}^{\mu\nu}(t,t')$ and $M_{\mu\nu}(t,t')$ (app.9.1)

$$\Delta^{\mu\nu}(t,t')= c\{\theta(t-t')\bar{J}^{\mu\lambda}(t,T)M_{\lambda\rho}(T,t_\circ)\bar{J}^{\rho\nu}(t_\circ,t')-\theta(t'-t)\bar{J}^{\mu\lambda}(t,t_\circ)M_{\lambda\rho}(t_\circ,T)\bar{J}^{\beta\nu}(T,t')\} \tag{44}$$

(38) and (40) give similar representations for $S_{\alpha}{}^{\beta}(t,t')$ and $D_{\alpha\beta}(t,t')$. The discontinuity of $S_{\alpha}{}^{\beta}(t,t')$ which forced us to use a discretization, follows from the one of the derivative of $\Delta^{\mu\nu}(t,t')$ (app.9.1(27))

$$\begin{aligned} S_{\alpha}{}^{\beta}(t,t+\delta)-S_{\alpha}{}^{\beta}(t,t-\delta)\Big|_{\delta\to 0^+} &= \frac{i}{c}\, g_{\alpha\mu}(t)\{\frac{d}{dt}\Delta^{\mu\beta}(t,t+\delta)- \frac{d}{dt}\Delta^{\mu\beta}(t,t-\delta)\}\Big|_{\delta\to 0^+} \\ &= \frac{i}{c}\, g_{\alpha\mu}(t)\; c\; g^{\mu\beta}(t) \\ &= i\, \delta_{\alpha}{}^{\beta} \end{aligned} \tag{45}$$

We remark that the discretization in (30) fixes in fact $S_{\alpha}{}^{\beta}(t,t)$ as $S_{\alpha}{}^{\beta}(t,t+\delta)\big|_{\delta\to 0^+}$ but for the perturbation expansion this is not relevant, since the interpretation of $S_{\alpha}{}^{\beta}(t,t)$ has to be determined from the discretization of the terms in the interaction Hamiltonian (Ch.4). In that sense, the ordering of (25) or the discretization of (23) does not play any role in the determination of the propagators (it is relevant for the normalization factor). In the operator method we used, this is reflected in the fact that the equations of motion (33) and (34) remain unchanged if we take a different ordering in (25).

## 3. The normalization factor

$\tilde{N} = \tilde{Z}_\circ(0,0)$ is given by (24) without source terms. It can be computed very easily, using the results of appendix 2. First , we have, by changing the discretization from postpoint to midpoint

$$- i \int_{\gamma_1(1)} dt\, \frac{\partial^2 \bar{H}_{cl}(y,x)}{\partial y_\mu\, \partial x^\nu}\, p_\mu q^\nu = - i \int_{\gamma_1(\frac{1}{2})} dt\, \frac{\partial^2 \bar{H}_{cl}(y,x)}{\partial y_\mu\, \partial x^\nu}\, p_\mu q^\nu + \frac{1}{2}\int dt\, \frac{\partial^2 \bar{H}_{cl}(y,x)}{\partial y_\mu\, \partial x^\nu} \tag{46}$$

which is a direct application of $\hat{q}^\nu\hat{p}_\mu = \frac{1}{2}(\hat{q}^\nu\hat{p}_\mu + \hat{p}_\mu\hat{q}^\nu) + \frac{i}{2}\,\delta^\nu{}_\mu$, and can also be obtained by using the replacements inthe corresponding configuration space functional integral.

After removal of the factor

$$\exp \frac{1}{2} \int dt \frac{\partial^2 \bar{H}_{cl}(y,x)}{\partial y_\mu \, \partial x^\nu} = \exp - \frac{1}{2} \int d\vartheta [\dot{\alpha}_\circ \partial_\mu A^\mu + ic\partial_\mu \, g^{\sigma\mu} y_\sigma] \tag{47}$$

the remaining part is a phase space integral discretized in the midpoint

$$\int_{\gamma_1(\frac{1}{2})} \mathcal{D}p\mathcal{D}q \quad \exp i \int_{t_\circ}^{T} d\tau [p_\mu \dot{q}^\mu - (\frac{1}{2} \frac{\partial^2 \bar{H}_{cl}(y,x)}{\partial y_\mu \, \partial y_\nu} p_\mu p_\nu + \frac{\partial^2 \bar{H}_{cl}(y,x)}{\partial y_\mu \, \partial x^\nu} p_\mu q_\nu + \frac{1}{2} \frac{\partial^2 \bar{H}_{cl}(y,x)}{\partial x^\mu \partial x^\nu} q^\mu q^\nu ] \tag{48}$$

giving the configuration space integral ($\bar{\mathcal{L}}_{cl} = \frac{i}{c} L_{cl}$), after integration over p.

$$\int_{\gamma_1(\frac{1}{2})} \mathcal{D}q \ (\frac{g}{c^d})^{1/2} \exp - \frac{1}{c} \int_{t_\circ}^{T} (\frac{1}{2} \frac{\partial^2 L_{cl}}{\partial \dot{x}^\mu \partial \dot{x}^\nu} \dot{q}^\mu \dot{q}^\nu + \frac{\partial^2 L_{cl}}{\partial \dot{x}^\mu \partial x^\nu} \dot{q}^\mu q^\nu + \frac{1}{2} \frac{\partial^2 L_{cl}}{\partial x^\mu \partial x^\nu} q^\mu q^\nu) \ \delta(q(T)) \ \delta(q(t_\circ)) \tag{49}$$

In appendix 9.2 we show that this integral has the value

$$(2\pi c)^{-d/2} \quad [\det K^{\mu\nu}(t_\circ)]^{-1/2} \tag{50}$$

where $K^{\mu\nu}(t)$ satisfies the Euler Lagrange equation derived from the quadratic Lagrangian,

$$L_\circ = \frac{1}{2} \frac{\partial^2 L_{cl}}{\partial \dot{x}^\mu \partial \dot{x}^\nu} \dot{q}^\mu \dot{q}^\nu + \frac{\partial^2 L_{cl}}{\partial \dot{x}^\mu \partial x^\nu} \dot{q}^\mu q^\nu + \frac{1}{2} \frac{\partial^2 L_{cl}}{\partial x^\mu \partial x^\nu} q^\mu q^\nu \tag{51}$$

(also called the Jacobi equation, corresponding to $L_{cl}$) with boundary conditions

$$K^{\mu\nu}(T) = 0 \tag{52}$$

$$\frac{d}{dt} K^{\mu\nu}(t) \Big|_{t=T} = - g^{\mu\nu}(T)$$

This is precisely the Jacobi field $\bar{J}^{\mu\nu}(t,T)$ [App. 9.1(15)], and we have also shown that [App. 9.1(19,21)]

$$[\det \bar{J}^{\mu\nu}(t_\circ,T)]^{-1/2} = [\det - \frac{\partial^2 A_{cl}(X,X_\circ)}{\partial X^\mu \partial X_\circ^\nu}]^{1/2} \tag{53}$$

the Van Vleck determinant. The complete normalization factor $\bar{N}$ (27) is then given by

$$\bar{N} = (2\pi\eta c)^{-d/2} (\det - \frac{\partial^2 A_{cl}(X,X_\circ)}{\partial X^\alpha \, \partial X_\circ^\beta})^{1/2} \exp \frac{1}{2} \int_{t_\circ}^{T} dt \frac{\partial^2 \bar{H}_{cl}(y,x)}{\partial y^\mu \, \partial x^\nu} \quad \times$$

$$\times \quad \exp i \int_{t_\circ}^{T} d\tau [a_1 A^\mu(x) y_\mu - \frac{i}{c} b_1 V(x)] \tag{54}$$

## 4. Remarks

It is not necessary to treat the problem (A.1) using functional integrals, discretized in the postpoint . In (A.1) we put the derivatives to the right, and the logical choice was to use the postpoint (or standard order in the operator formalism). Prepoint discretization is just as simple, if one makes the necessary commutators. Of course an arbitrary discretization can be used as well, but will lead to more complicated rules for the pq contractions at equal time. By this, we just want to stress the fact that this presentation, given in the postpoint , is not an argument in favour of this specific discretization. In [L11, L12,L17] use was made of discretizations giving the classical Lagrangian (Ch.6) (III.C) , in the functional integral representation.

Concerning the generality of the equation, it should be clear that it is the coefficient of $g^{\mu\nu}\partial_\mu\partial_\nu$ that plays the role of the expansion parameter. If $\eta$ is replaced by $f(\eta)$, one has to re express $\alpha(\eta)$ and $\beta(\eta)$ in powers of $f(\eta)$, which is then the new expansion parameter.

The form of the equation also admits some immediate generalizations which can be incorporated without problems. $\alpha(\eta)A^\mu(q)\partial_\mu + \beta(\eta)V(q)$ can be replaced by $\sum_i \alpha_i(\eta)A_i{}^\mu(q)\partial_\mu + \sum_j \beta_j(\eta)V_j(q)$. Another easy generalisation concerns the explicit time dependence of $\mathcal{L}(q, \partial/\partial q, t, \eta)$. The modifications, due to the time dependence of $g^{\mu\nu}$, $A^\mu$ and V, will be discussed in the context of the gauge invariant and covariant method (9.D)

Finally, the treatment given here can be extended to Hamiltonians which are more than quadratic in the momenta (app.9.3)

C. The covariant and gauge invariant method

As a special case of equation (A.1) , we study the invariant Fokker-Planck and Schrodinger equations[ D8,D9,D25,D26,G16]

$$\frac{\partial\psi}{\partial t} = \frac{\eta c}{2}\Delta_2\psi + \frac{1}{2}\left(A^\mu\partial_\mu\psi + \frac{1}{\sqrt{g}}\partial_\mu\sqrt{g}\,A^\mu\psi\right) + \frac{1}{\eta c}\left(\frac{1}{2}A_\mu A^\mu + V\right)\psi \qquad (1)$$

Writing this equation in the form of (A.1) , we can immediately use the results obtained in section B. We prefer however to do the calculation again, now in a more elegant way, in view of an expansion of the fundamental solution of (1), displaying in a manifest way the behaviour of the correction terms under coordinate and gauge transformations. What will be said about covariance under general coordinate transformations is valid for both Fokker-Planck and Schrodinger equations, while the gauge invariance of the correction terms will be studied in the case of the Schrodinger equations, when V is independent of $A_\mu$ .
The classical Lagrangian corresponding to (1) is (it can be found as in (B.9,12))

$$\mathcal{L}_{cl} = \frac{1}{2}g_{\mu\nu}(q)\dot{q}^\mu\dot{q}^\nu + \dot{q}^\mu A_\mu(q) - V(q) \qquad (2)$$

and the classical path x(t) satisfies the Euler-Lagrange equation

$$\nabla_t\dot{x}^\mu + g^{\mu\nu}(x)\dot{x}^\rho F_{\rho\nu}(x) + g^{\mu\nu}(x)\partial_\nu V(x) = 0 \qquad (3)$$

$$F_{\rho\nu} = \partial_\rho A_\nu - \partial_\nu A_\rho$$

1. Functional integral representation

We first do a time dependent coordinate transformation in (1) such that the new coordinates are normal coordinates around the classical path x(t). Defining the time dependent transformation

$$q^\mu = F^\mu(t,\bar{q}) \qquad (4)$$

we have, since $\psi$ is a scalar,

$$\bar{\psi}(t,\bar{q}) = \psi(t,F(t,\bar{q})) \qquad (5)$$

and

$$\frac{\partial}{\partial t}\bar{\psi}(t,\bar{q}) = \frac{\partial}{\partial t}\psi(t,q) + \frac{\partial\psi}{\partial q^\alpha}\frac{\partial F^\alpha(t,\bar{q})}{\partial t} \qquad (6)$$

which implies

$$\frac{\partial}{\partial t}\,\bar{\psi}\,(t,\bar{q}) = \frac{\partial}{\partial t}\,\psi(t,q) + \frac{\partial F^{\alpha}}{\partial t}\,\frac{\partial \bar{q}^{\sigma}}{\partial q^{\alpha}}\,\frac{\partial \bar{\psi}}{\partial \bar{q}^{\sigma}}\,(t,\bar{q}) \tag{7}$$

The term $\partial\psi/\partial t$ is computed using (1), and since the operator on the right hand side is a scalar, one obtains, indicating the quantities in the new coordinate system by a bar

$$\dot{\bar{\psi}}(t,\bar{q}) = \frac{1}{2}\,\eta c\,\bar{\Delta}_2\,\bar{\psi} + \frac{1}{2}\,\bar{A}^{\mu}\partial_{\mu}\bar{\psi} + \frac{1}{\sqrt{\bar{g}}}\partial_{\mu}\sqrt{\bar{g}}\,\bar{A}^{\mu}\bar{\psi} + \frac{1}{\eta c}[\tfrac{1}{2}\bar{A}_{\mu}\bar{A}^{\mu}+V] + \frac{\partial F^{\alpha}}{\partial t}\frac{\partial \bar{q}^{\sigma}}{\partial q^{\alpha}}\,\frac{\partial \bar{\psi}(t,\bar{q})}{\partial \bar{q}^{\sigma}} \tag{8}$$

If we transform $q_{\circ}$ as $q_{\circ}^{\mu} = F^{\mu}(t_{\circ},\bar{q}_{\circ})$ , the fundamental solution $I(q,t;q_{\circ},t_{\circ})$ satisfying $I\ (t=t_{\circ}) = \delta(q-q_{\circ})$ will transform as

$$I(q\ ,t;q_{\circ},t_{\circ}) = \frac{1}{\left|\det \dfrac{\partial F^{\alpha}}{\partial \bar{q}_{\circ}^{\beta}}(t_{\circ},\bar{q}_{\circ})\right|}\ \bar{I}(\bar{q},t;\bar{q}_{\circ},t_{\circ}) \tag{9}$$

Let $Q^{\mu}$ then indicate an arbitrary coordinate system, and let $I(X,T;X_{\circ},t_{\circ})$ be the propagator in this system of which we want to calculate the WKB-expansion $x(t)$ is the classical path in this coordinate system, satisfying the boundary conditions $x(T)=X$ , $x(t_{\circ})= X_{\circ}$ . We shall go to normal coordinates $q^{\mu}$ around this curve, by choosing $F^{\mu}(t,q)$ as

$$Q^{\mu}= F^{\mu}(t,q) = x^{\mu}(t) + \sqrt{\eta}q^{\mu} - \frac{\eta}{2}\Gamma^{\mu}_{\nu\rho}(x(t))q^{\nu}q^{\rho} - \frac{\eta^{3/2}}{3!}\,\Gamma^{\alpha}_{\nu\rho\sigma}q^{\nu}q^{\rho}q^{\sigma} + \ldots \tag{10}$$

The complete series, and the meaning of the terms is given in [V3] It is the transformation formula from the coordinate $Q$ to the normal coordinate $\sqrt{\eta}q$, centered at $x(t)$. $\sqrt{\eta}\,q$ is in fact the tangent in $x(t)$ to the geodesic of unit length from $x(t)$ to $Q$. We shall come back to this in (37)(38).
Owing to the boundary conditions of $x(t)$, and using (10) one has

$$X^{\mu} = F^{\mu}(T,0) \tag{11}$$

$$X^{\mu}_{\circ} = F^{\mu}(t_{\circ},0).$$

and also

$$\det\left|\frac{\partial F^{\alpha}(t_{\circ},q)}{\partial q^{\beta}}\right|_{q=o} = \eta^{d/2} \tag{12}$$

Using (9), we express the propagator as

$$I(X,T;X_\circ,t_\circ) = \eta^{-d/2} I^{\star}(0,T;0,t_\circ) \tag{13}$$

where $I^{\star}(q,t;q_\circ,t_\circ)$ is the new propagator in normal coordinates around $x(t)$, satisfying (8) in the variables $q$ and $t$.
Introducing the usual operator formalism, we can write

$$I^{\star}(q,T;q_\circ,t_\circ) = \langle q|U(T,t_\circ)|q_\circ\rangle \tag{14}$$

$$\text{with } i\,\dot U(t,t') = \hat H(t)U(t,t') \qquad U(t,t) = 1 \tag{15}$$

and $\hat H(t)$ given by (8)

$$\hat H(t) = -\frac{i\eta c}{2} g^{\star\alpha\beta}(x(t),\hat q)\hat p_\alpha \hat p_\beta + \frac{\eta c}{2} g^{\star\alpha\rho}(x(t),\hat q)\Gamma^{\star\beta}_{\alpha\rho}(x(t),\hat q)\hat p_\beta - A^{\star\mu}(x(t),\hat q)\hat p_\mu$$
$$+ \frac{i}{2}\nabla_\mu A^{\star\mu}(x(t),\hat q) + \frac{i}{\eta c}\,\frac{1}{2}A^{\star}_\mu A^{\star\mu}(x(t),\hat q) + V^{\star}(x(t)),\hat q) - \frac{\partial q^\sigma}{\partial Q^\alpha}\bigg|_t \frac{\partial Q^\alpha}{\partial x^\rho}\bigg|_t \dot x^\rho(t)\hat p_\sigma \tag{16}$$

if we choose to write it in standard order, and $\Gamma^{\star\beta}_{\alpha\rho}$ are the Christoffel symbols in the new coordinate system. In general, the star denotes quantities in normal coordinates $q^\mu$ around the curve, they are related to the old ones in the $Q^\mu$ coordinates by

$$g^{\star\mu\nu}(x(t),q) = \frac{\partial q^\mu}{\partial Q^\alpha}\bigg|_t \frac{\partial q^\nu}{\partial Q^\beta}\bigg|_t g^{\alpha\beta}(Q) \tag{17}$$

$$A^{\star\mu}(x(t),q) = \frac{\partial q^\mu}{\partial Q^\alpha}\bigg|_t A^\alpha(Q)$$

$$\Gamma^{\star\beta}_{\alpha\rho}(x(t),q) = \frac{\partial q^\beta}{\partial Q^\mu}\bigg|_t \frac{\partial Q^\nu}{\partial q^\rho}\bigg|_t \frac{\partial Q^\sigma}{\partial q^\alpha}\bigg|_t \Gamma^\mu_{\nu\sigma}(Q) + \frac{\partial^2 Q^\sigma}{\partial q^\alpha \partial q^\rho}\bigg|_t \frac{\partial q^\beta}{\partial Q^\sigma}\bigg|_t \tag{19}$$

$$V^{\star}(x(t),q) = V(Q(x(t),q)) \tag{20}$$

From (13) and (16) we know how to write down a phase space functional integral respresentation for $I(X,T;X_\circ,t_\circ)$ in the postpoint discretization (I.A)

$$I(X,T;X_\circ,t_\circ) = \frac{1}{\eta^{d/2}} \int_{\gamma_1(1)} \mathcal{D}q\mathcal{D}p \exp i\int d\tau\; p_\sigma\Big(\dot q^\sigma + \frac{\partial q^\sigma}{\partial Q^\alpha}\frac{\partial Q^\alpha}{\partial x^\rho}\dot x^\rho - \frac{\eta c}{2} g^{\star\alpha\rho}\Gamma^{\star\sigma}_{\alpha\rho} + A^{\star\sigma}\Big)$$
$$+ \frac{i\eta c}{2} g^{\star\alpha\beta} p_\alpha p_\beta - \frac{i}{\eta c}\Big(\frac{1}{2}A^{\star\mu}A^{\star}_\mu + V^{\star} + \frac{\eta c}{2}\nabla_\mu A^{\star\mu}\Big)\quad \delta(q(T))\delta(q(t_\circ)) \tag{21}$$

We can also obtain (21) starting from the functional integral representation of I in the original coordinates, doing a transformation of coordinates in the functional integral. This calculation is much more involved, but is interesting from the point of view of functional integral methods. It is presented in appendix 9.4.(21) is a good starting point to obtain a covariant expansion. We do one more manipulation now to make the gauge invariance clear. Integrating (21) over the p-variables we obtain a configuration space functional integral in the postpoint.

$$I = \eta^{-d/2} \int_{\gamma_1(1)} \mathcal{D}q \left( \frac{g^\star}{(\eta c)^d} \right)^{1/2} \exp \int_{t_\circ}^{T} E(\tau)d\tau \quad \delta(q(T))\, \delta(q(t_\circ)) \tag{22}$$

$$E(t) = -\frac{1}{2\eta c}\left(\frac{\dot{x}^\sigma}{\sqrt{\eta}} + \nabla_t q^\sigma + B^\sigma - \frac{\eta c}{2} g^{\star\alpha\gamma}\Gamma^{\star\sigma}_{\alpha\gamma}\right) g^\star_{\sigma\rho} \left(\frac{\dot{x}^\rho}{\sqrt{\eta}} + \nabla_t q^\rho + B^\rho - \frac{\eta c}{2} g^{\star\beta\nu}\Gamma^{\star\rho}_{\beta\nu}\right)$$

$$- \frac{1}{\eta c}\left(\frac{\dot{x}^\sigma}{\sqrt{\eta}} + \nabla_t q^\sigma + B^\sigma - \frac{\eta c}{2} g^{\star\alpha\gamma}\Gamma^{\star\sigma}_{\alpha\gamma}\right) A^\star_\sigma + \frac{1}{\eta c} V^\star + \frac{1}{2}\nabla_\mu A^{\star\mu} \tag{23}$$

It can immediately be verified that $B^\sigma$ defined by

$$\dot{q}^\sigma + \frac{\partial q^\sigma}{\partial \Omega^\alpha} \frac{\partial \Omega^\alpha}{\partial x^\rho} \dot{x}^\rho = \frac{\dot{x}^\sigma}{\sqrt{\eta}} + \nabla_t q^\sigma + B^\sigma \tag{24}$$

starts at order $q^2$. From (10) we have indeed

$$\frac{\partial q^\sigma}{\partial \Omega^\alpha} = \frac{1}{\sqrt{\eta}} \delta^\sigma{}_\alpha + \Gamma^\sigma_{\nu\alpha} q^\nu + \mathcal{O}(q^2)$$

$$\frac{\partial \Omega^\alpha}{\partial x^\rho} = \delta^\alpha{}_\rho + \mathcal{O}(q^2)$$

In (23) we change to midpoint discretization in the crossterm $\nabla_t q^\sigma A^\star_\sigma$ using the replacement (5.35)

$$\Delta q_j^\rho \, \Delta q_j^\sigma \doteq \varepsilon \eta c \, g^{\star\rho\sigma}$$

$$\int_{t_\circ}^{T} d\tau \, \nabla_t q^\sigma A^\star_\sigma \Big|_{\gamma_1(1)} = \int_{t_\circ}^{T} d\tau \nabla_t q^\sigma A^\star_\sigma \Big|_{\gamma_1(\frac{1}{2})} + \frac{\eta c}{2} \int_{t_\circ}^{T} d\tau \, g^{\star\rho\sigma} \partial_\rho A^\star_\sigma \tag{25}$$

The second term (25) combines with the term in $\Gamma^{\star\sigma}_{\alpha\gamma} A^\star_\sigma$ to cancel the last term $\frac{1}{2}\nabla_\mu A^{\star\mu}$, and one has

$$\int dt \, E(t) = -\frac{1}{2\eta c} \int dt \left(\frac{\dot{x}^\sigma}{\sqrt{\eta}} + \nabla_t q^\sigma + B^\sigma - \frac{\eta c}{2} g^{\star\alpha\gamma}\Gamma^{\star\sigma}_{\alpha\gamma}\right) g^\star_{\sigma\rho} \left(\frac{\dot{x}^\rho}{\sqrt{\eta}} + \nabla_t q^\rho + B^\rho - \frac{\eta c}{2} g^{\star\beta\nu}\Gamma^{\star\rho}_{\beta\nu}\right)\Big|_{\gamma_1(1)}$$

$$- \frac{1}{\eta c} \int dt \left(\frac{\dot{x}^\sigma}{\sqrt{\eta}} + \nabla_t q^\sigma + B^\sigma\right) A^\star_\sigma \Big|_{\gamma_1(\frac{1}{2})} + \frac{1}{\eta c} V^\star \tag{26}$$

It will be shown in app.9.5 that

$$\int_{t_\circ}^{T} dt \left(\frac{\dot{x}^\sigma}{\sqrt{\eta}} + \nabla_t q^\sigma + B^\sigma\right) A^\star_\sigma \Big|_{\gamma_1(1/2)} = \int_{t_\circ}^{T} dt \left[\nabla_t q^\rho K_\rho + G(q)\right]\Big|_{\gamma_1(1/2)} \tag{27}$$

where $K_\rho$ depends only on $F_{\mu\nu}$, and the same is valid for $G(q)$, except for the first term in the development in powers of q, which is just $\dot{x}^\mu A_\mu$. Having done this, we want to come back to an expression where everything is in the postpoint. Using (25) again, but now with $K_\rho$ instead of $A^\star_\rho$, we have

$$\int_{t_\circ}^{T} dt\ \nabla_t q^\rho\ K_\rho\Big|_{\gamma_1(1/2)} = \int_{t_\circ}^{T} dt\ \nabla_t q^\rho\ K_\rho\Big|_{\gamma_1(1)} - \frac{\eta c}{2}\int_{t_\circ}^{T} dt\ g^{\star\rho\sigma}\partial_\rho K_\sigma \tag{28}$$

Replacing now (27) in (26), and reintroducing the $\mathcal{D}p$ integration after using (28), one obtains finally the phase space functional integral representation as

$$\begin{aligned} I(X,T;X_\circ,t_\circ) = \eta^{-d/2}\int_{\gamma_1(1)}\mathcal{D}q\mathcal{D}p\ \exp i\int_{t_\circ}^{T} dt\ & p_\sigma\Big(\frac{\dot{x}^\sigma}{\sqrt{\eta}} + \nabla_t q^\sigma + B^\sigma - \frac{\eta c}{2}\, g^{\star\alpha\gamma}\Gamma^{\star\sigma}_{\alpha\gamma}\Big) + \frac{i\eta c}{2} g^{\star\alpha\beta} p_\alpha p_\beta \\ & + p_\rho\, g^{\star\rho\sigma} K_\sigma - \frac{i}{2\eta c} K_\rho K^\rho - \frac{i}{\eta c}\Big(\frac{\dot{x}^\rho}{\sqrt{\eta}} + B^\rho\Big)K_\rho - \frac{i}{2}\nabla_\mu K^\mu + \frac{i}{\eta c} G(q) - \frac{i}{\eta c} V^\star(q) \\ & \delta(q(T))\quad \delta(q(t_\circ)) \end{aligned} \tag{29}$$

Since we obtain $I(X,T;X_\circ,t_\circ)$ in the form

$$I(X,T;X_\circ,t_\circ) = \eta^{-d/2}\exp\Big[-\frac{1}{\eta c}\ A_{cl}\Big]\ \tilde{I} \tag{30}$$

where $A_{cl} = \int_{t_\circ}^{T} dt\ \mathcal{L}_{cl}(x,\dot{x})$, the gauge transformation $A_\mu \to A_\mu + \partial_\mu\phi$ must leave $\tilde{I}$ invariant. The factor $\exp - 1/\eta c\ [\phi(X)-\phi(X_\circ)]$ comes indeed from the crossterm $\dot{x}^\mu A_\mu(x)$ in the classical action. Because we expressed the exponential in (29) (except for the classical terms) in terms of $F_{\mu\nu}$, the invariance will also be clear in the perturbation expansion.

It is shown for instance in [V3] that the affine extension of a tensor $T^{\mu\nu\kappa\dots}_{\alpha\beta\gamma\dots}$ given in any coordinate system $Q$ at a point x, and which is defined by

$$\big(T^{\mu\nu\kappa\dots}_{\alpha\beta\gamma\dots,\lambda_1\dots\lambda_K}\big)_x = \left(\frac{\partial^K T^{\star\mu\nu\kappa\dots}_{\alpha\beta\gamma\dots}}{\partial q^{\lambda_1}\dots\partial q^{\lambda_K}}\right)\Bigg|_{q=0} \tag{31}$$

where q is the normal coordinate system centered at x, and related to $Q$ by (10) [ with x instead of x(t), and $\eta$= 1] and $T^\star$ the tensor T in these normal coordinates is itself a tensor. This implies that the coefficients of the series expansion of all quantities in (29) in powers of q are tensors. That this is also true for $\Gamma^{\star\sigma}_{\alpha\gamma}$ is also proved in [V3]. The affine extensions of the affine connection $\Gamma^{\star\sigma}_{\alpha\gamma}$ are called the affine normal tensors. The fact that all coefficients in the series expansions of $B^\sigma, K_\sigma, G, \Gamma^{\star\sigma}_{\alpha\gamma}$, $g^{\star\alpha\gamma}$, are tensors will give us correction terms to the WKB-approximation which are manifestly scalars under general coordinate transformations. In the next section, and in appendix 9.5 we give an easy method to compute most of these quantities.

## 2. The WKB-expansion

In order to generate the perturbation expansion it is convenient to make in (29) the translation $p_\sigma \to p_\sigma + y_\sigma/\sqrt{\eta}$, $y_\sigma = i/c\, g_{\sigma\alpha}\, \dot{x}^\alpha$ . Remark that the classical Hamiltonian , derived from (1), as done in (B.9) by taking $-i\eta\partial_\mu = \hat{p}_\mu$ and keeping terms in $1/\eta$ , is of the form

$$\bar{H}_{cl} = -\frac{ic}{2}\, g^{\alpha\beta} p_\alpha p_\beta - A^\mu p_\mu + \frac{i}{c}\left(\frac{1}{2} A_\mu A^\mu + V\right) \tag{32}$$

with Hamilton's equations

$$\begin{aligned} \tilde{y}_\mu &= \frac{i}{c}\, g_{\mu\nu}(x)(\dot{x}^\nu + A^\nu(x)) \\ \dot{\tilde{y}}_\mu &= \frac{ic}{2}\,\partial_\mu g^{\lambda\rho}(x) y_\lambda y_\rho - \frac{i}{c}\left[\frac{1}{2}\partial_\mu (A_\lambda A^\lambda)(x) + \partial_\mu V(x)\right] \end{aligned} \tag{33}$$

The fact that we don't have to translate by the classical momentum $\tilde{y}_\sigma$ is due to the manipulation we made to show the gauge invariance. The translation brings out the classical action, and the terms in $1/\sqrt{\eta}$ vanish owing to the equations of motion (3).

The expansion in powers of q(which is in fact an expansion in powers of $\sqrt{\eta}$ q) of the relevant quantities in (29) can be found, using the one parameter variation method [M12,DM4] . We take a two parameter family $\alpha^\mu(u,t)$ , such that

$$\text{(a)}\quad \alpha^\mu(0,t) = x^\mu(t) \tag{34}$$

$$\text{(b)}\quad \alpha^\mu(1,t) = Q^\mu(t)$$

where $Q^\mu$ is the original coordinate system and $x^\mu(t)$ the classical trajectory Moreover, we impose

$$\nabla_u \frac{\partial\alpha}{\partial u}(u,t) = 0 \tag{35}$$

and we define

$$\left.\frac{\partial\alpha^\mu}{\partial u}\right|_{u=0} = q^\mu(t) \tag{36}$$

In (34) and (35) we see that for varying u and fixed t the curves $\alpha^\mu(u,t)$ are geodesics, starting from x(t), and going to $Q^\mu(t)$. Since $\alpha^\mu(u,t)$ are geodesics for fixed t, they are given by [V3]

$$\alpha^\mu(u,t) = x^\mu(t) + uq^\mu(t) - \frac{u^2}{2!}\Gamma^\mu_{\nu\rho}(x(t))q^\nu q^\rho - \frac{u^3}{3!}\Gamma^\mu_{\nu\rho\sigma}(x(t))q^\nu q^\rho q^\sigma + \ldots \quad (37)$$

and setting $u = 1$

$$Q^\mu(t) = x^\mu(t) + q^\mu(t) - \frac{1}{2}\Gamma^\mu_{\nu\rho}(x(t))q^\nu q^\rho - \frac{1}{3!}\Gamma^\mu_{\nu\rho\sigma}(x(t))q^\nu q^\rho q^\sigma \quad \ldots \quad (38)$$

Comparing (38) to formula (10) we see that they coincide except for the powers of $\sqrt{\eta}$, but these can easily be included afterwards.
Consider the quantity

$$L(u,t) = \frac{1}{2} \langle\dot\alpha(u,t),\dot\alpha(u,t)\rangle = \frac{1}{2}\frac{\partial\alpha^\mu}{\partial t} g_{\mu\nu}(\alpha(u,t))\frac{\partial\alpha^\nu}{\partial t} \quad (39)$$

which is, for u=1,

$$\bar L(t) = L(1,t) = \frac{1}{2}\langle\dot Q,\dot Q\rangle = \frac{1}{2}\dot Q^\mu g_{\mu\nu}(Q)\dot Q^\nu \quad (40)$$

We denote by $\tilde g^\star_{\mu\nu}(x(t),q(t))$ the metric tensor in the new coordinates $q(t)$, to distinguish it from $g^\star_{\mu\nu}$ when the $\eta$ dependence is included and we shall use the tilde in the same sense for the other quantities.
Then

$$\bar L(t) = \frac{1}{2}\dot Q^\mu \left.\frac{\partial q^\rho}{\partial Q^\mu}\right|_t \tilde g^\star_{\rho\sigma}(x(t),q)\left.\frac{\partial q^\sigma}{\partial Q^\nu}\right|_t \dot Q^\nu \quad (41)$$

But $$\left.\frac{\partial q^\rho}{\partial Q^\mu}\right|_t \dot Q^\mu = \frac{\partial q^\rho}{\partial Q^\mu}\left(\frac{\partial Q^\mu}{\partial q^\sigma}\dot q^\sigma + \frac{\partial Q^\mu}{\partial x^\sigma}\dot x^\sigma\right) = \dot q^\rho + \frac{\partial q^\rho}{\partial Q^\mu}\frac{\partial Q^\mu}{\partial x^\sigma}\dot x^\sigma \quad (42)$$

Comparing to the definition (24), we have, dropppinq the $\eta$ dependence

$$\left.\frac{\partial q^\rho}{\partial Q^\mu}\right|_t \dot Q^\mu = \dot x^\rho + \nabla_t q^\rho + \tilde B^\rho(q) \quad (43)$$

where $\tilde B^\rho(q)$ is the development of (42) from the quadratic terms onwards, and where of course no new $\nabla_t q$ dependence appears. Then (41) is

$$\bar L(t) = \frac{1}{2}(\dot x^\rho + \nabla_t q^\rho + \tilde B^\rho(q))\,\tilde g^\star_{\rho\sigma}(x(t),q)\,(\dot x^\sigma + \nabla_t q^\sigma + \tilde B^\sigma(q)) \quad (44)$$

Consequently $$\frac{\partial^2 \bar L}{\partial(\nabla_t q^\rho)\partial(\nabla_t q^\sigma)} = \tilde g^\star_{\rho\sigma}(x(t),q) \quad (45)$$

$$\frac{\partial \bar L_\ell}{\partial(\nabla_t q^\rho)} = \tilde g^\star_{\rho\sigma}(\dot x^\sigma + \tilde B^\sigma(q)) = \tilde D_\rho \quad (46)$$

where the derivative in (46) is of the terms $\bar{L}_\ell$ in $\bar{L}$ which are linear in $\nabla_t q$. We see then from (45) and (46) that if we know the expansion of $\bar{L}(t)$ in powers of q, then we also have the expansion of $\tilde{g}^\star_{\rho\sigma}$ and $\tilde{B}^\sigma$. (formula (46) shows that $\tilde{B}^\sigma(q)$ is a contravariant vector). The calculation of the development of $\bar{L}(t)$ is explained in appendix 9.5 where we also give the expansions of $\tilde{g}^{\star\rho\sigma}$, $\tilde{B}^\sigma$, $K_\rho$ and G (computed expanding $\langle\dot{\Omega},A\rangle$) up to the order needed to exhibit the first order correction in $\eta$ to the WKB-approximation of $I(X,T;X_o,t_o)$. After reintroducing the powers of $\sqrt{\eta}$, the representation (29) becomes

$$I(X,T;X_o,t_o) = \eta^{-d/2} \int_{\gamma_1(1)} \mathcal{D}q\mathcal{D}p \exp \frac{i}{\eta} \int_{t_o}^{T} dt\, [\sqrt{\eta}p_\mu + y_\mu][\dot{x}^\mu + \sqrt{\eta}\nabla_t q^\mu + \tilde{B}^\mu(\sqrt{\eta}q)$$
$$- \frac{\eta c}{2} \tilde{g}^{\star\alpha\beta}(\sqrt{\eta}q)\tilde{\Gamma}^{\star\mu}_{\alpha\rho}(\sqrt{\eta}q)] + \frac{ic}{2} \tilde{g}^{\star\mu\rho}(\sqrt{\eta}q)(\sqrt{\eta}p_\mu + y_\mu)(\sqrt{\eta}p_\rho + y_\rho)$$
$$+ (\sqrt{\eta}p_\mu + y_\mu)\tilde{g}^{\star\mu\rho}\tilde{K}_\rho(\sqrt{\eta}q) - \frac{i\eta}{2} \tilde{g}^{\star\mu\rho}\tilde{\nabla}_\mu\tilde{K}_\rho(\sqrt{\eta}q) - \frac{i}{c}(\dot{x}^\alpha + \tilde{B}^\alpha(\sqrt{\eta}q))\tilde{K}_\alpha(\sqrt{\eta}q)$$
$$- \frac{i}{2c} \tilde{g}^{\star\rho\sigma}\tilde{K}_\rho\tilde{K}_\sigma(\sqrt{\eta}q) + \frac{i}{c} \tilde{G}(\sqrt{\eta}q) - \frac{i}{c}\tilde{V}^\star(\sqrt{\eta}q)] \quad \delta(q(T)) \quad \delta(q(t_o)) \tag{47}$$

The expansions of all quantities in (47) in powers of $\sqrt{\eta}$ q are computed in appendix 9.5. We write now the argument of the exponential as $\frac{i}{\eta}\int_{t_o}^{T} dt\, W(t)$ and we collect equal powers of $\sqrt{\eta}$ as $W(t) = \sum_{n=o} (\sqrt{\eta})^n W_n(t)$. In order to obtain the perturbation series we write as usual

$\bar{W}_3 = \sum_{n\geq 3} (\sqrt{\eta})^n W_n$, and

$$I(X,T,X_o,t_o) = \exp \frac{i}{\eta} \int_{t_o}^{T} dt\, \bar{W}_3(x(t),q(t),p(t))\Bigg|_{\substack{q^\mu(t) = \frac{1}{i}\frac{\delta}{\delta J_\mu(t)} \\ p_\mu(t) = \frac{1}{i}\frac{\delta}{\delta J^{\star\mu}(t)}}} \tilde{Z}_o(J,J^\star)\Bigg|_{J=J^\star=0}$$

with

$$\tilde{Z}_o(J,J^\star) = \eta^{-d/2} \int_{\gamma_1(1)} \mathcal{D}q\mathcal{D}p \exp \frac{i}{\eta} \int_{t_o}^{T} dt \sum_{n=o}^{2} (\sqrt{\eta})^n W_n(t) + J_\mu q^\mu + J^{\star\mu} p_\mu \quad \delta(q(T))\, \delta(q(t_o)) \tag{48}$$

## 3. The free generating functional

The free generating functional $\tilde{Z}_o(J,J^\star)$ reduces for zero sources to the WKB-approximation of the propagator $I(X,T;X_o,t_o)$. The dominant part in $\eta$ is determined by $W_o(t)$, given by

$$W_\circ(t) = y_\mu \dot{x}^\mu + \frac{ic}{2} g^{\mu\rho} y_\mu y_\rho + \frac{i}{c} \tilde{G}^{[0]} - \frac{i}{c} V(q)$$

$$= \frac{i}{c} [g_{\mu\rho} \dot{x}^\mu \dot{x}^\rho + \dot{x}^\mu A_\mu - V(x)] \tag{49}$$

which gives the classical action $\frac{i}{c} A_{cl}$ (2). We use the notation $E^{[n]}(q)$ to indicate the term $\sim(q^n)$ in the development in powers of q for any quantity E. All quantities we use, are given in appendix 9.5 , and from now on we don't indicate the x(t)-dependence of the quantities ($g_{\mu\nu}$ stands for $g_{\mu\nu}(x(t))$.
Next, one finds

$$W_1(t) = p_\mu \dot{x}^\mu + y_\mu \nabla_t q^\mu + ic\, g^{\mu\rho} p_\mu y_\rho + y_\mu g^{\mu\rho} \tilde{K}_\rho^{[1]}$$

$$- \frac{i}{c} \dot{x}^\alpha \tilde{K}_\alpha^{[1]} + \frac{i}{c} \tilde{G}^{[1]} - \frac{i}{c} \tilde{V}^{\star[1]}$$

$$= p_\mu (\dot{x}^\mu + ic\, g^{\mu\rho} y_\rho) + \frac{i}{c} (g_{\mu\sigma} \dot{x}^\mu \nabla_t q^\sigma + \dot{x}^\sigma F_{\sigma\rho} q^\rho - \nabla_\rho V q^\rho) \tag{50}$$

$\int_{t_\circ}^T dt\, W_1(t)$ will vanish, after partial integration, using the equations of motion(3).

For $W_2(t)$ one obtains

$$W_2(t) = p_\sigma \nabla_t q^\sigma + y_\sigma \tilde{B}^\sigma_{[2]} + \frac{ic}{2} g^{\rho\sigma} p_\rho p_\sigma + \frac{ic}{2} \tilde{g}^{\star\rho\sigma}_{[2]} y_\rho y_\sigma$$

$$+ p_\mu g^{\mu\rho} \tilde{K}_\rho^{[1]} - \frac{i}{2c} g^{\sigma\rho} \tilde{K}_\rho^{[1]} \tilde{K}_\sigma^{[1]} + \frac{i}{c} \tilde{G}^{[2]} - \frac{i}{c} \tilde{V}^{\star}_{[2]}$$

$$= p_\sigma (\nabla_t q^\sigma - \frac{1}{2} F_\alpha{}^\sigma q^\alpha) + \frac{ic}{2} g^{\lambda\rho} p_\lambda p_\rho + \frac{i}{2c} q^\rho q^\sigma (\dot{x}^\mu \dot{x}^\nu R_{\nu\rho\sigma\mu} - \nabla_\rho \partial_\sigma V +$$

$$+ \nabla_\sigma F_{\rho\nu} \dot{x}^\nu - \frac{1}{4} F_{\lambda\rho} F^\lambda{}_\sigma) \tag{51}$$

The sign convention of the curvature tensor $\tilde{R}_{\mu\nu\rho\sigma}$ appearing here, is also given in app.9.5. In $Z_\circ$ (J,J*) this expression is discretized in the postpoint, thus the term $p_\sigma \nabla_t q^\sigma$ has to be interpreted as

$$p_{j\sigma} \left( \frac{\Delta q_j^\sigma}{\varepsilon} + \dot{x}_j^\rho \Gamma^\sigma_{\rho\alpha} (x_j) q_j^\alpha \right)$$

It is convenient to change here again to $\gamma_1(\frac{1}{2})$ or midpoint discretization

$$\int_{t_\circ}^T dt\, p_\mu \nabla_t q^\mu \Big|_{\gamma_1(1)} = \int_{t_\circ}^T dt\, p_\mu \nabla_t q^\mu \Big|_{\gamma_1(1/2)} + \frac{i}{2} \int_{t_\circ}^T dt\, \delta^\alpha{}_\sigma \dot{x}^\rho \Gamma^\sigma_{\rho\alpha}(x(t))$$

We only have to consider this part since the extra contribution coming from $p_\sigma F^\sigma{}_\alpha q^\alpha$ vanishes due to the antisymmetry of $F_{\mu\rho}$ . The extra contribution is then equal to $i/2 \int_{t_\circ}^T dt\, \dot{x}^\rho \partial_\rho \ln \sqrt{g(x(t))}$ and does not depend on p(t) or q(t) , and consequently goes out of the functional integral as

$$\exp - \frac{1}{2} \int_{t_\circ}^{T} dt \frac{d}{dt} (\ell n \sqrt{g(x(t))}) = [\frac{g(X_\circ)}{g(X)}]^{1/4} \tag{52}$$

We have then $\tilde{Z}_\circ(J,J^\star) = \eta^{-d/2} [\frac{g(X_\circ)}{g(X)}]^{1/4} \exp - \frac{1}{\eta c} A_{cl}\, Z_\circ(J,J^\star)$ where (53)

$$Z_\circ(J,J^\star) = \int_{\gamma_1(\frac{1}{2})} \mathcal{D}q\mathcal{D}p \exp i \int_{t_\circ}^{T} dt\, [p_\mu(\nabla_t \delta^\mu{}_\rho - \frac{1}{2} F^\mu{}_\rho)q^\rho + \frac{ic}{2} g^{\lambda\rho} p_\lambda p_\rho$$

$$+ \frac{i}{2c} q^\rho q^\sigma (\dot{x}^\mu \dot{x}^\nu R_{\nu\rho\sigma\mu} - \nabla_\rho \partial_\sigma V + \nabla_\sigma F_{\rho\nu} \dot{x}^\nu - \frac{1}{4} F_{\lambda\rho} F^\lambda{}_\sigma)$$

$$+ J_\mu q^\mu + J^{\star\mu} p_\mu] \qquad \delta(q(T)) \quad \delta(q(t_\circ)) \tag{54}$$

As for the noncovariant method, the two remaining problems are the determination of the propagators $G^{\mu\nu}$, $T_\mu{}^\nu$, $D_{\mu\nu}$ and the normalization factor in the solution of (54), which we write as

$$Z_\circ(J,J^\star) = \bar{N} \exp - \int_{t_\circ}^{T} \int_{t_\circ}^{T} dt\, dt' \{\frac{1}{2} J_\alpha(t) G^{\alpha\beta}(t,t') J_\beta(t') + J^{\star\alpha}(t) T_\alpha{}^\beta(t,t') J_\beta(t') +$$

$$+ \frac{1}{2} J^{\star\alpha}(t) D_{\alpha\beta}(t,t') J^{\star\beta}(t')\} \tag{55}$$

$$\bar{N} = Z_\circ(0,0)$$

In terms of the discretized integral, these problems consist in finding the inverse and the determinant of the matrix giving the quadratic form in the exponential.

a. The Jacobi equation

Using the results of appendix 9.1, we compute the Jacobi operator corresponding to $\mathcal{L}_{cl}(2)$

$$S(x(t))_{\mu\rho} = A^{(2)}_{\mu\rho}(t) \frac{d^2}{dt^2} + A^{(1)}_{\mu\rho}(t) \frac{d}{dt} + A^{(0)}_{\mu\rho}(t) \tag{56}$$

Using app.9.1, 9.2 we have

$$A^{(2)}_{\mu\rho}(t) = - g_{\mu\rho}$$

$$A^{(1)}_{\mu\rho}(t) = F_{\mu\rho} - 2 g_{\mu\lambda} \Gamma^\lambda_{\kappa\rho} \dot{x}^\kappa$$

$$A^{(0)}_{\mu\rho}(t) = \dot{x}^\kappa \partial_\rho F_{\mu\kappa} - \partial_\mu \partial_\rho V - \partial_\rho g_{\mu\kappa} \ddot{x}^\kappa + \dot{x}^\kappa \dot{x}^\lambda [\frac{1}{2} \partial_\mu \partial_\rho g_{\kappa\lambda} - \partial_\lambda \partial_\rho g_{\mu\kappa}]$$

(57)

giving

$$S(x(t))_{\mu\rho} = [-g_{\mu\rho}\nabla_t\nabla_t + F_{\rho\mu}\nabla_t + \nabla_\rho F_{\lambda\mu}\dot{x}^\lambda - R_{\nu\mu\rho\sigma}\dot{x}^\nu\dot{x}^\sigma + \nabla_\mu\partial_\rho V] \tag{58}$$

which can also be obtained immediately as the Euler-Lagrange equations from the Lagrangian in the configuration space integral corresponding to (54)

$$Z_0(0,0) = \int_{\gamma_1(\frac{1}{2})} \mathcal{D}q \left(\frac{g}{c^d}\right)^{1/2} \exp -\frac{1}{2c}\int_{t_0}^{T} (\nabla_t q^\mu g_{\mu\nu}\nabla_t q^\nu - \nabla_t q^\mu F_{\mu\rho}q^\rho + q^\rho q^\sigma$$

$$(\dot{x}^\mu\dot{x}^\nu R_{\nu\rho\sigma\mu} - \nabla_\rho\partial_\sigma V + \nabla_\sigma F_{\rho\nu}\dot{x}^\nu) \tag{59}$$

This Lagrangian is the second variation of the original classical Lagrangian (2), since expansion of this Lagrangian with $Q^\mu = x^\mu + q^\mu$ or $Q^\mu = x^\mu + q^\mu + \ldots$ (38) leads to the same quadratic terms in q, if the first variation vanishes. The covariant form of the Jacobi equation (58) can also be obtained immediately if one makes use of the method of app.9.5. Given a set of classical solutions $\alpha(u,t)$ satisfying (3), we have

$$\nabla_t \dot{\alpha}^\mu - F^\mu(\dot{\alpha}) + V'^\mu = 0 \tag{60}$$

Applying $\nabla_u$, and using [ App.5.4 ]

$$0 = \nabla_u\nabla_t \dot{\alpha}^\mu - \nabla_u F^\mu(\dot{\alpha}) + \nabla_u(V'^\mu)$$

$$= \nabla_t\nabla_t \frac{\partial\alpha^\mu}{\partial u} + R^\mu(\dot{\alpha}, \frac{\partial\alpha}{\partial u})\dot{\alpha} - \frac{\partial\alpha^\rho}{\partial u}\nabla_\rho F^\mu(\dot{\alpha}) - F^\mu(\nabla_t \frac{\partial\alpha}{\partial u}) + V''^\mu \tag{61}$$

showing at once that $\frac{\partial\alpha}{\partial u}$ satisfies (58) and giving the operator in covariant form. The fact that derivatives of the classical solution with respect to some parameter, satisfy the corresponding Jacobi equation was also used in app.9.1.

b. The propagators of the free generating functional

We can determine $G^{\mu\nu}(t,t')$, $T_\mu{}^\nu(t,t')$ and $D_{\mu\nu}(t,t')$ using the operator formalism, as we did in (B) . Instead of repeating this procedure, we present here the method based on the discretized functional integral representation, leading to a set of finite difference equations, and in the limit $\varepsilon \to 0$ to the corresponding differential equations.

The discrete form of the integral (54) can be written in the form

$$\int dx \exp -[\tfrac{1}{2}(x,\tilde{A}x) - i(\tilde{J},x)] \tag{62}$$

where x and $\tilde{J}$ are vectors with d(2N+1) components

$$x = (p_{1\mu},\ldots,p_{N+1\mu}, q_1^\lambda,\ldots,q_N^\lambda) \tag{63}$$

$$\tilde{J} = (J^{\star\mu}_1,\ldots,J^{\star\mu}_{N+1}, J_{1\lambda},\ldots,J_{N\lambda})$$

Here $p_{1\mu}$ stands for $p_{11},\ldots p_{1\lambda},\ldots,p_{1d}$. We used the notation

$$(x,\tilde{A}x) = \sum_{jk} \varepsilon^2 \sum_{\mu\nu} x^\mu_j \tilde{A}^{\mu\nu}_{jk} x^\nu_k$$

$$(\tilde{J},x) = \sum_j \varepsilon \sum_\mu J^\mu_j x^\mu_j$$

The symmetric matrix $\tilde{A}$ has the structure

$$\tilde{A} = \begin{vmatrix} K^{\mu\nu}(i,j) & B^\mu{}_\nu(i,j) \\ C_\mu{}^\nu(i,j) & L_{\mu\nu}(i,j) \end{vmatrix} \tag{64}$$

with symmetry relations $B^\mu{}_\nu(i,j) = C_\nu{}^\mu(j,i)$ ; $A^{\mu\nu}(i,j) = A^{\nu\mu}(j,i)$ , $L_{\mu\nu}(i,j) = L_{\nu\mu}(j,i)$. The blocks have the following dimensions

K (N+1)d x (N+1)d

B (N+1)d x Nd

C Nd x (N+1)d

L Nd x Nd

The inverse matrix $\tilde{G}$, in the sense $\sum_{k,\nu} \varepsilon \tilde{A}^{\mu\nu}_{ik} \tilde{G}^{kj}_{\nu\lambda} = \frac{\delta_{ij}\delta^\mu{}_\lambda}{\varepsilon}$ has the same block structure

$$\tilde{G} = \begin{vmatrix} D_{\nu\lambda}(j,k) & T_\nu{}^\lambda(j,k) \\ T^\nu{}_\lambda(j,k) & G^{\nu\lambda}(j,k) \end{vmatrix} \tag{65}$$

The explicit form of the matrices in (64) is found from the discretized functional integral as

$$K^{\mu\nu}(i,j) = g^{\mu\nu}(i)\,\delta_{ij}\,\frac{c}{\varepsilon} \tag{a}$$

$$B^\mu{}_\nu(i,j) = -\,i\,\delta_{ij}\left(\frac{\delta^\mu{}_\nu}{\varepsilon^2} + \Gamma^\mu_{\lambda\nu}(i)\,\frac{\dot{x}^\lambda(i)}{2\varepsilon} - \frac{1}{4}\,\frac{F^\mu{}_\nu(i)}{\varepsilon}\right) - i\delta_{i,j+1}$$
$$\left(-\,\frac{\delta^\mu{}_\nu}{\varepsilon^2} + \Gamma^\mu_{\lambda\nu}(i)\,\frac{\dot{x}^\lambda(i)}{2\varepsilon} - \frac{1}{4}\,\frac{F^\mu{}_\nu(i)}{\varepsilon}\right) \tag{b}$$

$$C_\mu{}^\nu(i,j) = -i\delta_{i,j}\,(\delta_\mu^\nu/\epsilon^2 + \Gamma^\nu_{\lambda\mu}\frac{\dot x^\lambda(i)}{2\epsilon} - \frac{1}{4}\frac{F^\nu{}_\mu(i)}{\epsilon}) - i\delta_{i,j-1}(-\frac{\delta_\mu^\nu}{\epsilon^2} +$$

$$+\Gamma^\nu_{\lambda\mu}(i+1)\frac{\dot x^\lambda(i+1)}{2\epsilon} - \frac{1}{4}F^\nu{}_\mu\frac{(i+1)}{\epsilon}) \qquad (c) \qquad (66)$$

$$L_{\mu\nu}(i,j) = \frac{1}{c\epsilon}(\dot x^\rho \dot x^\sigma R_{\rho\mu\nu\sigma} - \nabla_\mu\partial_\nu V + \nabla_\mu F_{\nu\sigma}\dot x^\sigma + \frac{1}{2}\nabla_t F_{\mu\nu} - \frac{1}{4}F_{\lambda\mu}F^\lambda{}_\nu)_i\delta_{i,j} \qquad (d)$$

where we symmetrized $\nabla_\mu F_{\nu\sigma}\,\dot x^\sigma$ to $(\nabla_\mu F_{\nu\sigma}\,\dot x^\sigma + \frac{1}{2}\nabla_t F_{\mu\nu})$. The equation for the inverse $\tilde G$ now becomes a system of finite difference equations

$$c\, g^{\mu\nu}(i)D_{\nu\lambda}(i,j) - i\{\frac{T^\mu{}_\lambda(i,j)-T^\mu{}_\lambda(i-1,j)}{\epsilon} + (\Gamma^\mu_{\lambda\nu}(i)\dot x^\lambda(i) - \frac{1}{2}F^\mu{}_\nu(i))$$

$$\frac{T^\nu{}_\lambda(i,j)+T^\nu{}_\lambda(i-1,j)}{2}\} = \delta^\mu_\lambda\frac{\delta_{i,j}}{\epsilon} \qquad \begin{matrix} i=1,N+1 \\ j=1,N+1\end{matrix} \qquad (a)$$

$$c\, g^{\mu\nu}(i)T_\nu{}^\lambda(i,j) - i\{\frac{G^{\mu\lambda}(i,j) - G^{\mu\lambda}(i-1,j)}{\epsilon} + (\Gamma^\mu_{\lambda\nu}\,\dot x^\lambda - \frac{1}{2}F^\mu{}_\nu(i))$$

$$\frac{G^{\nu\lambda}(i,j) + G^{\nu\lambda}(i-1,j)}{2}\} = 0 \qquad \begin{matrix} i=1,..N+1 \\ j=1,..,N\end{matrix} \qquad (b)$$

$$-i\{\frac{D_{\mu\lambda}(i,j)-D_{\mu\lambda}(i+1,j)}{\epsilon} + \frac{1}{2}[(\Gamma^\nu_{\sigma\mu}\dot x^\sigma - \frac{1}{2}F^\nu{}_\mu)(i)D_{\nu\lambda}(i,j) + (\Gamma^\nu_{\sigma\mu}\dot x^\sigma - \frac{1}{2}F^\nu{}_\mu)(i+1)D_{\nu\lambda}(i+1,j)]\} \qquad \begin{matrix} i=1,N \\ j=1,..N+1\end{matrix} \qquad (c) \qquad (67)$$

$$+\frac{1}{c}(\dot x^\rho\dot x^\sigma R_{\rho\mu\nu\sigma} - \nabla_\mu\partial_\nu V + \nabla_\mu F_{\nu\sigma}\dot x^\sigma + \frac{1}{2}\nabla_t F_{\mu\nu} - \frac{1}{4}F_{\lambda\mu}F^\lambda{}_\nu)(i)T^\nu{}_\lambda(i,j) = 0$$

$$-i\{\frac{T_\mu{}^\lambda(i,j) - T_\mu{}^\lambda(i+1,j)}{\epsilon} + \frac{1}{2}[(\Gamma^\nu_{\sigma\mu}\dot x^\sigma - \frac{1}{2}F^\nu{}_\mu)(i)T_\nu{}^\lambda(i,j) + (\Gamma^\nu_{\sigma\mu}\dot x^\sigma - \frac{1}{2}F^\nu{}_\mu)(i+1)T_\nu{}^\lambda(i+1,j)]\} \qquad \begin{matrix} i=1,..N \\ j=1,..N\end{matrix} \qquad (d)$$

$$+\frac{1}{c}(\dot x^\rho\dot x^\sigma R_{\rho\mu\nu\sigma} - \nabla_\mu\partial_\nu V + \nabla_\mu F_{\nu\sigma}\dot x^\sigma + \frac{1}{2}\nabla_t F_{\mu\nu} - \frac{1}{4}F_{\lambda\mu}F^\lambda{}_\nu)(i)G^{\nu\lambda}(i,j) = \frac{\delta_{i,j}}{\epsilon}\,\delta_\mu{}^\lambda$$

These are the general forms of the equations, but at the boundaries($i,j=1$ or $N+1$) they are slightly different. For the first two, the correct finite difference equations are obtained by taking

$$T^\nu{}_\lambda(0,j) = T^\nu{}_\lambda(N+1,j) = G^{\nu\lambda}(0,j) = G^{\nu\lambda}(N+1,j) = 0 \qquad (68)$$

As can be seen from (66) and the dimensions of the matrices these matrix elements do not exist. In the last two equations, all quantities are defined.

Since we want to take the limit $\varepsilon \to 0$, and to interpret the equations (67) as differential equations in this limit, we have to add the quantities (10) in the equations, specifying at the same time that they are zero in the sense of boundary conditions. In the limit, we obtain the following set of equations

$$\text{(a)}\quad c\, g^{\mu\nu}(t) D_{\nu\lambda}(t,t') - i(\nabla_t T^{\mu}{}_{\lambda}(t,t') - \frac{1}{2} F^{\mu}{}_{\nu}(t) T^{\nu}{}_{\lambda}(t,t')) = \delta(t-t')\delta^{\mu}{}_{\lambda}$$

$$\text{(b)}\quad c\, g^{\mu\nu}(t) T_{\nu}{}^{\lambda}(t,t') - i(\nabla_t G^{\mu\lambda}(t,t') - \frac{1}{2} F^{\mu}{}_{\nu}(t) G^{\nu\lambda}(t,t')) = 0 \qquad (69)$$

$$\text{(c)}\quad i\,(\nabla_t D_{\mu\lambda}(t,t') + \frac{1}{2} F^{\nu}{}_{\mu}(t) D_{\nu\lambda}(t,t') + \frac{1}{c}(\dot{x}^{\rho}\dot{x}^{\sigma} R_{\rho\mu\nu\sigma} - \nabla_{\mu}\partial_{\nu} V + \nabla_{\mu} F_{\nu\sigma}\dot{x}^{\sigma} + \frac{1}{2}\nabla_t F_{\mu\nu} - \frac{1}{4} F_{\lambda\mu} F^{\lambda}{}_{\nu})(t) T^{\nu}{}_{\lambda}(t,t') = 0$$

$$\text{(d)}\quad i\,(\nabla_t T_{\mu}{}^{\lambda}(t,t') + \frac{1}{2} F^{\nu}{}_{\mu}(t) T_{\nu}{}^{\lambda}(t,t') + \frac{1}{c}(\dot{x}^{\rho}\dot{x}^{\sigma} R_{\rho\mu\nu\sigma} - \nabla_{\mu}\partial_{\nu} V + \nabla_{\mu} F_{\nu\sigma}\dot{x}^{\sigma} + \frac{1}{2}\nabla_t F_{\mu\nu} - \frac{1}{4} F_{\sigma\mu} F^{\sigma}{}_{\nu})(t)\, G^{\nu\lambda}(t,t') = \delta(t-t')\delta_{\mu}{}^{\lambda}$$

which have to be supplemented by $G^{\nu\lambda}(t_0,t) = G^{\nu\lambda}(T,t') = 0$ (70)

We only need these two, since the other two in (68) follow from (69b) using the symmetry properties of $G^{\nu\lambda}(t,t')$ and $T_{\nu}{}^{\lambda}(t,t')$

$$G^{\lambda\rho}(t,t') = G^{\rho\lambda}(t',t) \qquad (71)$$

$$T^{\nu}{}_{\lambda}(t,t') = T_{\lambda}{}^{\nu}(t',t)$$

The covariant derivatives in (69) have to be interpreted as acting on the index associated with the time

so $\nabla_t T^{\mu}{}_{\lambda}(t,t') = \frac{d}{dt} T^{\mu}{}_{\lambda}(t,t') + \Gamma^{\mu}_{\rho\sigma}(t)\dot{x}^{\sigma}(t) T^{\rho}{}_{\lambda}(t,t')$, etc...

Combination of (69b) and (69d) gives the equation for $G^{\mu\nu}(t,t')$

$$(g_{\mu\rho}\nabla_t\nabla_t + F_{\rho\mu}(t)\nabla_t + \nabla_{\rho} F_{\sigma\mu}\dot{x}^{\sigma} - \dot{x}^{\kappa}\dot{x}^{\sigma} R_{\kappa\mu\rho\sigma} + \nabla_{\mu}\partial_{\rho} V) G^{\rho\lambda}(t,t') = -c\,\delta_{\mu}{}^{\lambda}\,\delta(t-t') \qquad (72)$$

which means that $G^{\mu\nu}(t,t')$ is the Green's function of the Jacobi equation, satisfying zero boundary conditions. (69.b) fixes $T_{\nu}{}^{\lambda}(t,t')$ as function of $G^{\rho\lambda}(t,t')$

$$T_{\nu}{}^{\lambda}(t,t') = \frac{i}{c}(g_{\nu\mu}(t)\nabla_t - \frac{1}{2} F_{\nu\mu}(t))\, G^{\mu\lambda}(t,t') \qquad (73)$$

The combination of (69.a) and (69.b) yields $D_{\mu\nu}(t,t')$, if we use (71)

$$D_{\nu\lambda}(t,t') = -\frac{1}{c^2}(g_{\lambda\rho}(t')\nabla_{t'} - \frac{1}{2} F_{\lambda\rho}(t'))(g_{\nu\kappa}(t)\nabla_t - \frac{1}{2} F_{\nu\kappa}(t)) G^{\kappa\rho}(t,t') + \frac{1}{c} g_{\nu\lambda}(t)\,\delta(t-t') \qquad (74)$$

This completes the determination of the equations for the propagators, and the only remaining problem is to express $G^{\mu\nu}(t,t')$ in terms of the classical solution. Using the results of app.9.1 , we have

$$G^{\mu\lambda}(t,t') = c\{\theta(t-t')\bar{J}^{\mu\sigma}(t,T)M_{\sigma\rho}(T,t_\circ)\bar{J}^{\rho\lambda}(t_\circ,t') - \theta(t'-t)\bar{J}^{\mu\sigma}(t,t_\circ)M_{\sigma\rho}(t_\circ,T)\bar{J}^{\rho\lambda}(T,t')\} \quad (75)$$

or in terms of the Van Vleck matrix $\bar{M}$ and the explicit derivatives (app.9.1(30))

$$= c\{\theta(t-t')\frac{\partial x^\mu(t)}{\partial X_\circ{}^\gamma}(\bar{M}^{-1})^{\kappa\gamma}\frac{\partial x^\lambda(t')}{\partial X^\kappa} + \theta(t'-t)\frac{\partial x^\mu(t)}{\partial X^\kappa}(\bar{M}^{-1})^{\kappa\gamma}\frac{\partial x^\lambda(t')}{\partial X_\circ{}^\gamma}\} \quad (76)$$

The other propagators are then found applying (73) and (74). The discontinuity of $T_\nu{}^\lambda(t,t')$ is obtained using (app.9.1(27)), as

$$T_\nu{}^\lambda(t,t+\varepsilon) - T_\nu{}^\lambda(t,t-\varepsilon) = i\,\delta_\nu{}^\lambda \quad (77)$$

c. The normalization factor

The normalization factor $\bar{N} = Z_\circ(0,0)$ is given by the configuration space functional integral (59)

$$\int_{\gamma_1(\frac{1}{2})} \mathcal{D}q\ \left(\frac{g(x(t))}{c^d}\right)^{1/2} \exp -\frac{1}{2c}\int_{t_\circ}^{T} dt[\nabla_t q^\mu g_{\mu\nu}\nabla_t q^\nu - \nabla_t q^\mu F_{\mu\rho}q^\rho + q^\rho q^\sigma(\dot{x}^\mu\dot{x}^\nu R_{\nu\rho\sigma\mu} - \nabla_\rho\partial_\sigma V + \nabla_\sigma F_{\rho\nu}\dot{x}^\nu]\quad \delta(q(T))\ \delta(q(t_\circ)) \quad (78)$$

and it is computed using the results of appendix(9.2). The normalization factor is given by

$$[\,2\pi c]^{-d/2}\ [\det K^{\mu\nu}(t_\circ)]^{-1/2} \quad (79)$$

where $K^{\mu\nu}(t)$ satisfies the Euler Lagrange equations derived from the quadratic Lagrangian in the exponential (since this is the second variation of the Lagrangian, it is again the Jacobi equation).
The boundary conditions are

$$K^{\mu\nu}(T) = 0 \quad (80)$$

$$\frac{d}{dt}K^{\mu\nu}(t)\Big|_{t=T} = -g^{\mu\nu}(x(T))$$

Comparing to (app.9.1(15)) we conclude

$$K^{\mu\nu}(t) = \bar{J}^{\mu\nu}(t,T) \quad (81)$$

and using (app.9.1(19,22))

$$\det \bar{J}^{\mu\nu}(t_\circ,T) = [\det - \frac{\partial^2 A}{\partial X^\rho \partial X_\circ^\beta}]^{-1} \tag{82}$$

It would have been more in the spirit of b) to calculate the determinant of $\tilde{A}_{ij}$(62)or of the matrix giving the quadratic form in (59). An appropriate generalization of the method used in[F3,G6] for the 1-dimensional case, where the recursion relation for the Jacobi matrices is used to give the Jacobi equation for the determinant, should lead to the same result here.
Another method has been used by Dewitt in [D13] , using the form of $G^{\mu\nu}(t,t')$, and we shall comment on it in app.10.1.
Combining all factors we have for $\tilde{Z}_\circ(0,0)$ (53)

$$(2\pi\eta c)^{-d/2} \sqrt{g(X_\circ)}[ \frac{1}{\sqrt{g(X_\circ)}} \frac{1}{\sqrt{g(X)}} \det( \frac{-\partial^2 A_{cl}}{\partial X_\circ^\alpha \partial X^\rho}) ]^{1/2} \exp - \frac{1}{\eta c} A_{cl} \tag{83}$$

and we wrote it in this form since

$[g(X_\circ)]^{-1/2} \det ( - \frac{\partial^2 A}{\partial X_\circ^\alpha \partial X^\rho}) [g(X)]^{-1/2}$ is a scalar, to show that the WKB-approximation transform as $\sqrt{g(X_o)}$ which is the transformation property of $I(X,T;X_o,t_o)$.The corrections to this WKB-approximation should then be scalars.

4. The higher order corrections

Using the results of app.9.5 we calculate $W_3$ and $W_4$ in (47) , which are the terms one must know to compute the first order correction to the WKB approximation.
We easily find, collecting all terms

$$W_3 = \frac{2}{3}< p,R(\dot{x},q)q> + \frac{i}{6c} <R'(\dot{x},q)q,\dot{x}> + \frac{i}{3}<\mathcal{R}(q),\dot{x}> - \frac{1}{3}<p,F'(q)> + \frac{i}{6c}<R(\dot{x},q)q,F(q)>$$
$$- \frac{i}{6c} <F(q),F'(q)> - \frac{i}{6c}<\dot{x},F''(q)> - \frac{i}{6c} V''' + \frac{i}{6} q^\alpha \nabla_\mu F^\mu{}_\alpha \tag{84}$$

$$W_4 = \frac{1}{4}<p,R'(\dot{x},q)q> + \frac{i}{24c} <R''(\dot{x},q)q,\dot{x}> - \frac{i}{18c} <R\ R\ \dot{x},q)q,q)\dot{q},\dot{x}> - \frac{ic}{6} <p,R(p,q)q>$$
$$+ i [- \frac{ic}{3} <\mathcal{R}(q),p> + \frac{1}{4} <\mathcal{R}(q)\dot{x}> - \frac{1}{24}<\nabla_t \mathcal{R}(q),q>]$$
$$- \frac{1}{8} <p,F''(q)> + \frac{1}{8} <p,R(F(q)q)q> + \frac{i}{c}\frac{7}{72}<R(\dot{x},q)q,F'(q)>$$
$$+ \frac{i}{48c} <R(F(q),q)q,F(q)> - \frac{i}{16c} <F(q),F''(q)> + \frac{i}{12c} <\dot{x},R'(F(q),q)q >$$

$$- \frac{i}{18c} \langle F'(q), F'(q)\rangle - \frac{i}{24c} \langle \dot{x}, F'''(q)\rangle - \frac{i}{24c} V''''$$

$$+ i \left( \frac{5}{24} \langle R(q), F(q)\rangle + \frac{1}{8} \nabla_\rho \nabla_\beta F^\rho{}_\mu q^\beta q^\mu \right) \tag{85}$$

The structure of the correction terms is now clear. They have the form of multiple integrals $\int dt_1 \ldots dt_m$ of quantities with tensorial indices, evaluated at $x(t_i)$, completely contracted with the propagators $G^{\mu\nu}(t_i,t_j)$ , $D_{\mu\nu}(t_i,t_j)$, $T_\mu{}^\lambda(t_i,t_j)$ . The scalar character of the correction terms under a transformation $\bar{q}=\bar{q}(Q)$ such that $\bar{x}_0=\bar{q}(X_0)$, $\bar{x}=\bar{q}(X)$ is now clear for each term . If K is such a term, we should have $K(X(\bar{x}),X_0(\bar{x}_0),T-t_0) = \bar{K}(\bar{x},\bar{x}_0,T-t_0)$ where $\bar{K}$ is the corresponding term in the expansion of $\bar{I}(\bar{x},T,\bar{x}_0,t_0)$.
Since $x_{c\ell}\ (X(\bar{x}),X_0(\bar{x}_0),t) = \bar{q}(\bar{x}_{cl}(\bar{x},\bar{x}_0,t))$, the tensors at $x(t_i)$ transform with a number of

$$\left( \frac{\partial \bar{q}^\lambda}{\partial Q^\mu} \right)_{x_{cl}(t_i)} , \left( \frac{\partial Q^\sigma}{\partial \bar{q}^\rho} \right)_{\bar{x}_{cl}(t_i)}$$

The scalar character follows if the propagators satisfy analogous relations . For $G^{\mu\nu}(t_i,t_j)$ , we should have, indicating the $X_0,X$ dependence

$$G^{\mu\nu}(t_i,t_j,X_0(\bar{x}_0),X(\bar{x})\ ) = \left( \frac{\partial Q^\mu}{\partial \bar{q}^\lambda} \right)_{\bar{x}(t_i)} \left( \frac{\partial Q^\nu}{\partial \bar{q}^\rho} \right)_{\bar{x}(t_j)} \bar{G}^{\mu\nu}(t_i,t_j,\bar{x}_0,\bar{x})$$

and this follows immediately from the covariant form of the Jacobi equation. We then see that $G^{\mu\nu}(t_i,t_j)$ transforms in the index $\mu$ as a vector at $x(t_i)$, while in $\nu$ it is a vector at $x\,(t_j)$. In this sense $G^{\mu\nu}(t_i,t_j)$ can be called a two-point vector. The quantities $\bar{J},M$ occurring in the definition of G have the same properties [D14] . The other propagators $D_{\mu\nu}(t_i,t_j)$ $T_{\mu\lambda}(t_i,t_j)$ have similar transformation properties as indicated by the position of the indices and the times.

D. Explicit time dependence

It is possible to extend the treatment, given in B, or C to the case where the coefficients in the equation for the propagator are explicitly time dependent . As an example we treat the equation

$$\frac{\partial\psi}{\partial t} = \frac{\eta c}{2} \frac{1}{\sqrt{g}}\partial_\mu \sqrt{g}\; g^{\mu\nu}\, \partial_\nu \psi \tag{1}$$

describing free motion on a Riemannian manifold, in the case where $g_{\mu\nu}$ has an explicit time dependence, $g_{\mu\nu}(q,t)$. The explicit time dependence does not change the transformation properties of the equation under general (time independent) coordinate transformations and we can try to construct again a covariant expansion for the propagator $I(X,T,X_o,t_o)$. We use the covariant method and show in what points this method has to be changed.

The classical Lagrangian corresponding to (1) is

$$\mathcal{L}_{cl} = \frac{1}{2}\, g_{\mu\nu}(q,t)\dot{q}^\mu \dot{q}^\nu \tag{2}$$

corresponding the Hamiltonian ($\bar{\mathcal{L}}_{cl} = \frac{i}{c}\mathcal{L}_{cl}$)

$$\bar{H}_{cl} = -\frac{ic}{2}\, g^{\mu\nu} p_\mu p_\nu \tag{3}$$

with Hamiltons equations

$$\begin{aligned} \dot{x}^\mu &= -ic\, g^{\mu\nu} y_\nu \\ \dot{y}_\lambda &= \frac{ic}{2}\, \partial_\lambda g^{\mu\nu} y_\mu y_\nu \end{aligned} \tag{4}$$

The classical equation is $\nabla_t\, \dot{x}_\mu = 0$ (but not $\nabla_t\, \dot{x}^\mu$ , because of the explicit time dependence). As in (C.10) we can make now the time dependent transformation

$$Q^\mu = x^\mu(t) + \sqrt{\eta}\, q^\mu - \frac{\eta}{2}\, \Gamma^\mu_{\nu\rho}(x(t),t)\; q^\nu q^\rho \; \ldots \tag{5}$$

to arrive at the functional integral representation

$$I(X,T;X_o,t_o) = \frac{1}{\eta^{d/2}} \int_o^o \mathcal{D}q\mathcal{D}p \exp i \int d\tau\; p_\sigma [\dot{q}^\sigma + \frac{\partial q^\sigma}{\partial Q^\alpha}( \frac{\partial Q^\alpha}{\partial x^\rho}\dot{x}^\rho + \frac{\partial \tilde{Q}^\alpha}{\partial t}) - \frac{\eta c}{2}\, g^{\star\alpha\beta}\Gamma^\sigma_{\alpha\beta}] + \frac{i\eta c}{2}\, g^{\star\alpha\rho} p_\alpha p_\rho \tag{6}$$

where $\frac{\partial Q^\alpha}{\partial t}$ now includes a part $\frac{\partial \tilde{Q}^\alpha}{\partial t}$ , obtained by differentiating the explicit time dependence in the coefficients of (5).

From (5) and (6), we see that after the translation $p_\sigma \to p_\sigma + y_\sigma/\sqrt{\eta}$ , we obtain an extra contribution to the free generating functional. It is of the form

$$\exp\; i \int d\tau\; y_\sigma \left(-\frac{1}{2}\, \frac{\partial}{\partial t}\, \Gamma^\sigma_{\lambda\rho}(x(t),t) q^\lambda q^\rho\right) \tag{7}$$

and $\tilde{Z}_\circ(J,J^\star)$ (see C.54) will be proportional to

$$\tilde{Z}_\circ(J,J^\star) \sim \int_{\gamma(1)} \mathcal{D}p\mathcal{D}q \exp i\int d\tau[p_\mu \nabla_t q^\mu - \frac{1}{2} y_\mu \frac{\tilde{\partial}}{\partial t}\Gamma^\mu_{\lambda\rho} q^\lambda q^\rho + \frac{ic}{2} g^{\lambda\rho} p_\lambda p_\rho + \frac{i}{2c} q^\rho q^\sigma \dot{x}^\mu \dot{x}^\nu R_{\nu\rho\sigma\mu}$$

$$+ J_\mu q^\mu + J^{\star\mu} p_\mu ] \tag{8}$$

In the time independent case, we computed the terms in $\tilde{Z}_\circ(J,J\star)$ and the higher order corrections using the one parameter variation method. The same method can be used here since (C.46)

$$\frac{\partial \bar{L}_\ell}{\partial(\nabla_t q^\rho)} = \tilde{g}^\star_{\rho\sigma} (\dot{x}^\sigma + \tilde{B}^\sigma(q)) = \tilde{D}_\rho(q). \tag{9}$$

now includes the terms with an explicit time derivative. However, the series $\bar{L}(t)$ changes . $\alpha(u,t)$ is again a geodesic with respect to the parameter u, but now in the metric $g_{\mu\nu}(\alpha(u,t),t)$ and therefore the expansion of $\bar{L}(t)$ will change from the second term onwards. Indeed, $<\nabla_u\nabla_t \frac{\partial\alpha}{\partial u}, \frac{\partial\alpha}{\partial t}> = <(\nabla_u\nabla_t - \nabla_t\nabla_u)\frac{\partial\alpha}{\partial u}, \frac{\partial\alpha}{\partial t}>$ is no longer [app.9.5] $<R(\dot{\alpha}, \frac{\partial\alpha}{\partial u}) \frac{\partial\alpha}{\partial u}, \dot{\alpha}>$

but has an extra contribution

$$-\frac{\tilde{\partial}}{\partial t}\Gamma^\lambda_{\mu\rho}(\alpha(u,t),t) \frac{\partial\alpha^\mu}{\partial u}\frac{\partial\alpha^\rho}{\partial u} g_{\lambda\sigma}(\alpha(u,t),t) \frac{\partial\alpha^\sigma}{\partial t} \tag{10}$$

These changes lead to a more complicated series, and will give the extra term (7) in the free generating functional, as the first change.

Restricting ourselves now to the WKB-approximation, we have for $\tilde{Z}_\circ(0,0)$, by changing (8) to the midpoint, the extra term

$$\exp -\frac{1}{2}\int_{t_\circ}^{T} \dot{x}^\rho \partial_\rho \ln \sqrt{g(x(t),t)} \tag{11}$$

which is different from (C.52) because of the explicit time dependence.

The Green's functions are constructed in the same manner as before. Indeed, if $\alpha^\mu(u,t)$ satisfies

$$\nabla_t g_{\rho\lambda} \dot{\alpha}^\lambda = 0$$

then

$$\nabla_u\nabla_t g_{\rho\lambda} \dot{\alpha}^\lambda = \nabla_t g_{\rho\lambda}\nabla_t \frac{\partial\alpha^\lambda}{\partial u} + (\nabla_u\nabla_t - \nabla_t\nabla_u) g_{\rho\lambda} \frac{\partial\alpha^\lambda}{\partial u} \tag{12}$$

and as in (10) one has an extra contribution

$$\nabla_t g_{\rho\lambda} \nabla_t \frac{\partial\alpha^\lambda}{\partial u} + R_{\rho\sigma\lambda\mu} \frac{\partial\alpha^\sigma}{\partial t}\frac{\partial\alpha^\lambda}{\partial u}\frac{\partial\alpha^\mu}{\partial t} + \frac{\tilde{\partial}}{\partial t}\Gamma^\kappa_{\rho\mu} g_{\kappa\lambda}\frac{\partial\alpha^\lambda}{\partial u}\frac{\partial\alpha^\mu}{\partial u} \tag{13}$$

In fact we have that in spite of its appearance the term $\frac{\tilde{\partial}\Gamma^{\kappa}_{\mu\rho}}{\partial t}(x(t),t)$ transforms as a tensor under (time independent) coordinate transformations. Indeed

$$\bar{\Gamma}^{\lambda}_{\mu\rho}\frac{\partial\bar{x}^{\mu}}{\partial x^{\kappa}}\frac{\partial\bar{x}^{\rho}}{\partial x^{\sigma}}\frac{\partial x^{\tau}}{\partial\bar{x}^{\lambda}} = \Gamma^{\tau}_{\kappa\sigma} + \frac{\partial^2 x^{\tau}}{\partial\bar{x}^{\mu}\partial\bar{x}^{\nu}}\frac{\partial\bar{x}^{\mu}}{\partial x^{\kappa}}\frac{\partial\bar{x}^{\rho}}{\partial x^{\sigma}}$$

implies $$\frac{\tilde{\partial}}{\partial t}(\bar{\Gamma}^{\lambda}_{\mu\rho})\frac{\partial\bar{x}^{\mu}}{\partial x^{\kappa}}\frac{\partial\bar{x}^{\rho}}{\partial x^{\sigma}}\frac{\partial x^{\tau}}{\partial\bar{x}^{\lambda}} = \frac{\tilde{\partial}}{\partial t}\Gamma^{\tau}_{\kappa\sigma} \tag{14}$$

Finally, there is no change in the computation of the normalization factor and we obtain for $\tilde{Z}_{\circ}(0,0)$

$$\tilde{Z}_{\circ}(0,0) = [2\pi\eta c]^{-d/2}\exp\{-\frac{1}{2}\int_{t_\circ}^{T}\dot{x}^{\rho}\partial_{\rho}\ln\sqrt{g(x(t),t)}\}\,[\det(-\frac{\partial^2 A_{cl}}{\partial x^{\alpha}\partial x^{\beta}_{\circ}})]^{1/2}\exp -\frac{1}{\eta c}A_{cl} \tag{15}$$

which has the correct transformation properties.

## E. Applications

We present two applications of the phase space formalism, developed to compute the higher order corrections to the WKB-approximation. We look at the short time propagator obtained as the short time limit of the propagator, given by its perturbation expansion, where we keep the relevant terms when $T-t_o=\varepsilon$ . Because of the form of the WKB-approximation, we expect to find this short time propagator in the form of the Feynman definition and using the formulas (C84,85) for the first order correction, we shall show explicitly how to obtain this.
A second application concerns the calculation of the propagator up to first order in $\eta$ of a free particle, constrained on the sphere in 4 dimensions ($S_3$). Since everything can be explicitly calculated, and since the full propagator is known, it serves as a nice check of the calculations in (C).

### 1. The short time propagator

To compute the short time propagator corresponding to the equation (C.1), from the semiclassical expansion, one must retain, apart from the dominating exponential, terms up to order $\varepsilon$ , using $\Delta^\mu = X^\mu - X_o^\mu = \mathcal{O}(\sqrt{\varepsilon})$.
From our calculation of the short time propagator (Ch.5) , we expect to find, in addition to the terms given by the WKB-approximation, a term proportional to R/12, coming from the first order correction and contributing to the short time propagator. The WKB-expansion is an expansion in powers of $\eta$ and does not coincide with an expansion in powers of $(T-t_o)=\varepsilon$ , except in the free case (A=V=0) where the expansion is effectively in $\eta(T-t_o)$. In general, we have to use therefore some other method to see which terms contribute up to order $\varepsilon$ .
We shall use a rescaled time $t=\varepsilon u$, so that the correction terms are expressed as integrals in the interval [0,1] , and we shall also use rescaled propagators [C.72-75]

$$\tilde{G}^{\mu\lambda}(u,u') = G^{\mu\lambda}(\varepsilon u,\varepsilon u') \qquad \text{(a)}$$

$$\tilde{T}_\nu{}^\lambda(u,u') = T_\nu{}^\lambda(\varepsilon u,\varepsilon u') \qquad \text{(b)} \qquad (1)$$

$$\tilde{D}_{\nu\lambda}(u,u') = D_{\nu\lambda}(\varepsilon u,\varepsilon u') \qquad \text{(c)}$$

of which we only retain the leading term as a power series in $\varepsilon$ and $\Delta^\mu = X^\mu - X_o^\mu$ . To obtain this leading term we take the Euler-Lagrange equations for the classical path x(t) (C.3)

$$\ddot{x}^\mu + \Gamma^\mu_{\lambda\rho}\dot{x}^\lambda\dot{x}^\rho - F^\mu{}_\rho(x)\dot{x}^\rho + g^{\mu\lambda}(x)\,\partial_\lambda V = 0 \qquad (2)$$

and compute from it the series for x(t)

$$x^\mu(t) = X_\circ{}^\mu + t\,\dot{x}^\mu(t_\circ) + \frac{t^2}{2}\,\ddot{x}^\mu(t_\circ)$$

imposing $x^\mu(\varepsilon) = X^\mu$. For the scaled path $x^\mu(\varepsilon u)$ we obtain a systematic expansion in powers of $\varepsilon$ and $\Delta$

$$\tilde{x}^\mu(u) = x^\mu(\varepsilon u) = X_\circ^\mu + \Delta^\mu u + O(\Delta^2,\varepsilon) \tag{3}$$

The terms we kept in (3) are the only ones we need to find the leading behaviour of the scaled propagators . We find, using (C.75) for the q-q propagator $\tilde{G}^{\mu\lambda}$

$$\text{(a)}\quad \frac{\partial\tilde{x}^\mu}{\partial X_\circ^\alpha} = (1-u)\delta^\mu{}_\alpha + \ldots.$$

$$\text{(b)}\quad \frac{\partial\tilde{x}^\mu}{\partial X^\alpha} = u\,\delta^\mu{}_\alpha + \ldots \tag{4}$$

and since the leading term in the action $A$ is $\int_{t_\circ}^{T} dt\,\dot{x}^\mu g_{\mu\nu}(x)\dot{x}^\nu \to \frac{1}{2\varepsilon}(\Delta^\mu g_{\mu\nu}(X_\circ)\Delta^\nu)$ (III.C.19) we have for $(\bar{M}^{-1})^{\tau\alpha}$

$$(\bar{M}^{-1})^{\tau\alpha} = \varepsilon g^{\tau\alpha}(X_\circ) \tag{5}$$

Here we can choose of course also $g^{\tau\alpha}(X)$ differing only in a term proportional to $\Delta$ . Using then (C.75) for $\tilde{G}^{\mu\lambda}$ , and (C.73)(C.74) for the other propagators, we find

$$\text{(a)}\quad \tilde{G}^{\mu\lambda}(u,u') = c\,\varepsilon\,\{\theta(u-u')\,(1-u)u' + \theta(u'-u)\,(1-u')u\,\}\,g^{\mu\lambda}(X_\circ) + \ldots$$

$$\text{(b)}\quad \tilde{T}_\nu{}^\lambda(u,u') = i\,\{\theta(u-u')\,(-u') + \theta(u'-u)\,(1-u')\,\}\delta_\nu{}^\lambda + \ldots \tag{6}$$

$$\text{(c)}\quad \tilde{D}_{\nu\lambda}(u,u') = \frac{1}{\varepsilon c}\,g_{\nu\lambda}(X_\circ) + \ldots.$$

For the leading behaviour of $\tilde{T}_\nu{}^\lambda(u,u')$ , $\tilde{D}_{\nu\lambda}(u,u')$ , we only had to take

$$\tilde{T}_\nu{}^\lambda(u,u') = \frac{i}{c}\,g_{\nu\mu}(X_\circ)\,\frac{1}{\varepsilon}\,\frac{d}{du}\,\tilde{G}^{\mu\lambda}(u,u') + \ldots$$

and $$\tilde{D}_{\nu\lambda}(u,u') = -\frac{1}{c^2}\,g_{\lambda\rho}(X_\circ)g_{\nu\kappa}(X_\circ)\,\frac{1}{\varepsilon}\frac{d}{du}\,\frac{1}{\varepsilon}\frac{d}{du'}\,\tilde{G}^{\kappa\rho}(u,u') + \ldots$$

From (6) we see that the resulting propagatorshave as leading term, the exact scaled propagators obtained from the classical Lagrangian $\mathcal{L} = \frac{1}{2}\,\dot{q}^\mu g_{\mu\nu}(X_\circ)\dot{q}^\nu$. It is also clear from (6) that each p in the interaction Hamiltonian will give a contribution proportional to $1/\sqrt{\varepsilon}$, while each q contributes as $\sqrt{\varepsilon}$ . In this way it is possible to asign a definite power of $\varepsilon$ an $\Delta$ to each interaction term. Replacing $\dot{x}^\mu$ by $\Delta^\mu/\varepsilon$ we find as leading terms in $\int_{t_o}^{T} dt\ W_3$

$$\varepsilon\frac{i}{3}\int_0^1 du \langle \mathcal{R}(q), \frac{\Delta}{\varepsilon}\rangle \text{ coming from } \int_{t_o}^{T} dt\, \frac{i}{3}\langle \mathcal{R}(q), \dot{x}\rangle$$

and

$$\varepsilon\,\frac{2}{3}\int_0^1 du \langle p, R(\frac{\Delta}{\varepsilon}, q)q\rangle \text{ from } \int_{t_o}^{T} dt\, \frac{2}{3}\langle p, R(\dot{x}, q)q\rangle$$

which are both of order $\sqrt{\varepsilon\Delta}$.

In $\int_{t_o}^{T} dt\, W_4$ there are two terms of order $\varepsilon$ coming from

$$\int_{t_o}^{T} dt[\frac{-ic}{6}\langle p, R(p,q)q\rangle + \frac{c}{3}\langle \mathcal{R}(q), p\rangle] \tag{7}$$

while the others are of higher order.

It can best be seen from (C.47), since $p_\mu$ and $\dot{x}^\mu$ which lower the order are explicit there, that in $\int_{t_o}^{T} W_n dt$ the leading terms come from

$$\int_{t_o}^{T} dt\, \frac{ic}{2}\, \tilde{g}^{\star\mu\rho}(q)^{[n-2]}\, p_\mu p_\rho - \frac{c}{2}\, p_\mu[\tilde{g}^{\star\alpha\rho}(q)\tilde{\Gamma}^{\star\mu}_{\alpha\rho}(q)]^{[n-3]} \tag{8}$$

which are both of order $\varepsilon^{\frac{n-2}{2}}$. These terms do not contribute in $W_3$, which explains the higher order we have found there. Since we must take at least $\int_{t_o}^{T} dt \int_{t_o}^{T} dt'\, W_3(t)W_3(t')$ and this is of order $\varepsilon\Delta^2$, we conclude that the only contribution of order $\varepsilon$ and which therefore plays a role in the short time propagator, is given by (7). In terms of the rescaled propagators, the leading contribution of these terms to $K_1$ is given by

$$K_1^\varepsilon = i\eta\varepsilon[-\frac{ic}{6}\int_0^1 du\, R^\nu{}_{\sigma\rho}{}^\lambda(X_o)(\tilde{D}_{\nu\lambda}(u,u)\tilde{G}^{\sigma\rho}(u,u) + \tilde{T}_\nu{}^\sigma(u,u)\tilde{T}_\lambda{}^\rho(u,u) + \tilde{T}_\nu{}^\rho(u,u)\tilde{T}_\lambda{}^\sigma(u,u))$$
$$+ \frac{c}{3}\int_0^1 du\, R^\nu{}_\lambda(X_o)\, \tilde{T}_\nu{}^\lambda(u,u)] \quad \ldots \tag{9}$$

The curvature and Ricci-tensor can be taken in $X_o$ (or in any intermediate point $X_o + \Delta\alpha$), and the undefined $\tilde{T}_\nu{}^\sigma(u,u)$ has to be interpreted as $\lim_{\delta\to o^+} \tilde{T}_\nu{}^\sigma(u,u+\delta)$, as follows from the postpoint discretization, used in (C).

Using (6) we find

$$K_1^\varepsilon = i\eta\varepsilon\,\{\int_0^1 du\, R(X_o)(-\frac{ic}{6})\,[(1-u)u + (1-u)^2] + \int_0^1 du\, R(X_o)(\frac{ic}{3})(1-u)\}$$
$$= -\frac{\varepsilon\eta c}{6}R(X_o)\int_0^1 du(1-u) = -\frac{\varepsilon\eta c}{12}R(X_o) \tag{10}$$

which confirms what we said in section 5.5.

2. The sphere $S_3$

As a second example of the formalism developed in (C) , we treat free motion on the sphere $S_3$, the group manifold of SU(2). By free, we mean that the propagator $I(X,T,X_o,t_o)$ satisfies the equation

$$\frac{\partial I}{\partial T} = \frac{\eta c}{2} \Delta_2 I = \frac{\eta c}{2} \frac{1}{\sqrt{g}} \partial_\mu \sqrt{g}\, g^{\mu\lambda} \partial_\lambda I \tag{11}$$

$\Delta_2$ is the Laplace-Beltrami operator for the sphere, and it will be specified if we give the metric for the sphere. Instead of giving it immediately, we start from the fact that for a sphere $S_n$ of radius $1/\sqrt{K}$ in (n+1) dimensions, one has the following relation between the curvature tensor and the metric [W2 ]

$$R_{ijkl} = K(g_{ik}\, g_{jl} - g_{il}\, g_{jk}) \tag{12}$$

which implies that the Ricci tensor and the scalar curvature are respectively given by

$$R_{jk} = R^a_{jka} = g^{al}(g_{lk}g_{ja} - g_{la}\, g_{jk})K = g_{jk}(1-n)K \tag{13}$$

$$R = g^{jk}R_{jk} = n(1-n)K \tag{14}$$

In particular we shall be interested in the case n=3.
Assume now thatthere is given a curve x(t) on the sphere, which we describe in an arbitrary coordinate system. Using the expansion method of app.9.5 we express $\bar{L} = \frac{1}{2} \dot{Q}^\mu(t) g_{\mu\nu}(Q(t))\dot{Q}^\nu(t)$ in normal coordinates q(t) of Q(t), centered at x(t). Using only relation (12), which implies $R'(.,.). = 0$ we arrive at the following expansion

$$\bar{L} = \sum_{n=o}^{\infty} S^n \frac{1}{n!} \tag{15}$$

$$S^o = \frac{1}{2} \langle \dot{x},\dot{x}\rangle$$
$$S^1 = \langle \nabla_t q,\dot{x}\rangle$$
$$S^2 = \langle \nabla_t q,\nabla_t q\rangle - K[\langle q,q\rangle\langle \dot{x},\dot{x}\rangle - (\langle \dot{x},q\rangle)^2]$$
$$S^3 = 4K[\langle \dot{x},q\rangle \langle q,\nabla_t q\rangle - \langle q,q\rangle S^1]$$
$$S^4 = -4K \langle q,q\rangle S^2 + 4K (\langle \nabla_t q,q\rangle)^2$$
$$S^5 = -4K \langle q,q\rangle S^3$$
$$\vdots$$
$$S^n = -4K \langle q,q\rangle S^{n-2} \tag{16}$$

We recall that all contractions are with respect to $g_{\mu\nu}(x(t))$, in particular

$$<q,q> = q^{\mu}(t)\, g_{\mu\nu}(x(t))q^{\nu}(t)$$

We see that (12) determines all terms in (16), and hence the metric which can be obtained from the summed series (15). We obtain, using the recursion relation in (16)

$$\begin{aligned}\bar{L} = \frac{1}{2}|\dot{x}|^2 &+ \{<\nabla_t q,\nabla_t q> - K|q|^2\,|\dot{x}|^2 + K(<\dot{x},q>)^2\}\,\frac{1-\cos 2\sqrt{K}|q|}{4K\,|q|^2} \\ &+ \frac{4K}{(4K|q|^2)^2}(\cos(2\sqrt{K}\,|q|)+2K|q|^2-1)\,(<q,\nabla_t q>)^2 + <\nabla_t q,\dot{x}>\left(\frac{\sin 2\sqrt{K}|q|}{2\sqrt{K}|q|}\right) \\ &+ 4K<\dot{x},q><q,\nabla_t q>\left(\frac{2\sqrt{K}\,|q| - \sin 2\sqrt{K}\,|q|}{(4K\,|q|^2)^{3/2}}\right)\end{aligned} \qquad (17)$$

with the notation $|q| = (q^{\mu}(t)\, g_{\mu\nu}(x(t))q^{\nu}(t))^{1/2}$. From (C.45) we obtain an expression for the metric $\bar{g}_{\mu\nu}\,(q(t),x(t))$ in normal coordinates centered at $x(t)$

$$\begin{aligned}\bar{g}_{\mu\nu}(q(t),x(t)) &= \frac{\partial^2\bar{L}}{\partial(\nabla_t q^{\mu})\partial(\nabla_t q^{\nu})} \\ &= 2\, g_{\mu\nu}(x(t))\left(\frac{1-\cos 2\sqrt{K}|q|}{4K|q|^2}\right) + 2\, g_{\mu\lambda}(x(t))q^{\lambda}q^{\rho}g_{\rho\nu}(x(t)) \\ &\qquad (\cos 2\sqrt{K}|q| + 2K|q|^2 - 1)\,\frac{4K}{(4K|q|^2)^2} \\ &= g_{\mu\nu}(x(t))\,\frac{\sin^2\sqrt{K}\,q}{K|q|^2} + g_{\mu\lambda}\,(x(t))q^{\lambda}q^{\rho}g_{\rho\nu}(x(t))\,\frac{K|q|^2 - \sin^2\sqrt{K}|q|}{K|q|^4}\end{aligned} \qquad (18)$$

To obtain the normal coordinates centered at an arbitrary point $\Omega_0$, we can take $g_{\mu\nu}(\Omega_0)$ in (18) instead of $g_{\mu\nu}(x(t))$ , and because the original metric can always be brought to a form satisfying $g_{\mu\nu}(\Omega_0) = \delta_{\mu\nu}$ , we finally obtain that we can take the metric in normal coordinates around an arbitrary point of the sphere as ($q(t)$ now becomes $q$)

$$\bar{g}_{\mu\nu}\,(q) = \left(\delta_{\mu\nu} - \frac{q^{\mu}q^{\nu}}{q^2}\right)\frac{\sin^2(\sqrt{K}\,q)}{Kq^2} + \frac{q^{\mu}q^{\nu}}{q^2} \equiv X^{\mu\nu}D + Y^{\mu\nu} \qquad (19)$$

where we use the notation

(a) $Y^{\mu\nu} = \dfrac{q^{\mu}q^{\nu}}{q^2}$ (c) $D = \dfrac{\sin^2[\sqrt{K}\,q\,]}{Kq^2}$

(b) $X^{\mu\nu} = \delta_{\mu\nu} - Y^{\mu\nu}$ (d) $q = (\sum_{\mu} q^{\mu}q^{\mu})^{1/2}$ (20)

The propagator $I(X,T,X_o,t_o)$ in an arbitrary coordinate system can now easily be related to the propagator in the normal coordinate system centered at $X_o$, with metric given by (19). If $\sigma_{\mu j}\sigma_{\lambda j} = g_{\mu\lambda}(X_o)$ and $\bar{x}^\mu$ is related to $X^\mu$ by (C.38) where $x(t)$ is replaced by $X_o, \Omega \to X, q \to \bar{x}$, we have

$$I(X,T-t_o,X_o,0) = \sqrt{g(X_o)}\; I^\star(\sigma_{\mu j}\, x^\mu,\; T-t_o,0,0) \tag{21}$$

What remains to be done is the construction of the propagator $I^\star(x,T-t_o,0,0)$ in the coordinate system with metric (19). We have seen in (C) that the perturbation expansion of $I^\star$ will be completely determined by the knowledge of the classical trajectory, and the solutions of the Jacobi equation. Scaling the time (we shall reintroduce the explicit time dependence afterwards), we have as classical path the geodesic $\gamma^\mu(t) = x^\mu t$, $t \in [0,1]$.
The solutions of the Jacobi equation, here of the form

$$\nabla_t\nabla_t\, K^\mu + \bar{R}^\mu\,(\dot{\bar{\gamma}},K)\,\dot{\bar{\gamma}} = 0 \tag{22}$$

needed to construct the $\overline{qq}$ propagator (C.75), are found by differentiation of $\bar{\gamma}$ with respect to the boundary conditions. The first set of solutions is then, taking the derivative with respect to the final point,

$$K^{(1)\mu}{}_\rho = t\delta^\mu{}_\rho \tag{23}$$

We don't have an explicit initial condition. However, if $\gamma^\mu(t)$ is the same geodesic from $X_o$ to $X$ in a coordinate system $\Omega$ with metric $g_{\mu\nu}(\Omega)$ (we suppose $g_{\mu\nu}(X_o) = \delta_{\mu\nu}$), then the second set of solutions $K^{(2)\mu}{}_\beta$ is found by transforming $\partial\gamma^\mu(t)/\partial X_o{}^\lambda$ to the normal coordinates centered at $X_o$.

Writing the coordinate transformation as $\Omega(X_o,\bar{q})$

$$K^{(2)\lambda}{}_\rho = \left(\frac{\partial\bar{q}^\lambda}{\partial\Omega^\mu}\right)_{\Omega^\mu=\gamma^\mu(t)} \frac{\partial\gamma^\mu(t)}{\partial X_o{}^\sigma} \left(\frac{\partial\Omega^\sigma}{\partial\bar{q}^\rho}\right)_{\bar{q}=0} = \left(\frac{\partial\bar{q}^\lambda}{\partial\Omega^\mu}\frac{\partial\gamma^\mu}{\partial X_o^\sigma}\right)_{\Omega^\mu=\gamma^\mu(t)} \tag{24}$$

In the expression for the geodesic (C.37)

$$\gamma^\mu(t) = X_o{}^\mu + x^\mu t - \frac{1}{2}\,(\Gamma^\mu_{\lambda\rho})_{X_o}\, x^\lambda x^\rho t^2 - \frac{1}{3!}\,\cdots \tag{25}$$

not all the $X_o$ dependence is explicit, since we have $x\,(X_o)$ by the relation

$$X^\mu = X_o^\mu + x^\mu - \frac{1}{2}\,(\Gamma^\mu_{\lambda\rho})_{X_o}\, x^\lambda x^\rho - \frac{1}{3!}\,\cdots\cdot \tag{26}$$

(26) is just the relation $X^\mu = \Omega^\mu(X_o,x)$. The derivative of $\gamma^\mu(t)$ with respect to $X_o$ is then found in the form

$$\frac{\partial\gamma^\mu}{\partial X_\circ^\rho} = \left(\frac{\partial Q^\mu(X_\circ,\bar{q})}{\partial X_\circ^\rho}\right)_{\bar{q}=xt} + \frac{\partial\gamma^\mu}{\partial x^\sigma}\frac{\partial x^\sigma}{\partial X_\circ^\rho} \tag{27}$$

but from (25) and (26) we see that

$$\frac{\partial\gamma^\mu}{\partial x^\sigma} = t\left(\frac{\partial Q^\mu(X_\circ,\bar{q})}{\partial\bar{q}^\sigma}\right)_{\bar{q}=xt} \tag{28}$$

and therefore the second set of solutions of the Jacobi equation, $K^{(2)\lambda}{}_\rho$, takes the form

$$K^{(2)\lambda}{}_\rho = \left(\frac{\partial\bar{q}^\lambda}{\partial Q^\mu}\frac{\partial Q^\mu}{\partial X_\circ^\rho}\right)_{\bar{q}=xt} + t\left(\frac{\partial\bar{q}^\lambda}{\partial X_\circ^\rho}\right)_{\bar{q}=x} . \tag{29}$$

The first term in (29) is a power series in t, the second term is linear in t (and satisfies the Jacobi equation because of (23)). We check the boundary conditions of (29). At $t = 0$, we have

$$K^{(2)\lambda}{}_\rho(0) = \left(\frac{\partial\bar{q}^\lambda}{\partial Q^\mu}\frac{\partial Q^\mu}{\partial X_\circ^\rho}\right)_{\bar{q}=0} = \delta^\lambda{}_\rho .$$

In $t = 1$ we should find $K^{(2)\lambda}{}_\rho(1) = 0$, and for this we have to check in (29)

$$\left(\frac{\partial\bar{q}^\lambda}{\partial Q^\mu}\frac{\partial Q^\mu}{\partial X_\circ^\rho}\right)_{\bar{q}} = -\left(\frac{\partial\bar{q}^\lambda}{\partial X_\circ^\rho}\right)_{\bar{q}} \tag{30}$$

From (26), we see that the inverse relation $\bar{q}(Q)$ is of the form $\bar{q}(Q - X_\circ, X_\circ)$, where the second $X_\circ$ indicates the dependence on $X_\circ$ in the coefficients of the series. By differentiation of the identity $\bar{q}^\mu(Q(X_\circ,\bar{q}) - X_\circ, X_\circ) = \bar{q}^\mu$, we obtain ($\tilde{\partial}$ indicates differentiation of the last argument)

$$\left(\frac{\partial\bar{q}^\lambda}{\partial Q^\rho}\frac{\partial Q^\rho}{\partial X_\circ^\sigma} - \frac{\partial\bar{q}^\lambda}{\partial Q^\sigma} + \frac{\tilde{\partial}\bar{q}^\lambda}{\partial X_\circ^\sigma}\right) = 0$$

or

$$\left(\frac{\partial\bar{q}^\lambda}{\partial Q^\rho}\frac{\partial Q^\rho}{\partial X_\circ^\sigma}\right)_{\bar{q}} = -\left(\frac{\tilde{\partial}\bar{q}^\lambda}{\partial X_\circ^\sigma} - \frac{\partial\bar{q}^\lambda}{\partial Q^\sigma}\right) = -\frac{\partial\bar{q}^\lambda}{\partial X_\circ^\sigma}$$

which verifies (30) and leads to another form of (29).

$$K^{(2)\lambda}{}_\rho(t) = \frac{\partial\bar{q}^\lambda}{\partial Q^\mu}\frac{\partial Q^\mu}{\partial X_\circ^\rho}(X_\circ,\bar{q})_{\bar{q}^\mu=x^\mu t} - t\left[\frac{\partial\bar{q}^\lambda}{\partial Q^\mu}\frac{\partial Q^\mu}{\partial X_\circ^\rho}(X_\circ,\bar{q})\right]_{\bar{q}^\mu=x^\mu} \tag{31}$$

To obtain the explicit form of $K^{(2)\lambda}{}_\rho(t)$, we use again (17). Indeed (compare to (C.41) and (C.42))

$$\frac{\partial^2\bar{L}}{\partial\dot{x}^\sigma\partial(\nabla_t q)^\rho} = \bar{g}_{\rho\lambda}(q(t),\ x(t))\left(\frac{\partial\bar{q}^\lambda}{\partial Q^\mu}\frac{\partial Q^\mu}{\partial x^\sigma}\right)$$

and hence

$$\left(\frac{\partial\bar{q}^\lambda}{\partial Q^\mu}\frac{\partial Q^\mu}{\partial X_\circ^\rho}\right)_{\bar{q}} = \left[\bar{g}^{\lambda\rho}(q(t),\ x(t))\frac{\partial^2\bar{L}}{\partial\dot{x}^\sigma\partial(\nabla_t q)^\rho}\right]_{\substack{x(t)\to X_\circ \\ q(t)\to\bar{q}}} \tag{32}$$

Here we need the contravariant metric tensor which has the form

$$\bar{g}^{\lambda\rho}(q) = X^{\lambda\rho}\frac{1}{D} + Y^{\lambda\rho}. \tag{33}$$

This can be verified from (19) and the simple rules $Y^{\mu\nu}Y^{\nu\lambda} = Y^{\mu\lambda}$, $X^{\mu\nu}X^{\nu\lambda} = X^{\mu\lambda}$, $X^{\mu\nu}Y^{\nu\lambda} = Y^{\mu\nu}X^{\nu\lambda} = 0$ which follow from (20). From (17) we obtain then the result (using $g_{\mu\nu}(X_\circ) = \delta_{\mu\nu}$, and $\Gamma_\mu{}^\lambda{}_\sigma$ are the Christoffel symbols at $X_\circ$)

$$\left[\frac{\partial^2\bar{L}}{\partial\dot{x}^\lambda\partial(\nabla_t q)^\rho}\right]_{\substack{x(t)\to X_\circ \\ q(t)\to q}} = X^{\lambda\rho}\frac{\sin\sqrt{K}\,q\,\cos\sqrt{K}\,q}{\sqrt{K}\,q} + Y^{\lambda\rho} + \Gamma_\mu{}^\kappa{}_\lambda\bar{g}_{\rho\kappa} \tag{34}$$

and finally

$$\left(\frac{\partial\bar{q}^\lambda}{\partial Q^\mu}\frac{\partial Q^\mu}{\partial X_\circ^\rho}\right) = X^{\lambda\rho}(\mathrm{cotg}\sqrt{K}\,q)\sqrt{K}\,q + Y^{\lambda\rho} + \Gamma_\mu{}^\lambda{}_\rho q^\mu \tag{35}$$

Let us indicate by $\tilde{X}^{\mu\nu}$, $\tilde{Y}^{\mu\nu}$ the quantities obtained from $X^{\mu\nu}$, $Y^{\mu\nu}$ under the substitution $q^\lambda = x^\lambda t$ (or under $q^\lambda = x^\lambda$, because they are independent of t). (31) and (35) give us an explicit formula for $K^{(2)\lambda}{}_\rho(t)$

$$K^{(2)\lambda}{}_\rho(t) = \tilde{X}^{\lambda\rho}\sqrt{K}\,x\,t[\mathrm{cotg}(\sqrt{K}\,x\,t) - \mathrm{cotg}\,(\sqrt{K}\,x)] + \tilde{Y}^{\lambda\rho}(1-t) \tag{36}$$

The Van Vleck matrix $\bar{M}_{\lambda\rho}(0,1)$ ( app.9.1(23))is also immediately found as (app.9.1(18))

$$\bar{M}_{\lambda\rho}(0,1) = \frac{\partial P_\lambda(0)}{\partial x^\rho} = \frac{\partial\dot{\bar{\gamma}}^\mu(0)}{\partial x^\rho}\bar{g}_{\mu\lambda}(0) = \delta_{\lambda\rho} \tag{37}$$

(23), (31) and (37) enable us to compute the q-q propagator , while for the other propagators we need the covariant derivatives of $K^{(1)\sigma}{}_\lambda(t)$ and $K^{(2)\sigma}{}_\lambda(t)$. Calculating the Christoffel symbols from (19)

$$\bar{\Gamma}^{\mu}_{\lambda\rho} = (X^{\mu\rho}q^{\lambda} + X^{\mu\lambda}q^{\rho})\ K\ \left(\frac{\cot g\sqrt{K}q}{\sqrt{K}\ q} - \frac{1}{Kq^2}\right) - X^{\lambda\rho}\left(\frac{\sin\sqrt{K}q}{\sqrt{K}\ q}\ \cos\sqrt{K}\ q - 1\right)\frac{1}{q^2} \tag{38}$$

which lead to

$$\nabla_t\ K^{(1)\sigma}{}_{\lambda}(t) = [x\ t\sqrt{K}\ \cot g\ (xt\sqrt{K})]\tilde{X}^{\sigma\lambda} + \tilde{Y}^{\sigma\lambda} \tag{39}$$

$$\nabla_t\ K^{(2)\sigma}{}_{\lambda}(t) = -[Kx^2t(1 + \cot g\sqrt{K}\ x\ t\ \cot g\sqrt{K}x)]\tilde{X}^{\sigma\lambda} - \tilde{Y}^{\sigma\lambda} \tag{40}$$

Applying (C.75) for $G^{\mu\lambda}(t, t')$, where $K^{(2)\lambda}{}_{\rho}(t)$ now replaces $\dfrac{\partial x^{\lambda}(t)}{\partial X^{\rho}_{\circ}}$ and $K^{(1)\lambda}{}_{\rho}(t)$ is $\dfrac{\partial x^{\lambda}(t)}{\partial X^{\rho}}$, while $\bar{M}_{\lambda\rho} = \delta_{\lambda\rho}$

$$G^{\mu\lambda}(t, t') = c\{\theta(t - t')\ K^{(2)\mu}{}_{\gamma}(t)K^{(1)\lambda}{}_{\gamma}(t') + \theta(t' - t)K^{(1)\mu}{}_{\kappa}(t)K^{(2)\lambda}{}_{\kappa}(t')\} \tag{41}$$

From (C.73), (C.74)

$$T_{\nu}{}^{\lambda}(t,t') = i\bar{g}_{\nu\mu}(\bar{\gamma}(t))\{\theta(t-t')\nabla_t K^{(2)\mu}{}_{\gamma}(t)K^{(1)\lambda}{}_{\gamma}(t') - \theta(t'-t)\nabla_t K^{(1)\mu}{}_{\gamma}(t)K^{(2)\lambda}{}_{\gamma}(t')\} \tag{42}$$

$$D_{\nu\lambda}(t,t') = -\frac{1}{c}\ \bar{g}_{\lambda\rho}(\bar{\gamma}(t'))\bar{g}_{\nu\kappa}(\bar{\gamma}(t))\{\theta(t-t')\nabla_t K^{(2)\kappa}{}_{\gamma}(t)\nabla_{t'}K^{(1)\lambda}{}_{\gamma}(t') - \theta(t'-t)\nabla_t K^{(1)\kappa}{}_{\gamma}(t)\nabla_{t'}K^{(2)\lambda}{}_{\gamma}(t')\} \tag{43}$$

and explicitly

$$G^{\mu\lambda}(t,t') = c\{\theta(t-t')[\tilde{X}^{\mu\lambda}A(t)t' + \tilde{Y}^{\mu\lambda}(1-t)t'] + \theta(t'-t)[\tilde{X}^{\mu\lambda}A(t')t + \tilde{Y}^{\mu\lambda}(1-t')t]\} \tag{44}$$

$$T_{\alpha}{}^{\beta}(t,t') = -i\{\theta(t-t')[\tilde{X}^{\alpha\beta}\ \frac{B(t)t'}{t} + \tilde{Y}^{\alpha\beta}t'] - \theta(t'-t)[\tilde{X}^{\alpha\beta}C(t)A(t') + \tilde{Y}^{\alpha\beta}(1-t)]\} \tag{45}$$

$$D_{\alpha\beta}(t,t') = \frac{1}{c}\{\theta(t-t')[\tilde{X}^{\alpha\beta}\frac{B(t)C(t')}{t} + \tilde{Y}^{\alpha\beta}] + \theta(t'-t)[\tilde{X}^{\alpha\beta}\frac{B(t')C(t)}{t'} + \tilde{Y}^{\alpha\beta}]\} \tag{46}$$

where

$$\text{(a)}\quad A(t) = \sqrt{K}\,xt\;(\mathrm{cotg}\sqrt{K}\,xt - \mathrm{cotg}\sqrt{K}\,x) \tag{47}$$

$$\text{(b)}\quad B(t) = (1 + \mathrm{cotg}\sqrt{K}\,xt\;\mathrm{cotg}\sqrt{K}\,x)\sin^2\sqrt{K}\,xt$$

$$\text{(c)}\quad C(t) = \frac{\cos\sqrt{K}\,xt\;\sin\sqrt{K}\,xt}{\sqrt{K}\,xt}$$

These are the explicit formulas we need to do the perturbation expansion. To compute the first order correction, we copy the interaction Hamiltonian from (C.84)(C.85) (for F = V = 0), using (12) for the curvature tensor.

$$i\int dt\,\sqrt{\eta}W_3 + \eta W_4 = \int_0^1 dt\; i\sqrt{\eta}\,\{\frac{2}{3}K[\langle\dot{x},q\rangle\langle p,q\rangle - \langle p,\dot{x}\rangle\langle q,q\rangle] - \frac{2}{3}i\langle q,\dot{x}\rangle\}$$

$$+ i\eta\{-\frac{i}{18c}K^2[\langle q,q\rangle^2\langle\dot{x},\dot{x}\rangle - \langle q,q\rangle\langle\dot{x},q\rangle^2] - \frac{ic}{6}K(\langle p,q\rangle^2 - \langle p,p\rangle\langle q,q\rangle) - \frac{2c}{3}K\langle p,q\rangle \tag{48}$$

It should be clear that in this formula the contractions <.,.> are done with respect to $\bar{g}_{\mu\nu}(xt)$. The remaining calculations are straightforward, and lead to a first order correction given by

$$K^{(1)} = c\,\frac{K}{2} = -\frac{c}{12}R \tag{49}$$

The WKB approximation is given by (C.83). In our case, since

$$\bar{g}(q) = \det(X^{\mu\nu}D + Y^{\mu\nu}) = D^2$$

we obtain up to first order in $\eta$

$$I^{\star}(x,1;0,0) = \frac{1}{(2\pi\eta c)^{3/2}}\;\frac{\sqrt{K}x}{\sin\sqrt{K}x}\;\exp\{-\frac{1}{2\eta c}x^2\}\;(1 + \frac{\eta c K}{2} + \ldots) \tag{50}$$

The classical action in (50) is found as

$$\frac{1}{2}\int_0^1 dt\;\dot{\bar{\gamma}}^{\mu}\bar{g}_{\mu\nu}(\bar{\gamma}(t))\dot{\bar{\gamma}}^{\nu} = \frac{1}{2}\int_0^1 dt\;x^{\mu}\bar{g}_{\mu\nu}(xt)x^{\nu} = \frac{1}{2}x^2$$

where we used $x^{\mu}\tilde{X}^{\mu\nu}x^{\nu} = 0$, $x^{\mu}\tilde{Y}^{\mu\nu}x^{\nu} = x^2$

Finally, we reintroduce the time $T - t_o$, and this is most easily done, remarking that $I^\star(x, T-t_o, 0, 0)$ is in fact a function of $\eta c(T-t_o)$ (see (11). So

$$I^\star(x, T-t_o; 0,0) = \frac{1}{[2\pi\eta c(T-t_o)]^{3/2}} \left(\frac{\sqrt{K}\, x}{\sin\sqrt{K}\, x}\right) \exp\left\{-\frac{1}{\eta c}\frac{x^2}{2(T-t_o)}\right\}$$

$$\left(1 + \frac{\eta c K}{2}(T-t_o) + \mathcal{O}(\eta^2)\right) \tag{51}$$

In fact the higher order corrections sum up to give

$$I^\star(x,T-t_o; 0,0) = [2\pi\eta c(T-t_o)]^{-3/2} \left(\frac{\sqrt{K}\, x}{\sin\sqrt{K}\, x}\right) \exp\left\{-\frac{1}{\eta c}\frac{x^2}{2(T-t_o)} + \frac{\eta c K}{2}(T-t_o)\right\} \tag{52}$$

The complete solution will be treated in Ch.12, were functional integrals will be used to obtain this global result. There we shall make clear the meaning of (42), since in fact the manifold of the sphere $S_3$ is compact, and there are an infinity of classical paths connecting 0 and x,two facts which do not fit in the formalism of (C).

Appendix 9.1

If $\mathcal{L}(q,\dot{q},t)$ is a Lagrangian, and $z(t)$ any solution of the Euler-Lagrange equations of $\mathcal{L}$, we denote by $S(z(t))_{\mu\rho}$ the Jacobi-operator constructed with this solution, i.e.

$$S(z(t))_{\mu\rho} = A^{(2)}_{\mu\rho}(t)\frac{\partial^2}{\partial t^2} + A^{(1)}_{\mu\rho}(t)\frac{\partial}{\partial t} + A^{(0)}_{\mu\rho}(t) \tag{1}$$

with
$$A^{(2)}_{\mu\rho}(t) = -\frac{\partial^2\mathcal{L}}{\partial\dot{q}^\mu\partial\dot{q}^\rho}\Big|_{q = z(t)} \quad \text{(a)}$$

$$A^{(1)}_{\mu\rho}(t) = \frac{\partial^2\mathcal{L}}{\partial q^\mu\partial\dot{q}^\rho} - \frac{\partial^2\mathcal{L}}{\partial\dot{q}^\mu\partial q^\rho} - \frac{\partial}{\partial t}\left(\frac{\partial^2\mathcal{L}}{\partial q^\mu\partial\dot{q}^\rho}\right)\Big|_{q = z(t)} \quad \text{(b)} \quad (2)$$

$$A^{(0)}_{\mu\rho}(t) = \frac{\partial^2\mathcal{L}}{\partial q^\mu\partial\dot{q}^\rho} - \frac{\partial}{\partial t}\left(\frac{\partial^2\mathcal{L}}{\partial\dot{q}^\mu\partial q^\rho}\right)\Big|_{q = z(t)} \quad \text{(c)}$$

(2) is obtained [D10] by evaluating the second variation of the action $\int\mathcal{L}(q,\dot{q},t)dt$ along the path $z(t)$ (neglecting boundary terms)

$$\int dtdt'q^\alpha(t)\frac{\delta^2 S}{\delta q^\alpha(t)\delta q^\beta(t')}\Big|_{z(t)} q^\beta(t') = \int dtq^\alpha(t)\, S(z(t))_{\alpha\beta}q^\beta(t) \tag{3}$$

Alternatively one can say that $S(z(t))$ is obtained as the operator in the Euler-Lagrange equations corresponding to the quadratic Lagrangian $L_o$

$$L_o(t) = \frac{1}{2}\dot{q}^\mu\frac{\partial^2\mathcal{L}}{\partial\dot{q}^\mu\partial\dot{q}^\rho}\Big|_{z(t)}\dot{q}^\rho + \dot{q}^\mu\frac{\partial^2\mathcal{L}}{\partial\dot{q}^\mu\partial q^\rho}\Big|_{z(t)} q^\rho + \frac{1}{2}q^\mu\frac{\partial^2\mathcal{L}}{\partial q^\mu\partial q^\rho}\Big|_{z(t)} q^\rho \tag{4}$$

It's easily shown, that if $u$ is any of the 2d constants of integration of $z(t)$, then [D14,D17,M13]

$$S(z(t))_{\mu\rho}\frac{\partial z^\rho(t)}{\partial u} = 0 \tag{5}$$

Let $q(t)$ be the solution with boundary conditions $q(t_a) = a$, $p(t_a) = p_a$ where $p_\mu = \frac{\partial\mathcal{L}}{\partial\dot{q}^\mu}$ is the conjugate momentum. We consider the function $\tilde{J}^{\alpha\beta}(t,s)$, defined by

$$\tilde{J}^{\alpha\beta}(t,s) = \frac{\partial q^\alpha(t)}{\partial a^\gamma}\frac{\partial q^\beta(s)}{\partial p_{a\gamma}} - \frac{\partial q^\alpha(t)}{\partial p_{a\gamma}}\frac{\partial q^\beta(s)}{\partial a^\gamma} = \{q^\alpha(t),\, q^\beta(s)\}_{(a,p_a)} \tag{6}$$

where $\{\ \}_{(a,p_a)}$ stands for the Poisson-bracket with respect to $(a,p_a)$[G10].

Because of (5) $\tilde{J}^{\alpha\beta}(t,s)$ satisfies the Jacobi equation

$$S(q(t))_{\mu\rho}\tilde{J}^{\rho\sigma}(t,s) = 0 \tag{7}$$

The boundary conditions are

$$\tilde{J}^{\alpha\beta}(s,s) = \{q^{\alpha}(s),\ q^{\beta}(s)\}_{(a,p_a)} = 0 \tag{8}$$

$$\frac{\partial}{\partial t}\tilde{J}^{\alpha\beta}(t,s) = \{\dot{q}^{\alpha}(t),\ q^{\beta}(s)\}_{(a,p_a)} = +\{\frac{\partial H(t)}{\partial p_a(t)},\ q^{\beta}(s)\}_{(a,p_a)}$$

Here $H(t)$ is the Hamiltonian corresponding to the original Lagrangian $\mathcal{L}$, taken at $(q(t),\ p(t))$.

At equal times $t = s$, we obtain

$$\frac{\partial}{\partial t}\tilde{J}^{\alpha\beta}(t,s)\Big|_{t=s} = \{\frac{\partial H(s)}{\partial p_{\alpha}(s)},\ q^{\beta}(s)\}_{(q(s),p(s))} = -\frac{\partial^2 H(s)}{\partial p_{\beta}(s)\partial p_{\alpha}(s)} = -g^{\alpha\beta}(q(s)) \tag{9}$$

We used here $\{f,g\}_{(a,p_a)} = \{f,g\}_{(q(s),p(s))}$ since $(a,\ p_a)$ and $(q(s),\ p(s))$ are related by a canonical transformation. Since $\dot{q}^{\mu} = \frac{\partial H}{\partial p_{\mu}}$ and $p_{\alpha} = \frac{\partial \mathcal{L}}{\partial \dot{q}^{\alpha}}$, we have by differentiating $\dot{q}^{\mu} = \frac{\partial H}{\partial p_{\mu}}$ with respect to $\dot{q}^{\nu}$

$$\delta^{\mu}_{\nu} = \frac{\partial^2 H}{\partial p_{\mu}\partial p_{\alpha}}\ \frac{\partial p_{\alpha}}{\partial \dot{q}^{\nu}} = \frac{\partial^2 H}{\partial p_{\mu}\partial p_{\alpha}}\ \frac{\partial^2 \mathcal{L}}{\partial \dot{q}^{\alpha}\partial \dot{q}^{\nu}}$$

and $g^{\alpha\beta}(q(s))$ is the inverse of $g_{\beta\rho}(q(s)) = \frac{\partial^2 \mathcal{L}}{\partial \dot{q}^{\beta}\partial \dot{q}^{\rho}}$

From (6) we see that $\tilde{J}^{\alpha\beta}(t,s) = -\tilde{J}^{\beta\alpha}(s,t)$ and hence $\tilde{J}^{\alpha\beta}(t,s)$ also satisfies the Jacobi equation in the variable s

$$S(q(s))_{\mu\rho}\ \tilde{J}^{\alpha\rho}(t,s) = 0 \tag{10}$$

Taking now $s = t_a$, one has from (6)

$$\tilde{J}^{\alpha\beta}(t,t_a) = -\frac{\partial q^{\alpha}(t)}{\partial p_{a\beta}} = -\frac{\partial q^{\alpha}(t)}{\partial p_{\beta}(t_a)} \tag{11}$$

Let $Q(t)$ be the solution of the Euler-Lagrange equations of $\mathcal{L}$ for boundary conditions $Q(t_a) = a$, $Q(t_b) = b$, and let $P_{\mu}(t) = \frac{\partial \mathcal{L}}{\partial \dot{Q}^{\mu}}$ be the conjugate momentum. The relation between b and $p_a$ is given by

$$b^\mu = q^\mu(t_b, t_a, a, p_a) \tag{12}$$

which can be inverted as

$$p_{a_\mu} = f_\mu(t_b, t_a; b, a) = \frac{\partial \mathcal{L}(t_a)}{\partial \dot{Q}^\mu(t_a)} = P_\mu(t_a) \tag{13}$$

In(6) we used $\tilde{J}^{\alpha\beta}(t,s)$ as a function of $(t_a, a, p_a)$, i.e., $\tilde{J}^{\alpha\beta}(t,s) = \tilde{J}^{\alpha\beta}(t,s|t_a, a, p_a)$. Now we define the function $\bar{J}^{\alpha\beta}(t,s)$ as the function obtained from $\tilde{J}^{\alpha\beta}(t,s)$ replacing $p_{a_\mu} = f_\mu(t_b, t_a; b, a)$. So $\bar{J}^{\alpha\beta}(t,s)$ satisfies the Jacobi equation

$$S(Q(t))_{\mu\rho}\, \bar{J}^{\rho\nu}(t,s) = 0 \tag{14}$$

with the same boundary conditions as $\tilde{J}^{\alpha\beta}(t,s)$

$$\bar{J}^{\alpha\beta}(s,s) = 0 \qquad \text{(a)} \tag{15}$$

$$\frac{\partial}{\partial t}\bar{J}^{\alpha\beta}(t,s)\Big|_{t=s} = -\frac{\partial^2 H(P,Q,s)}{\partial P_\alpha(s)\,\partial P_\beta(s)} = -g^{\alpha\beta}(Q(s)) \qquad \text{(b)}$$

The equivalent of (11) is now

$$\bar{J}^{\alpha\beta}(t,t_a) = -\frac{\partial q^\alpha(t)}{\partial p_{a_\beta}}\Big|_{p_{a_\mu} = f_\mu(t_b, t_a, b, a)} \tag{16}$$

From (12) and (13) we obtain

$$\delta_\mu^\rho = \frac{\partial p_{a_\mu}}{\partial b^\sigma}\,\frac{\partial b^\sigma}{\partial p_{a_\rho}} = \frac{\partial P_\mu(t_a)}{\partial b^\sigma}\,\frac{\partial q^\sigma(t_b)}{\partial p_{a_\rho}}\Big|_{p_{a_\mu} = f_\mu(t_b, t_a; b, a)} \tag{17}$$

which implies

$$M_{\mu\sigma}(t_a, t_b) = -\frac{\partial p_\mu(t_a)}{\partial b^\sigma} \tag{18}$$

if we define the inverse $M_{\mu\nu}(t,t')$ of $\bar{J}^{\nu\rho}(t',t)$ as

$$M_{\mu\nu}(t,t')\, \bar{J}^{\nu\rho}(t',t) = \delta_\mu^\rho \tag{19}$$

In view of (6), defined by the Poisson-bracket, we have in terms of the Lagrange-bracket $[\,,\,]_L(a,p_a)$

$$\sum_\alpha [q^\alpha(t), q^\rho(s)]_{L(a,p_a)} \{q^\alpha(t), q^\beta(s)\}_{(a,p_a)} = \delta^{\rho\beta} \tag{20}$$

which gives for $\tilde{M}_{\mu\nu}(t,t')\, \tilde{J}^{\nu\rho}(t',t) = \delta^\rho_\mu$

$$\tilde{M}_{\mu\nu}(t,t') = [q^\mu(t), q^\nu(t')]_{L(a,p_a)} = \frac{\partial a^\gamma}{\partial q^\mu(t)} \frac{\partial p_{a_\gamma}}{\partial q^\nu(t')} - \frac{\partial p_{a_\gamma}}{\partial q^\mu(t)} \frac{\partial a^\gamma}{\partial q^\nu(t')} \tag{21}$$

and $M_{\mu\nu}$ is obtained from $\tilde{M}_{\mu\nu}$, using $p_{a_\mu} = f_\mu(t_b,t_a;b,a)$. This also leads to (18). In general, the sum in (20) is over a set of (2d) independent variables $u^\alpha(p_a,a)$. It will be valid in this form if $(q^\alpha(s), q^\beta(t))$ is a set of 2d independent variables. This assumption is similar to the one we made inverting (12)(13).

Let $A(b,t_b|a,t_a) = A(Q(t)) = \int_{t_a}^{t_b} dt\mathcal{L}(Q, \dot{Q},t)$ be the action for the solution $Q(t)$, then one has

$$P_\mu(t_a) = -\frac{\partial A(b,t_b|a,t_a)}{\partial a^\mu} \quad \text{and consequently}$$

$$M_{\mu\sigma}(t_a,t_b) = \frac{\partial^2 A(b,t_b|a,t_a)}{\partial a^\mu \partial b^\sigma} \tag{22}$$

which shows that $M_{\mu\sigma}(t_a,t_b)$ is the negative of the Van Vleck matrix $\bar{M}_{\mu\sigma}(t_a,t_b)$

$$\bar{M}_{\mu\sigma}(t_a,t_b) = -\frac{\partial^2 A(b,t_b|a,t_a)}{\partial a^\mu \partial b^\sigma} \tag{23}$$

In terms of the Jacobi fields and the matrix $M_{\sigma\rho}$, it is easy to express the Green's function of the Jacobi equation

$$S(Q(t))_{\mu\rho} G^{\rho\lambda}(t,t') = \delta^\lambda_\mu \delta(t-t') \tag{24}$$

We have

$$G^{\rho\lambda}(t,t') = \theta(t-t')\, \bar{J}^{\rho\mu}(t,T) M_{\mu\sigma}(T,t_o) \bar{J}^{\sigma\lambda}(t_o,t') - \theta(t'-t) \bar{J}^{\rho\mu}(t,t_o)$$
$$M_{\mu\sigma}(t_o,T) \bar{J}^{\sigma\lambda}(T,t') \tag{25}$$

The continuity of $G^{\rho\lambda}(t,t')$ at $t = t'$ follows from

$$\bar{J}^{\rho\mu}(r,t)M_{\mu\lambda}(t,t')\,\bar{J}^{\lambda\sigma}(t',s) + \bar{J}^{\rho\mu}(r,t')M_{\mu\lambda}(t',t)\bar{J}^{\lambda\sigma}(t,s) = \bar{J}^{\rho\sigma}(r,s) \tag{26}$$

which is proved by remarking that both sides satisfy the Jacobi equation in r, and coincide for r=t, and r=t'. From this formula we can immediately compute the discontinuity of $\frac{d}{dt}\,G^{\rho\lambda}(t,t')$ at t=t'. Indeed

$$\frac{d}{dt}\,G^{\rho\lambda}(t,t+\varepsilon) - \frac{d}{dt}\,G^{\rho\lambda}(t,t-\varepsilon)\Big|_{\varepsilon=0} = -(\frac{d}{dt}\,\bar{J}^{\rho\mu}(t,t_o)M_{\mu\sigma}(t_o,T)\bar{J}^{\sigma\lambda}(T,t) +$$

$$+ \frac{d}{dt}\,\bar{J}^{\rho\mu}(t,T)M_{\mu\sigma}(T,t_o)\bar{J}^{\sigma\lambda}(t_o,t)) = -\frac{d}{dt}\,\bar{J}^{\rho\lambda}(t,t')\Big|_{t=t'} =$$

$$= g^{\rho\lambda}(Q(t)) \tag{27}$$

Finally, we give the representation of $G^{\rho\lambda}(t,t')$ in terms of derivatives of the classical solution with respect to the endpoints and the Van-Vleck matrix $\bar{M}_{\mu\sigma}(t_a,t_b)$ (23).
We have, expressing $\bar{J}$ in terms of b,a

$$\bar{J}^{\alpha\beta}(t,t_a) = -\frac{\partial Q^{\alpha}(t)}{\partial b^{\lambda}}\,\frac{\partial b^{\lambda}}{\partial p_{a_\beta}} \tag{28}$$

$$\bar{J}^{\alpha\beta}(t_b,t) = -\frac{\partial b^{\alpha}}{\partial p_{a_\gamma}}\,\frac{\partial Q^{\beta}(t)}{\partial a^{\gamma}} + \frac{\partial Q^{\beta}(t)}{\partial b^{\lambda}}\,(\frac{\partial b^{\lambda}}{\partial p_{a_\gamma}}\,\frac{\partial b^{\alpha}}{\partial a^{\gamma}} - \frac{\partial b^{\lambda}}{\partial a^{\gamma}}\,\frac{\partial b^{\alpha}}{\partial p_{a_\gamma}})$$

$$= -\frac{\partial b^{\alpha}}{\partial p_{a_\gamma}}\,\frac{\partial Q^{\beta}(t)}{\partial a^{\gamma}} \tag{29}$$

In the first expression for $\bar{J}^{\alpha\beta}(t_b,t)$ the second part vanished since it was proportional to $\{q^{\alpha}(T),\, q^{\lambda}(T)\}_{(a,p_a)} = 0$. Using (18) we rewrite $G^{\rho\lambda}(t,t')$ as

$$G^{\rho\lambda}(t,t') = \theta(t-t')\,\frac{\partial b^{\mu}}{\partial p_{a_\gamma}}\,\frac{\partial Q^{\rho}(t)}{\partial a^{\gamma}}\,\frac{\partial P_{\sigma}(t_a)}{\partial b^{\mu}}\,\frac{\partial Q^{\lambda}(t')}{\partial b^{\kappa}}\,\frac{\partial b^{\kappa}}{\partial p_{a_\sigma}} +$$

$$\theta(t'-t)\,\frac{\partial Q^{\rho}(t)}{\partial b^{\kappa}}\,\frac{\partial b^{\kappa}}{\partial p_{a_\tau}}\,\frac{\partial P_{\tau}(t_a)}{\partial b^{\sigma}}\,\frac{\partial b^{\sigma}}{\partial p_{a_\mu}}\,\frac{\partial Q^{\lambda}(t')}{\partial a^{\mu}}$$

and noting $\bar{M}_{\mu\sigma}(t_a,t_b)$ as $\bar{M}_{\mu\sigma}$

$$= \theta(t-t')\,\frac{\partial Q^{\rho}(t)}{\partial a^{\gamma}}\,(\bar{M}^{-1})^{\kappa\gamma}\,\frac{\partial Q^{\lambda}(t')}{\partial b^{\kappa}} + \theta(t'-t)\,\frac{\partial Q^{\rho}(t)}{\partial b^{\kappa}}\,(\bar{M}^{-1})^{\kappa\mu}\,\frac{\partial Q^{\mu}(t')}{\partial a^{\mu}} \tag{30}$$

Most of the material, given here, can be found in a similar form in [D17]

Appendix 9.2

In this section we compute a general Gaussian functional integral of the form

$$\int_{\gamma_1(1/2)} \mathcal{D}u(t)\ \exp\{-\frac{1}{2c}\int_{t_o}^{T} (\dot{u})^2 + u^i A_{ij}(t)\dot{u}^j(t) + u^i V_{ij}(t)u^j\}$$

$$\delta(u(t_o))\,\delta(u(T)) \tag{1}$$

with zero boundary condition. $V_{ij}$ is symmetric in (i,j) and $A_{ij}$ is antisymmetric. We shall relax this condition on $A_{ij}$ afterwards, since a nonvanishing symmetric part in $A_{ij}$ can always be partially integrated and added to the potential $V_{ij}$. In fact the specification $\gamma_1$ (1/2) is not necessary in (1) since any other discretization differs only in a term proportional to

$$\Delta u^i A_{ij}\Delta u^j \doteq c\varepsilon A_{ii} = 0.$$

In order to compute (1), we do a change of variables $u(t) \to y(t)$ of the Cameron-Martin type[C2,G6,K5] in such a way that (1) is reduced to the standard form, which is well known

$$\int \mathcal{D}y(t)\ \exp\{-\frac{1}{2c}\int_{t_o}^{T} (\dot{y})^2\}\delta(y(t_o))\,\delta(y(T)). \tag{2}$$

More precisely, we make the transformation

$$u^i(t) = y^i(t) - f^{i\ell}(t)\int_{t_o}^{t} ds\ g^{\ell k}(s)y^k(s) \tag{3}$$

and we impose $f^{i\ell}(T) = 0$ which implies $y^i(T) = 0$ because $u^i(T) = 0$. We do this to have zero boundary conditions in the transformed integral (2): $y(T) = y(t_o) = 0$. Since (1) is defined by its discrete form, (3) also has to be specified in its discrete form, or, which is equivalent, as a stochastic integral. For convenience, we take it in the midpoint as

$$u^i(k) = y^i(k) - f^{i\ell}(k)\varepsilon\sum_{j=1}^{k} g^{\ell m}(j)\,\frac{1}{2}\,[y^m(j) + y^m(j-1)] \tag{4}$$

$$u^i(0) = y^i(0) = 0$$

This change of variables is to be used in (1) and all terms up to order $\varepsilon$ should be retained, taking into account $\Delta y = \mathcal{O}(\sqrt{\varepsilon})$. It turns out that one can work in fact directly with (3), and that partial integrations can be done with the normal rules of calculus, due to the midpoint discretization of $y^j(s)$. The problem is then to determine $f^{i\ell}$ and $g^{\ell k}$ such that $\int dtL_o$, with $L_o$ the

Lagrangian in (1), transforms into $\int_{t_o}^{T} dt(\dot{y})^2$.

From (4) we obtain, using the notation $\bar{y}^m_{j-1} = \frac{1}{2}[y^m(j) + y^m(j-1)]$

$$u^i(k) - u^i(k-1) = y^i(k) - y^i(k-1) - f^{i\ell}(k)\varepsilon \sum_1^k g^{\ell m}(j)\bar{y}^m_{j-1} +$$

$$+ f^{i\ell}(k-1)\varepsilon \sum_1^{k-1} g^{\ell m}(j)\bar{y}^m_{j-1}$$

or

$$\Delta u^i(k) = \Delta y^i(k) - [f^{i\ell}(k) - f^{i\ell}(k-1)]\varepsilon \sum_{j=1}^{k-1} g^{\ell m}(j)\bar{y}^m_{j-1} - f^{i\ell}(k)\varepsilon\, g^{\ell m}(k)\bar{y}^m_{k-1} \tag{5}$$

corresponding to

$$\dot{u}^i(t) = \dot{y}^i(t) - \dot{f}^{i\ell}(t)\int_{t_o}^{t} ds g^{\ell m}(s)y^m(s) - f^{i\ell}(t)g^{\ell m}(t)y^m(t) \tag{6}$$

We use formally (6), while keeping in mind that the last term in (5) is discretized in the midpoint.

Using (3) and (6), we obtain from (1)

$$\int_{t_o}^{T} dt(\dot{u}^2 + u^i V_{ij}u^j + u^i A_{ij}\dot{u}^j) - \int_{t_o}^{T} dt\dot{y}^2$$

$$= \int_{t_o}^{T} dt\ y^i(t)y^j(t)(V_{ij} + 2\dot{f}^{ik}g^{kj} + f^{\ell k}g^{ki}f^{\ell m}g^{mj} - A_{i\ell}f^{\ell k}g^{kj} + f^{\ell k}A_{\ell j}g^{ki})$$

$$+\int_{t_o}^{T} dt(\int_{t_o}^{t} ds\ g^{k\ell}(s)y^{\ell}(s))y^j(t)[2\ddot{f}^{jk}(t) - 2f^{ik}V_{ij} + g^{mj}(f^{\ell m}\dot{f}^{\ell k} - f^{\ell k}\dot{f}^{\ell m})$$

$$- A_{jm}\dot{f}^{mk} + f^{ik}A_{i\ell}f^{\ell m}g^{mj} + \frac{d}{dt}(f^{ik}A_{ij})]$$

$$+ \int_{t_o}^{T} dt \int_{t_o}^{t} ds\ g^{k\ell}(s)y^{\ell}(s) \int_{t_o}^{t} ds'\ g^{m\ell'}(s')y^{\ell'}(s')(V_{ij}f^{ik}f^{jm} - f^{ik}\ddot{f}^{im} +$$

$$+ f^{ik}A_{ij}\dot{f}^{jm})(t)+\int_{t_o}^{T} dt\ \dot{y}^i(-2f^{ik}g^{k\ell}+ A_{\ell i}y^{\ell}) \tag{7}$$

The partial integrations needed to obtain this result are all allowed. In fact one can easily check that in a term like

$$\int_{t_o}^{T} dt\ \dot{y}^i(t)\ \dot{f}^{ik}(t)\left(\int_{t_o}^{t} ds\ g^{k\ell}(s)y^\ell(s)\right) = -\int_{t_o}^{T} dt\ y^i(t)$$

$$[\ddot{f}^{ik}(t)\int_{t_o}^{t} ds\ g^{k\ell}(s)y^\ell(s) + \dot{f}^{ik}(t)g^{k\ell}(t)y^\ell(t)] \tag{8}$$

the discretization of the term $\int g^{k\ell}(s)y^\ell(s)$ doesn't play a role.
The last term in (7) can be integrated by parts if $-2f^{ik}g^{k\ell} + A_{\ell i}$ is symmetric in $(i,\ell)$, or, defining

$$K^{i\ell} = f^{ij}g^{j\ell} \tag{9}$$

if $$A_{i\ell} = K^{\ell i} - K^{i\ell} \tag{10}$$

We impose this condition on $K^{i\ell}$, and using the fact that the last term in (7) is discretized in the midpoint (4,6), we do the partial integration to obtain

$$\int_{t_o}^{T} dt\ \dot{y}^i(-2K^{i\ell} + A_{\ell i})y^\ell = -\frac{1}{2}\int_{t_o}^{T} dt\ y^i(t)\ \frac{d}{dt}\ (-2f^{ik}(t)g^{k\ell}(t) + A_{\ell i}(t))y^\ell(t) \tag{11}$$

Using this in (7), and requiring that the terms on the right vanish, we obtain the set of equations (we use now matrix notation $f^{ij} \to f$, $\tilde{B}$ stands for the transposed of B)

$$(V + \dot{K} + 2\dot{f}g + (\tilde{K}^2) - AK - \frac{1}{2}\dot{A})^s = 0 \tag{12}$$

$$\ddot{f} - Vf - A\dot{f} - \frac{1}{2}\dot{A}f + \frac{1}{2}\tilde{g}(\tilde{\dot{f}}f - \tilde{f}\dot{f} - \tilde{f}Af) = 0 \tag{13}$$

$$[\tilde{f}(-\ddot{f} + Vf + A\dot{f})]^s = 0 \tag{14}$$

where we used $A = \tilde{K} - K$, $\tilde{A} = -A$, and the notation $(B)^s$ stands for the symmetric part of the matrix inside the bracket, $(B)^s = \frac{1}{2}(B+\tilde{B})$. These three equations come from putting equal to zero in the integral of (7) the coefficients of $y^iy^j$, $y^j\int ds\ y^{k\ell}(s)y^\ell(s)$, and $(\int ds\ g^{k\ell}y^\ell)(\int ds' g^{m\ell'}y^{\ell'})$ respectively.
Since $A = -\tilde{A}$, the last term in (12) can be dropped.
We shall show now that (9) and (12-14) are satisfied when we determine f(t) as solution of the differential equation

$$\ddot{f} = A\dot{f} + (V + \frac{1}{2}\dot{A})f \tag{15}$$

$$f(T) = 0$$

$$\dot{f}(T) = -1$$

and, putting $S = \dot{f}f^{-1}$, $L = \dot{\ell}\ell^{-1}$ with $\ell(t) = 1\,(T-t)$ (1 is the unit matrix), we determine g by

$$K = fg = L - S \tag{16}$$

Since L is symmetric, $\tilde{K} - K = S - \tilde{S}$. (10) will be satisfied if $A = \dot{f}f^{-1} - \tilde{f}^{-1}\dot{\tilde{f}}$ or $\tilde{f}Af = \tilde{f}\dot{f} - \dot{\tilde{f}}f$. Considering $F(t) = \tilde{f}\dot{f} - \dot{\tilde{f}}f - \tilde{f}Af$, one has $F(T) = 0$ since $f(T) = 0$ and computing $\dot{F}$ using (15), one obtains

$$\dot{F}(t) = \tilde{f}\ddot{f} - \ddot{\tilde{f}}f - \frac{d}{dt}(\tilde{f}Af)$$

$$= \tilde{f}A\dot{f} + \tilde{f}(V + \frac{1}{2}\dot{A})f + \dot{\tilde{f}}Af - \tilde{f}(V - \frac{1}{2}\dot{A})f - \frac{d}{dt}(\tilde{f}Af) = 0 \tag{17}$$

Consequently $F(t) = 0$ and $A = \tilde{K} - K$

(14) reduces to $(-\frac{1}{2}\tilde{\dot{f}}Af)^s = 0$, using (15) and the antisymmetry of $\dot{A}$. (13) reduces to $\frac{1}{2}\tilde{g}\,F(t) = 0$.

In order to check (12) we compute

$$\dot{K} = \dot{L} - \dot{S} = -L^2 - V - AS - \frac{1}{2}\dot{A} + S^2 \quad \text{which can be written as}$$

$$\dot{K} = -V + AK - K^2 - 2SK + (SK - KS) - AL - \frac{1}{2}\dot{A}$$

One can easily check that $(SK - KS - AL - \frac{1}{2}\dot{A})^s = 0$ while

$$(\dot{K} + V - AK + K^2 + 2SK)^s = 0$$

is exactly (12).

This completes the determination of the functions f and g, which, by (3) or (4) give the transformation bringing the Gaussian functional integral (1) to the standard form (2). The only part missing is the Jacobian of (4)

$$J = \det\left(\frac{\partial u^i(k)}{\partial y^j(\ell)}\right) = \prod_{j=1}^{N} \det \frac{\partial u^i(k)}{\partial y^j(k)} \tag{18}$$

since the transformation matrix is blocktriangular. But $\frac{\partial u^i(k)}{\partial y^j(k)} = \delta^i_j - \frac{\varepsilon}{2}K^{ij}(k)$

and using $\det(1 + B) = \exp \operatorname{tr} \ln(1 + B)$, we have, keeping terms up to order $\varepsilon$

$$J = \exp\{-\frac{\varepsilon}{2}\sum_{j=1}^{N} \operatorname{tr} K(t_j)\} \to \exp\{-\frac{1}{2}\int_{t_o}^{T} dt\, \operatorname{tr}(\dot{\ell}\ell^{-1} - \dot{f}f^{-1})\} \tag{19}$$

but $\int_{t_o}^{T} dt \operatorname{tr} K(t) = \operatorname{tr} \int_{t_o}^{T} dt \frac{\partial}{\partial t} \ln \ell(t) f^{-1}(t) = \operatorname{tr}[\ln \ell(T) f^{-1}(T) -$

$$\ln \ell(t_o) f^{-1}(t_o)]$$

$\ell(T)f^{-1}(T)$ is undetermined $(f(T) = 0)$, but from (19) we see that it must be taken as $\ell(T-\varepsilon)f^{-1}(T-\varepsilon)\big|_{\varepsilon\to 0^+}$ and this gives the value 1, and therefore one has

$$J = [\det \ell(t_o) f^{-1}(t_o)]^{1/2} = \frac{(T-t_o)^{d/2}}{\sqrt{\det f(t_o)}} \tag{20}$$

(1) then reduces to

$$I = \frac{(T - t_o)^{d/2}}{(\det f(t_o))^{1/2}} \int \mathcal{D}y \exp\{-\frac{1}{2c}\int_{t_o}^{T} dt\, \dot{y}^2\}\delta(y(T))\delta(y(t_o)) \tag{21}$$

$$= (2\pi c)^{-d/2}(\det f(t_o))^{-1/2} \tag{22}$$

If $A_{\mu\nu}$ in (1) is $A_{\mu\nu}(t) = A^{(1)}_{\mu\nu}(t) + A^{(2)}_{\mu\nu}(t)$ where $A^{(1)}_{\mu\nu}$ is symmetric and $A^{(2)}_{\mu\nu}$ is antisymmetric we add $\int dt\, u^\mu A^{(1)}_{\mu\nu}\dot{u}^\nu = -\frac{1}{2}\int dt\, u^\mu \dot{A}^{(1)}_{\mu\nu} u^\nu$ to the potential and we are in the case (1) with $V_{\mu\nu} \to V_{\mu\nu} - \frac{1}{2}\dot{A}^{(1)}_{\mu\nu}$. This partial integration is allowed since we are in the midpoint $(\gamma_1(1/2))$ discretization. It is easily checked that the equation for f (15) is the Euler-Lagrange equation derived from the quadratic Lagrangian in (1), $L_o$

$$L_o = \frac{1}{2}(\dot{u}^2 + u^\mu A_{\mu\nu}\dot{u}^\nu + u^\mu V_{\mu\nu} u^\nu) \tag{23}$$

The same property remains valid if A has a symmetric part i.e. I is given by (22) where f satisfies the Euler-Lagrange equations derived from (23) with boundary conditions $f(T) = 0$, $\dot{f}(T) = -1$.

As a final generalization we calculate the most general Gaussian in d dimensions

$$I = \int_{\gamma_1(1/2)} \mathcal{D}q\sqrt{g}\, \exp\{-\frac{1}{c}\int_{t_o}^{T} dt\, L_1(q,\dot{q},t)\}\delta(q(T))\delta(q(t_o)) \tag{24}$$

with

$$L_1 = \frac{1}{2} g_{\mu\nu}(t)\dot{q}^\mu\dot{q}^\nu + B_{\mu\nu}(t)q^\mu\dot{q}^\nu + C_{\mu\nu}(t)q^\mu q^\nu \tag{25}$$

The discretization defines I as $I = \lim I_N, N \to \infty (t_j = t_o + j\varepsilon, t_{N+1} = T, \Delta q_j = q_j - q_{j-1}, \bar{q}_{j-1} = \frac{1}{2}(q_{j-1} + q_j), g(t) = \det g_{\mu\nu}(t))$ with

$$I_N = \int_{q_o=0}^{q_{N+1}=0} \prod_1^N dq_i \prod_{j=1}^{N+1} \frac{\sqrt{g(t_j)}}{(2\pi\varepsilon c)^{d/2}} \exp - \frac{\varepsilon}{2c} \sum_{j=1}^{N+1} g_{\mu\nu}(t_j) \frac{\Delta q_j^\mu}{\varepsilon} \frac{\Delta q_j^\nu}{\varepsilon}$$

$$+ B_{\mu\nu}(t_j)\bar{q}_{j-1}^\mu \frac{\Delta q_j}{\varepsilon} + C_{\mu\nu}(t_j) q_j^\mu q_j^\nu \tag{26}$$

Let $g^{\mu\nu}(t)$ be the inverse matrix of $g_{\mu\nu}(t)$. One can always find a $\tilde{\sigma}_{\mu i}$, such that $\tilde{\sigma}_{\mu i} g^{\mu\nu}(t_o)\tilde{\sigma}_{\nu j} = \delta_{ij}$. We define then $\sigma_{\mu i}(t)$ as the solution of

$$\dot{\sigma}_{\mu i}(t) + \frac{1}{2} g_{\mu\nu}(t)\, \dot{g}^{\nu\lambda}(t)\sigma_{\lambda i}(t) = 0$$

$$\sigma_{\mu i}(t_o) = \tilde{\sigma}_{\mu i} \tag{27}$$

From this we conclude

$$\frac{d}{dt}(\sigma_{\mu i} g^{\mu\nu}\sigma_{\nu j}) = (\frac{d}{dt}\sigma_{\mu i} + \frac{1}{2}\sigma_{\lambda i}\frac{d}{dt}(g^{\lambda\rho})g_{\rho\mu})g^{\mu\nu}(t)\sigma_{\nu j}(t)$$

$$+ \sigma_{\mu i} g^{\mu\nu}(\frac{d}{dt}\sigma_{\nu j} + \frac{1}{2} g_{\nu\rho}\frac{d}{dt}(g^{\rho\lambda})\sigma_{\lambda j} = 0$$

or $\sigma_{\mu i} g^{\mu\nu}\sigma_{\nu j} = \delta_{ij}$ for all t. Defining $\sigma^\mu{}_i(t) = g^{\mu\alpha}(t)\sigma_{\alpha i}(t)$ one has $\sigma^\mu{}_i(t) g_{\mu\nu}(t) \sigma^\nu{}_j(t) = \delta_{ij}$, or $\sigma^\mu{}_i(t)\sigma_{\mu j}(t) = \delta_{ij}$ and $\sigma^\mu{}_i(t)\sigma_{\lambda i}(t) = \delta^\mu_\lambda$

In (26) we change variables $q_j^\mu = \sigma^\mu{}_i(t_j)u^i(t_j)$. This transformation has Jacobian

$$J = \prod_{j=1}^N \det[\sigma^\rho{}_i(t_j)] = \prod_{j=1}^N [g(t_j)]^{-1/2} \tag{28}$$

and one obtains

$$I_N = \sqrt{g(T)} \exp\{\frac{1}{2}\int_{t_o}^T dt\sigma_{\mu i}(t)\dot{\sigma}^\mu{}_i(t)\} \int_{u_o=0}^{u_{N+1}=0} \prod_{i=1}^N du_i \prod_{j=1}^{N+1} (2\pi\varepsilon c)^{-d/2}$$

$$\exp\{-\frac{\varepsilon}{2c}\sum_{j=1}^{N+1} (\frac{\Delta u_j^i}{\varepsilon})^2 + (B_{\mu\nu}(t_j)\sigma^\mu{}_k(t_j)\sigma^\nu{}_\ell(t_j) + 2\dot{\sigma}^\mu{}_k(t_j)\sigma_{\mu\ell}(t_j))\bar{u}_{j-1}^k \frac{\Delta u_j^\ell}{\varepsilon}$$

$$+ (C_{\mu\nu}(t_j)\sigma^\mu{}_k(t_j)\sigma^\nu{}_\ell(t_j) + g_{\mu\nu}(t_j)\dot{\sigma}^\mu{}_k(t_j)\dot{\sigma}^\nu{}_\ell(t_j) + B_{\mu\nu}(t_j)\sigma^\mu{}_k(t_j)\dot{\sigma}^\nu{}_\ell(t_j))$$

$$u_j^k u_j^\ell]\} \tag{29}$$

where we have kept all terms up to order $\varepsilon$ and used the replacement $\Delta u^k_j \Delta u^\ell_j \doteq \varepsilon c \delta^{k\ell}$. Evaluating $\exp\{\frac{1}{2}\int_{t_o}^T dt\,\sigma_{\mu i}\dot{\sigma}^\mu{}_i\} = [g(t_o)]^{1/4}[g(T)]^{-1/4}$, we have from (29)

$$I = [g(t_o)\ g(T)]^{1/4}\ \tilde{I}$$

$$\tilde{I} = \int_{\gamma_1(1/2)} \mathcal{D}u \ \exp\{-\frac{1}{c}\int_{t_o}^T dt\ L_2(u,\dot{u},t)\}\delta(u(T))\delta(u(t_o)) \tag{30}$$

with $L_2(u,\dot{u},t) = L_1(q^\mu(t) = \sigma^\mu{}_k(t)u^k(t))$, using the normal rules of calculus, $\tilde{I}$ is of type (1) and consequently its value is $(2\pi c)^{-d/2}\ [\det f(t_o)]^{-1/2}$ where $f^{ij}(t)$ satisfies the Euler-Lagrange equation of $L_2$ with boundary conditions $f^{ij}(T) = 0$ $\dot{f}^{ij}(T) = -\delta^i_j$. Because of the relation between $L_2$ and $L_1$, $J^{\mu\rho}(t) = \sigma^\mu{}_i(t)\ f^{ij}(t)\sigma^\rho{}_j(T)$ satisfies the Euler-Lagrange equations derived from $L_1(q,\dot{q},t)$ for the boundary conditions $J^{\mu\rho}(T) = 0$, $\dot{J}^{\mu\rho}(T) = -\sigma^\mu{}_i(T)\sigma^\rho{}_i(T) = -g^{\mu\rho}(T)$. Since $f^{ij}(t) = \sigma_{\mu i}(t)J^{\mu\rho}(t)\sigma_{\rho j}(T)$, $\det f(t) = \det(J(t))\ [g(t)g(T)]^{1/2}$ which gives $I = (2\pi c)^{-d/2}\ [g(t_o)\ g(T)]^{-1/4}\ \det[J^{\mu\nu}(t_o)]^{-1/2}$, and finally (24) is given by

$$I = (2\pi c)^{-d/2}\ \det[J^{\mu\nu}(t_o)]^{-1/2} \tag{31}$$

where $J^{\mu\nu}(t)$ satisfies the Euler-Lagrange equations derived from $L_1$, with boundary conditions

$$J^{\mu\nu}(T) = 0$$
$$\dot{J}^{\mu\nu}(T) = -g^{\mu\nu}(T) \tag{32}$$

Appendix 9.3

As a generalization of the problem (9.A(1)), we consider the equation ($\mu = 1, \ldots, d$) [L26]

$$\dot{I}(q,t) = \mathcal{L}\left(\frac{\partial}{\partial q^\mu}, q^\mu, t, \eta\right) I \tag{1}$$

in which the differential operator $\mathcal{L}$ (that can be time dependent) depends on $\eta$ as follows: $\mathcal{L} = \mathcal{L}_1 + \mathcal{L}_2$

$$\mathcal{L}_1 = \frac{1}{\eta} \sum_{n \geq 0} \eta^n A^{\mu_1 \ldots \mu_n}(q,t)\partial\mu_1 \ldots \partial\mu_n \tag{2}$$

$$\mathcal{L}_2 = \sum_{n \geq 0} \eta^n \sum_j b_n^j(\eta)\, B_1^{\mu_1 \ldots \mu_n}(q,t)\partial\mu_1 \ldots \partial\mu_n \tag{3}$$

with $b_n^j(\eta) = \sum_{k \geq 0} b_{nk}^j \eta k$. The construction of the Hamiltonian $\hat{H} = i\eta\mathcal{L}(q \to \hat{Q}, -i\eta\partial\mu \to \hat{P}_\mu) = \hat{H}_1 + \hat{H}_2$ in standard order leads to a phase space functional integral in postpoint discretization for the propagator $I(X,T,X_o,t_o)$

$$I(X,T;X_o,t_o) = \int_{\gamma_1(1)} \mathcal{D}q\mathcal{D}_\eta p \exp\left\{\frac{i}{\eta}\int_{t_o}^T dt\; P_\mu \dot{Q}^\mu - H_1(P,Q,t) - \eta H_2(P,Q,t,\eta)\right\}$$

$$\delta(Q(T) - X)\delta(Q(t_o) - X_o) \tag{4}$$

where $H_1$, $H_2$ are the functions obtained from $\hat{H}_1, \hat{H}_2$ in standard order, replacing $\hat{Q}^\mu \to Q^\mu, \hat{P}_\mu \to P_\mu$.
The associated classical problem is determined by $H_1$, and we take $(x(t), y(t))$ as the solutions of Hamiltons equations

$$\dot{x}^\mu(t) = \frac{\partial H_1}{\partial y^\mu}(y,x,t) \qquad \dot{y}_\mu(t) = \frac{-\partial H_1}{\partial x^\mu}(y,x,t) \tag{5}$$

with boundary conditions $x^\mu(T) = X^\mu$, $x^\mu(t_o) = X_o^\mu$. In (4) we then make the change of variables $Q^\mu(t) = x^\mu(t) + \sqrt{\eta}q^\mu(t)$, $P_\mu(t) = y_\mu(t) + \sqrt{\eta}p_\mu(t)$, and expand all quantities in powers of $\sqrt{\eta}$. In this way, we obtain

$$I = \eta^{-d/2}\exp\{\tfrac{1}{\eta}A_{cl}\}\int_{\gamma_1(1)} \mathcal{D}q\mathcal{D}p \exp\left\{i\int_{t_o}^T dt\left[p_\mu \dot{q}^\mu - \frac{1}{2}\frac{\partial^2 H_1}{\partial y_\mu \partial y_\nu} p_\mu p_\nu - \right.\right.$$

$$\left.\left. - \frac{\partial^2 H_1}{\partial y_\mu \partial x^\nu} p_\mu q^\nu - \frac{1}{2}\frac{\partial^2 H_1}{\partial x^\mu \partial x^\nu} q^\mu q^\nu - F(t)\right]\right\}\delta(q(t_o))\delta(q(T)) \tag{6}$$

Here $F(t) = H_3(t) + H_2(t)$, with

$$H_3(t) = \sum_{n\geq 3} \frac{(\sqrt{\eta})^{n-2}}{n!} \left(p_\mu \frac{\partial}{\partial y^\mu} + q^\mu \frac{\partial}{\partial x^\mu}\right)^n H_1(y,x,t)$$

The first term in the development of $H_1$ gave the classical action $A_{cl} = \int dt\, L(x,\dot{x},t) = \int dt\, y_\mu \dot{x}^\mu - H_1$, while the term in $\eta^{-1/2}$ vanished owing to the equations of motion

From (6) it is possible to compute the WKB-approximation and the higher order corrections as

$$I = I_{WKB}\, I_c = I_{WKB}(1 + \sum_{n\geq 1} \eta^n J_n) \tag{7}$$

with

$$I_{WKB} = \eta^{-d/2}\, \bar{N} \exp\{\frac{i}{\eta} A_{cl}\} \exp\{\frac{1}{2}\int_{t_o}^{T} dt \frac{\partial^2 H_1}{\partial y_\mu \partial x^\mu} - i\int_{t_o}^{T} dt\, H_2^{(o)}(y(t),x(t))\} \tag{8}$$

$$I_c = \exp\{-i\int_{t_o}^{T} dt\, K(x,y,p(t),q(t+\delta),t)\}\Big|_{q = \frac{1}{i}\frac{\delta}{\delta J},\ p = \frac{1}{i}\frac{\delta}{\delta J^\star}} Z_o(J,J^\star) \tag{9}$$

and

$$\bar{N}\, Z_o(J,J^\star) = \int_{\gamma_1(1/2)} \mathcal{D}q\mathcal{D}p \exp\{i\int_{t_o}^{T} dt[p_\mu \dot{q}^\mu - \frac{1}{2}\frac{\partial^2 H_1}{\partial y_\mu \partial y_\nu} p_\mu p_\nu - \frac{\partial^2 H_1}{\partial y_\mu \partial x^\nu} p_\mu q^\nu - \frac{1}{2}\frac{\partial^2 H_1}{\partial x^\mu\, \partial x^\nu} q^\mu q^\nu + J_\mu q^\mu + J^{\star\mu} p_\mu]\}\delta(q(t_o))\delta(q(T)) \tag{10}$$

$$= \bar{N} \exp\{-\int_{t_o}^{T} dtdt'[\tfrac{1}{2}J_\mu(t)\Delta^{\mu\nu}(t,t')J_\nu(t') + J^{\star\mu}(t)S_\mu{}^\nu(t,t')J_\nu(t') + \frac{1}{2} J^{\star\mu}(t)\, D_{\mu\nu}(t,t')J^{\star\nu}(t')]\} \tag{11}$$

The discussion is completely parallel to the one given in sect.9B. Setting up the perturbation expansion from (6), we remark that F(t) contains terms of order $\eta^o$, coming from $H_2(P,Q,t) = \sum_{k\geq 0} \eta^k H_2^{(k)}(P,Q,t)$ and we write $F(t) = H_2^{(o)}(y,x,t) + K(t)$. The free generating function obtained from (6) is discretized in the postpoint, and changing it to the midpoint we obtain the extra factor $\exp\{\frac{1}{2}\int dt \frac{\partial^2 H_1}{\partial y_\mu \partial x^\mu}\}$ (9.B,(46)) allowing us to redefine the free functional integral as in (16) and giving the expansion in the form (8),(9). The only remaining task is to compute the normalization $\bar{N}$ and the propagators in (11), and we shall show that also this task has already been done in ( 9.B).

As before, we indicated in (9) $q(t+\delta)$ which means that the $p_\mu(t)\, q^\nu(t)$ contraction has to be interpreted as $S^\nu_\mu\,(t,t+\delta), \delta \to 0^+$. It is a consequence of the postpoint discretization that the ambiguity has to be resolved in this way. The expansion will contain only powers of $\eta$, since the terms proportional to $(\sqrt{\eta})^n$, n odd, will vanish due to the form of $Z_o(J,J^\star)$ and to the fact that they have an odd number of $p(\tau)$ and $q(\tau)$.

Doing the p-integration in (10), we obtain for $J = J^\star = 0$:

$$\bar{N} = \int_{\gamma_1(1/2)} \mathcal{D}q\left(\frac{\sqrt{G}}{i^d}\right) \exp\left\{i \int_{t_o}^{T} dt\bar{L}(q,\dot{q},t)\right\}\delta(q(T))\delta(q(t_o)) \tag{12}$$

$$\bar{L} = \frac{1}{2}\frac{\partial^2 L}{\partial \dot{x}^\mu \partial \dot{x}^\nu}\dot{q}^\mu \dot{q}^\nu + \frac{\partial^2 L}{\partial \dot{x}^\mu \partial x^\nu}\dot{q}^\mu q^\nu + \frac{1}{2}\frac{\partial^2 L}{\partial x^\mu \partial x^\nu} q^\mu q^\nu \tag{13}$$

That this relation (see (9B.48-43)) is valid for arbitrary Hamiltonians can easily be checked. $G = \det G_{\mu\nu}(t)$ where $G_{\mu\nu}(t) = \frac{\partial^2 L}{\partial \dot{x}^\mu \partial \dot{x}^\nu}$ is the inverse of $\frac{\partial^2 H_1}{\partial y_\mu \partial y_\nu}$, i.e. $G_{\mu\alpha}(t)\left(\frac{\partial^2 H_1}{\partial y_\alpha \partial y_\rho}\right) = \delta^\rho_\mu$.

$$\frac{\partial^2 L}{\partial \dot{q}^\mu \partial q^\alpha} = -\frac{\partial^2 L}{\partial \dot{q}^\mu \partial \dot{q}^\beta}\,\frac{\partial^2 H}{\partial q^\alpha \partial p_\beta}$$

and

$$\frac{\partial^2 L}{\partial q^\alpha \partial q^\beta} = -\frac{\partial^2 H}{\partial q^\alpha \partial q^\beta} + \frac{\partial^2 H}{\partial q^\alpha \partial p_\gamma}\,\frac{\partial^2 L}{\partial \dot{q}^\gamma \partial \dot{q}^\rho}\,\frac{\partial^2 H}{\partial p_\rho \partial q^\beta}$$

As before it is the Jacobi equation [App.9.1 (1),(2)] for the Lagrangian L that enters into the computation The propagators will satisfy, using the notation of [App. 9.1 (2)]:

$$\text{(a)}\quad \left[A^{(2)}_{\mu\rho}\frac{\partial^2}{\partial t^2} + A^{(1)}_{\mu\rho}\frac{\partial}{\partial t} + A^{(0)}_{\mu\rho}\right]\Delta^{\rho\nu}(t,t') = i\delta^\nu_\mu \delta(t-t')$$

$$\Delta^{\rho\nu}(t_o,t) = \Delta^{\rho\nu}(T,t) = 0$$

$$\text{(b)}\quad S_\rho{}^\nu(t,t') = \left[\left(\frac{\partial^2 L}{\partial \dot{x}^\rho \partial \dot{x}^\mu}\right)\frac{\partial}{\partial t} + \frac{\partial^2 L}{\partial \dot{x}^\rho \partial x^\mu}\right]\Delta^{\mu\nu}(t,t') \tag{14}$$

$$\text{(c)}\quad D_{\rho\sigma}(t,t') = \left[\left(\frac{\partial^2 L(t)}{\partial \dot{x}^\rho \partial \dot{x}^\mu}\frac{\partial}{\partial t} + \frac{\partial^2 L(t)}{\partial \dot{x}^\rho \partial x^\mu}\right)\left(\frac{\partial^2 L(t')}{\partial \dot{x}^\sigma \partial \dot{x}^\nu}\frac{\partial}{\partial t'} + \frac{\partial^2 L(t')}{\partial \dot{x}^\sigma \partial x^\nu}\right)\right]\Delta^{\mu\nu}(t,t') - i\,\frac{\partial^2 L(t)}{\partial \dot{x}^\rho \partial \dot{x}^\sigma}\,\delta(t-t')$$

while the normalization factor is given by

$$\bar{N} = (2\pi i)^{-d/2} (\det \bar{M})^{1/2}$$
$$\bar{M}_{\alpha\beta} = -\frac{\partial^2 A_{cl}(X,T;X_o,t_o)}{\partial X_o^\alpha \, \partial X^\beta} \tag{15}$$

They can be found doing exactly the same calculation as in (9.B). The WKB - approximation for general Hamiltonianswas also treated in [M19].

Appendix 9.4

We want to illustrate here how one can obtain representation (9.C(21)) for the propagator I

$$I((X,T;X_o,t_o) = \frac{1}{\eta^{d/2}} \int_{\gamma_1(1)} \mathcal{D}q\mathcal{D}p \exp\{i\int_{t_o}^{T} d\tau p_\sigma(\dot{q}^\sigma + \frac{\partial q^\sigma}{\partial Q^\alpha}\frac{\partial Q^\alpha}{\partial x^\rho}\dot{x}^\rho - \frac{\eta c}{2} g^{\star\alpha\rho}\Gamma^{\star\sigma}_{\alpha\rho} + A^{\star\sigma})\}$$

$$\exp\{i\int_{t_o}^{T} d\tau[\frac{i\eta c}{2} g^{\star\alpha\beta}p_\alpha p_\beta - \frac{i}{\eta c}(\frac{1}{2}A^{\star\mu}A^\star_\mu + V^\star + \frac{\eta c}{2}\nabla_\mu A^{\star\mu})]\}$$

$$\delta(q(T))\delta(q(t_o)) \qquad (1)$$

starting from an ordinary functional integral representation, by making an explicit change of variable in the functional integral. Since this involves a time dependent transformation which has not been treated in (app.6.1), it is worthwhile to present the calculation [L18]. We treat the case $A^\mu = V = 0$, but these can be included without additional difficulties.
The standard ordered form of the operator corresponding to the equation

$$\frac{\partial I}{\partial t} = \frac{\eta c}{2}\Delta I \qquad (2)$$

is
$$\hat{H} = -\frac{i\eta c}{2} g^{\mu\nu}\hat{P}_\mu\hat{P}_\nu + \frac{\eta c}{2} g^{\alpha\rho}\Gamma^\beta_{\alpha\rho}\hat{P}_\beta \qquad (3)$$

and provides us with a functional integral representation of the propagator I in postpoint discretization $[Q_o = X_o,\ Q_{N+1} = X]$

$$I = \int_{\gamma_1(1)} \mathcal{D}Q\mathcal{D}P \exp\{i\int_{t_o}^{T} d\tau[P_\sigma(\dot{Q}^\sigma - \frac{\eta c}{2} g^{\alpha\rho}\Gamma^\sigma_{\alpha\rho}) + \frac{i\eta c}{2} P_\alpha g^{\alpha\beta} P_\beta]\}$$

$$= \lim_{N\to\infty} I_N = \lim_{N\to\infty} \int \prod_1^N dQ_i \prod_{j=1}^{N+1}\frac{dP_j}{(2\pi)^d} \exp\{i\varepsilon\sum_1^{N+1} P_{j\sigma}[\frac{Q^\sigma_j - Q^\sigma_{j-1}}{\varepsilon} - \frac{\eta c}{2} g^{\alpha\rho}\Gamma^\sigma_{\alpha\rho}] + \frac{i\eta c}{2} P_{\alpha j} g^{\alpha\beta}_j P_{\beta j}\} \qquad (4)$$

Here we do the time dependent canonical tranformation from $Q_j$ to the normal coordinates $q_j$ centered at $x_j = x(t_j)$ (IV.C.10)

$$Q_j^\mu = x_j^\mu + \sqrt{\eta}\, q_j^\mu - \frac{\eta}{2}\Gamma^\mu_{\nu\rho}(x_j) q_j^\nu q_j^\rho + \ldots .$$
$$P_{j\nu} = p_{j\mu}\frac{\partial q_j^\mu}{\partial Q_j^\nu} \tag{5}$$

and we find $(q_o = q_{N+1} = 0)$

$$I_N = \eta^{-d/2}\int \prod_1^N dq_i \prod_1^{N+1} \frac{dp_i}{(2\pi)^d} \exp\{ i\varepsilon \sum_1^{N+1} p_{j\mu}\frac{\partial q_j^\mu}{\partial Q_j^{\nu}} [\frac{Q_j^\nu - Q_{j-1}^\nu}{\varepsilon} +$$
$$\frac{\eta c}{2}\frac{\partial^2 Q_j^\nu}{\partial q_j^\tau \partial q_j^\sigma} g^{\star\sigma\tau}(x_j,q_j)] - p_{j\mu}(g^{\star\sigma\tau}\Gamma^{\star\mu}_{\sigma\tau})_j \frac{\eta c}{2}\}$$
$$\exp\{i\varepsilon \sum_1^{N+1} \frac{i\eta c}{2} p_{j\alpha} g^{\star\alpha\beta}(x_j,q_j) p_{j\beta}\} \tag{6}$$

The quantities in the normal coordinate system are indicated by stars. To obtain this result we used the transformation rule for the Christoffel symbols

$$\Gamma^{\star\mu}_{\sigma\tau} = \Gamma^\lambda_{\rho\nu}\frac{\partial Q^\rho}{\partial q^\sigma}\frac{\partial Q^\nu}{\partial q^\tau}\frac{\partial q^\mu}{\partial Q^\lambda} + \frac{\partial^2 Q^\lambda}{\partial q^\sigma \partial q^\tau}\frac{\partial q^\mu}{\partial Q^\lambda} \tag{7}$$

and we did not yet transform $Q_j^\nu - Q_{j-1}^\nu$. Comparing to (1) for $A^\mu = V = 0$, we should prove that in the representation (6) the first two terms

$$p_{j\mu}\frac{\partial q_j^\mu}{\partial Q_j^\nu}(\frac{Q_j^\nu - Q_{j-1}^\nu}{\varepsilon} + \frac{\eta c}{2}\frac{\partial^2 Q_j^\nu}{\partial q_j^\tau \partial q_j^\sigma} g^{\star\sigma\tau}(x_j,q_j)) \tag{8}$$

can effectively be replaced by $(\Delta_j^\mu = q_j^\mu - q_{j-1}^\mu)$

$$p_{j\mu}(\frac{\Delta_j^\mu}{\varepsilon} + \frac{\partial q_j^\mu}{\partial Q_j^\rho}\frac{\partial Q_j^\rho}{\partial x_j^\lambda}\dot{x}^\lambda(j)) \tag{9}$$

We first develop $Q^\mu(x_j,q_j) - Q^\mu(x_{j-1},q_{j-1})$ as a Taylor series around $Q^\mu(x_j,q_j)$, and we obtain

$$\frac{Q_j^\mu - Q_{j-1}^\mu}{\varepsilon} = \frac{\Delta_j^\lambda}{\varepsilon}(\frac{\partial Q^\mu}{\partial q^\lambda})(x_j,q_j) + \dot{x}_j^\lambda(\frac{\partial Q^\mu}{\partial x^\lambda})_j - \Delta_j^\rho \dot{x}_j^\lambda(\frac{\partial^2 Q^\mu}{\partial x^\lambda \partial q^\rho})_j -$$
$$\frac{1}{2!}\frac{\Delta_j^\rho \Delta_j^\sigma}{\varepsilon}(\frac{\partial^2 Q^\mu}{\partial q^\rho \partial q^\sigma})_j - \frac{1}{3!}\frac{\Delta^\rho\Delta^\sigma\Delta^\tau}{\varepsilon}(\frac{\partial^3 Q^\mu}{\partial q^\rho \partial q^\sigma \partial q^\tau})_j \tag{10}$$

where we retained all terms up to order $\sqrt{\varepsilon}$ and $(\ )_j$ indicates evaluation at the postpoint $q_j$. Only in the first term the discretization of x(t) is important. The difference between the choice $x_j$ or $x_{j-1}$ is in fact given by the third term.

In order to see how these terms contribute, one has to do the integration over $dp_j$ in (6) which leads to a configuration space functional integral of the form:

$$\left(\frac{g^\star(Q)}{(2\pi\varepsilon\eta c)^d}\right)^{1/2} \int \prod_1^N \frac{dq_j}{(2\pi\varepsilon\eta c)^{d/2}} \sqrt{g^\star(q_j)}\, \exp - \frac{1}{\eta c} \sum_1^{N+1} \frac{\Delta_j^\mu g^\star_{\mu\nu} \Delta_j^\nu}{2\varepsilon} + K_{j_1} + K_{j_2} \tag{11}$$

where $K_{j_1}$ is a term of order ε including all terms without $\Delta_j$ and $K_{j_2}$ contains all the dependence in $\Delta_j$, i.e. the terms in $\Delta_j$, $(\Delta_j)^2$, $(\Delta_j)^3$, $(\Delta_j)^4$. We then develop $\exp - \frac{K_{j2}}{\eta c} = 1 - \frac{K_{j2}}{\eta c} + \frac{1}{2}\frac{(K_{j2})^2}{(\eta c)^2}$, and we use the replacements determined by the form of the dominating exponential, in order to eliminate the powers $(\Delta_j)^n$ (see (5.35))

For the quadratic terms $\frac{\Delta_j^\mu \Delta_j^\nu}{\varepsilon} \doteq \eta c g^{\mu\nu}(j)$.

Doing all this one arrives at

$$I_N = \frac{\sqrt{g^\star(Q)}}{(2\pi\varepsilon\eta c)^{d/2}} \int \prod_1^N \frac{dq_j \sqrt{g^\star(q_j)}}{(2\pi\varepsilon\eta c)^{d/2}} \exp\{-\frac{\varepsilon}{\eta c} \sum_1^{N+1} [\frac{\Delta_j^\mu}{\varepsilon} - \frac{\eta c}{2} g_j^{\star\sigma\tau}\Gamma^{\star\mu}_{j\sigma\tau} + (\frac{\partial q^\mu}{\partial Q^\rho}\frac{\partial Q^\rho}{\partial x^\sigma}\dot{x}^\sigma)_j] \frac{g^\star_{\mu\nu}(j)}{2} [\frac{\Delta_j^\nu}{\varepsilon} - \frac{\eta c}{2} g^{\star\lambda\rho}\Gamma^{\star\nu}_{\lambda\rho} + \frac{\partial q^\nu}{\partial Q^\rho}\frac{\partial Q^\rho}{\partial x^\sigma}\dot{x}^\sigma]\}$$

$$\times \exp\{\sum_1^{N+1} \frac{\partial q^\nu}{\partial Q_j^\rho} [\dot{x}_j^\kappa \frac{\partial}{\partial x^\kappa}\frac{\partial Q^\rho}{\partial q^\nu} + \frac{\Delta^\kappa}{\varepsilon}\frac{\partial}{\partial q^\kappa}\frac{\partial Q^\rho}{\partial q^\nu}]_j + \frac{\eta c}{2}[\frac{\partial q^\nu}{\partial Q^\rho}\frac{\partial^2 Q^\rho}{\partial q^\lambda \partial q^\tau} g^{\star\tau\rho} \frac{\partial q^\lambda}{\partial Q^\sigma}\frac{\partial^2 Q^\sigma}{\partial q^\nu \partial q^\rho} - g^{\star\lambda\sigma}\frac{\partial q^\nu}{\partial Q^\rho}\frac{\partial^3 Q^\rho}{\partial q^\nu \partial q^\lambda \partial q^\sigma}]\} \tag{12}$$

The terms $\varepsilon \sum_1^{N+1} (\frac{\partial q^\nu}{\partial Q^\rho})_j [\dot{x}^\kappa \frac{\partial}{\partial x^\kappa}(\frac{\partial Q^\rho}{\partial q^\nu}) + \frac{\Delta^\kappa}{\varepsilon}\frac{\partial}{\partial q^\kappa}(\frac{\partial Q^\rho}{\partial q^\nu})]_j$ can be integrated to give (13)

$$\int_{t_o}^{T} dt' \frac{d}{dt'} \ln\det(\frac{\partial Q^\rho}{\partial q^\nu}) = \ln\left(\frac{g^\star(q(T),x(T)}{g(Q(T))}\frac{g(Q(t_o))}{g^\star(q(t_o),x(t_o))}\right) = 0 \tag{14}$$

due to the boundary conditions $q(T) = q(t_o) = 0$ since $g^\star(o,x(T) = \eta^d g(Q(T))$ and $g^\star(o,x(t_o)) = \eta^d g(Q(t_o))$. The integration in (13) can only be done in midpoint discretization [app.4.3], and the terms resulting from this shift in (13) cancel the last two terms in (12). Reintroducing then the p-integration in (12), we obtain the result (1).

Appendix 9.5

We compute explicitly the first few terms in the series expansion of quantities introduced in (9.C) up to the order we need to give the expressions for the first order correction (9.C(84,85);10.D(2,3)). For the definitions of the Riemann-Christoffel tensor, the Ricci tensor and the curvature scalar, we follow [ 84 ], more precisely

$$R^a_{ijk} = \partial_j \Gamma^a_{ik} - \partial_k \Gamma^a_{ij} + \Gamma^a_{bj}\Gamma^b_{ik} - \Gamma^a_{bk}\Gamma^b_{ij} \tag{1}$$

$$R_{ij} = R^a_{ija} \tag{2}$$

$$R = g^{ij} R_{ij} \tag{3}$$

and as a consequence $(\nabla_j\nabla_k - \nabla_k\nabla_j)V_i = R^a_{ikj}\, V_a$ (4)
If possible we shall make use of the notation

$$\langle a,b\rangle = a^\mu g_{\mu\nu} b^\nu \tag{5}$$

$$R^\mu(a,b)c = R^\mu{}_{\nu\rho\sigma} a^\sigma b^\rho c^\nu \tag{6}$$

$$\mathcal{R}^\mu(a) = R^\mu_{\cdot\nu} a^\nu \tag{7}$$

$$F^\mu(a) = F^\mu_{\cdot\nu}\, a^\nu \tag{8}$$

a) $\tilde{g}^\star_{\rho\sigma}$, $\tilde{g}^{\star\rho\sigma}$, $\tilde{B}^\sigma$

The expansion of $\tilde{g}^{\star\rho\sigma}$ and $\tilde{B}^\sigma$ (9.C.(43)) follows from the expansion of $L(u,t) = \frac{1}{2} \langle \frac{\partial\alpha\,(u,t)}{\partial t}, \frac{\partial\alpha(u,t)}{\partial t} \rangle$ in powers of u (this is also an expansion in powers of q) which is given by [D14,D17 ]

$$L(u,t) = \sum_{n=o}^{\infty} \frac{u^n}{n!} \left.\frac{\partial^n L(u,t)}{\partial u^n}\right|_{u=o} \tag{9}$$

What we need is $\bar{L}(t) = L(1,t) = \frac{1}{2} \langle \dot{Q}, \dot{Q}\rangle$, which has the form

$$\bar{L}(t) = \sum_{n=o}^{\infty} \frac{1}{n!} \left.\frac{\partial^n L(u,t)}{\partial u^n}\right|_{u=o} \tag{10}$$

Using the notation $\nabla_u = \frac{\partial\alpha^\mu}{\partial u} \nabla_\mu$ for covariant differentiation along the u-curves, $\nabla^2_u = \frac{\partial\alpha^\mu}{\partial u}\frac{\partial\alpha^\nu}{\partial u} \nabla_\mu\nabla_\nu$, etc. since $\nabla_u \frac{\partial\alpha}{\partial u} = 0$. At u = 0 we then find $\nabla_u \to q^\mu\nabla_\mu$

$\nabla_u^2 \to q^\mu q^\nu \nabla_\mu \nabla_\nu$ , etc... On the other hand

$\nabla_t = \frac{\partial\alpha^\mu}{\partial t}\nabla_\mu$ which for $t = 0$ becomes $\nabla_t \to \dot{x}^\mu \nabla_\mu$. Using

$\nabla_t \frac{\partial\alpha}{\partial u} = \nabla_u \frac{\partial\alpha}{\partial t}$ we obtain successively

$$\frac{\partial L}{\partial u} = \langle \nabla_u \frac{\partial\alpha}{\partial t} , \frac{\partial\alpha}{\partial t} \rangle = \langle \nabla_t \frac{\partial\alpha}{\partial u} , \frac{\partial\alpha}{\partial t} \rangle \tag{11}$$

$$\frac{\partial^2 L}{\partial u^2} = \langle \nabla_u \nabla_t \frac{\partial\alpha}{\partial u}, \frac{\partial\alpha}{\partial t} \rangle + \langle \nabla_t \frac{\partial\alpha}{\partial u} , \nabla_t \frac{\partial\alpha}{\partial u} \rangle \tag{12}$$

Using here $\nabla_u \frac{\partial\alpha}{\partial u} = 0$, one has

$$\nabla_u \nabla_t \frac{\partial\alpha}{\partial u} = (\nabla_u \nabla_t - \nabla_t \nabla_u) \frac{\partial\alpha}{\partial u} = R(\frac{\partial\alpha}{\partial t} , \frac{\partial\alpha}{\partial u}) \frac{\partial\alpha}{\partial u}$$

and therefore

$$\frac{\partial^2 L}{\partial u^2} = \langle R (\frac{\partial\alpha}{\partial t} , \frac{\partial\alpha}{\partial u}) \frac{\partial\alpha}{\partial u}, \frac{\partial\alpha}{\partial t} \rangle + \langle \nabla_t \frac{\partial\alpha}{\partial u}, \nabla_t \frac{\partial\alpha}{\partial u} \rangle \tag{13}$$

For the higher derivatives, we obtain

$$\frac{\partial^3 L}{\partial u^3} = \langle R' (\frac{\partial\alpha}{\partial t}, \frac{\partial\alpha}{\partial u})\frac{\partial\alpha}{\partial u} , \frac{\partial\alpha}{\partial t} \rangle + 4 \langle R(\frac{\partial\alpha}{\partial t}, \frac{\partial\alpha}{\partial u}) \frac{\partial\alpha}{\partial u} , \nabla_t \frac{\partial\alpha}{\partial u} \rangle \tag{14}$$

where

$$R'(\frac{\partial\alpha}{\partial t} , \frac{\partial\alpha}{\partial u}) \frac{\partial\alpha}{\partial u} = (\nabla_u R)(\frac{\partial\alpha}{\partial t}, \frac{\partial\alpha}{\partial u}) \frac{\partial\alpha}{\partial u} = \frac{\partial\alpha^\mu}{\partial u} (\nabla_\mu R^\nu_{\rho\sigma\gamma}) \frac{\partial\alpha^\gamma}{\partial t} \frac{\partial\alpha^\sigma}{\partial u} \frac{\partial\alpha^\rho}{\partial u}.$$

For higher derivatives (and also for other quantities) we use the same notation: $R'' = \nabla_u^2 R$, $R''' = \nabla_u^3 R$, etc.

$$\frac{\partial^4 L}{\partial u^4} = \langle R'' (\frac{\partial\alpha}{\partial t}, \frac{\partial\alpha}{\partial u}) \frac{\partial\alpha}{\partial u}, \frac{\partial\alpha}{\partial t} \rangle + 6\langle R'(\frac{\partial\alpha}{\partial t}, \frac{\partial\alpha}{\partial u})\frac{\partial\alpha}{\partial u}, \nabla_t \frac{\partial\alpha}{\partial u} \rangle + 4\langle R(\nabla_t \frac{\partial\alpha}{\partial u}, \frac{\partial\alpha}{\partial u}) \frac{\partial\alpha}{\partial u},$$

$$\nabla_t \frac{\partial\alpha}{\partial u} \rangle + 4\langle R(\frac{\partial\alpha}{\partial t}, \frac{\partial\alpha}{\partial u}) \frac{\partial\alpha}{\partial u}, R(\frac{\partial\alpha}{\partial t}, \frac{\partial\alpha}{\partial u}) \frac{\partial\alpha}{\partial u} \rangle \tag{15}$$

$$\frac{\partial^5 L}{\partial u^5} = \langle R''' (\frac{\partial\alpha}{\partial t} , \frac{\partial\alpha}{\partial u}) \frac{\partial\alpha}{\partial u}, \frac{\partial\alpha}{\partial t} \rangle + 8\langle R''(\nabla_t \frac{\partial\alpha}{\partial u}, \frac{\partial\alpha}{\partial u}) \frac{\partial\alpha}{\partial u}, \frac{\partial\alpha}{\partial t} \rangle + 10\langle R'(\nabla_t \frac{\partial\alpha}{\partial u}, \frac{\partial\alpha}{\partial u})$$

$$\frac{\partial\alpha}{\partial u}, \nabla_t \frac{\partial\alpha}{\partial u} \rangle + 14\langle R'(\frac{\partial\alpha}{\partial t}, \frac{\partial\alpha}{\partial u}) \frac{\partial\alpha}{\partial u}, R(\frac{\partial\alpha}{\partial t}, \frac{\partial\alpha}{\partial u})\frac{\partial\alpha}{\partial u} \rangle + 16\langle R(\nabla_t \frac{\partial\alpha}{\partial u} , \frac{\partial\alpha}{\partial u})\frac{\partial\alpha}{\partial u},$$

$$R(\frac{\partial\alpha}{\partial t}, \frac{\partial\alpha}{\partial u}) \frac{\partial\alpha}{\partial u} \rangle \tag{16}$$

It is clear now that one can continue computing the successive derivatives of $L(u,t)$, but for our purpose here, which is to give the explicit expression of the first correction to the WKB-approximation, we only need $g^{\star\mu\nu}(x(t),q(t))$ up to order $O(q^4)$, so it is enough to calculate only the quadratic terms in $\nabla_t \frac{\partial\alpha}{\partial u}\Big|_{u=o} = \nabla_t q$ of $\frac{\partial^6 L}{\partial u^6}$ which are

$$\frac{\partial^6 L}{\partial u^6}\Big|(\text{quadratic } \nabla_t q) = 18\langle R''(\nabla_t \frac{\partial\alpha}{\partial u}, \frac{\partial\alpha}{\partial u}) \frac{\partial\alpha}{\partial u}, \nabla_t \frac{\partial\alpha}{\partial u}\rangle + 16\langle R(\nabla_t \frac{\partial\alpha}{\partial u}, \frac{\partial\alpha}{\partial u}) \frac{\partial\alpha}{\partial u}, R(\nabla_t \frac{\partial\alpha}{\partial u}, \frac{\partial\alpha}{\partial u}) \frac{\partial\alpha}{\partial u}\rangle \qquad (17)$$

$\bar{L}(t)$ can now be calculated from (10) to the required order, recalling that for $u=0$, one has $\frac{\partial\alpha}{\partial u} \to q$, $\nabla_t \frac{\partial\alpha}{\partial u} \to \nabla_t q$, $\frac{\partial\alpha}{\partial t} \to \dot{x}$, and that the contraction $\langle,\rangle$ is to be interpreted with $g_{\mu\nu}(x(t))$. Using (9.C(45)) and (9.C(46)) we determine $\tilde{g}^{\star}_{\rho\sigma}(x(t)q)$ and $\tilde{D}_\tau(x(t),q)$ as

$$\tilde{g}^{\star}_{\alpha\beta}(x(t),q) = g_{\alpha\beta}(x) + \frac{1}{3} R_{\alpha\nu\rho\beta}(x) q^\nu q^\rho + \frac{1}{6} R'_{\alpha\nu\rho\beta}(x) q^\nu q^\rho + \frac{1}{20} R''_{\alpha\nu\rho\beta} q^\nu q^\rho + \frac{2}{45} R_{\mu\nu\rho\alpha}(x) R^{\mu}_{\gamma\delta\beta}(x) q^\nu q^\rho q^\gamma q^\delta + O(q^5) \qquad (18)$$

$$\tilde{D}_\tau(x(t),q(t)) = g_{\tau\beta}(x)\dot{x}^\beta + \frac{2}{3} R_{\tau\nu\rho\sigma}(x)\dot{x}^\sigma q^\rho q^\nu + \frac{1}{4} R'_{\tau\nu\rho\sigma}(x)\dot{x}^\sigma q^\rho q^\nu + \frac{1}{15} R''_{\mu\nu\rho\tau}(x) q^\nu q^\rho \dot{x}^\mu + \frac{2}{15} R_{\mu\nu\rho\tau}(x) R^{\mu}_{\alpha\beta\gamma}(x) q^\nu q^\rho q^\alpha q^\beta \dot{x}^\gamma + O(q^5) \qquad (19)$$

From $\tilde{g}^{\star\alpha\beta}\tilde{g}^{\star}_{\beta\gamma} = \delta^\alpha_\gamma$, we obtain the expansion of $\tilde{g}^{\star\alpha\beta}$ as

$$\tilde{g}^{\star\gamma\tau}(x(t),q(t)) = g^{\gamma\tau}(x) - \frac{1}{3} R^{\gamma}{}_{\nu\rho}{}^{\tau} q^\nu q^\rho - \frac{1}{6} R'^{\gamma}{}_{\nu\rho}{}^{\tau} q^\nu q^\rho - \frac{1}{20} R''^{\gamma}{}_{\nu\rho}{}^{\tau} q^\nu q^\rho + \frac{1}{15} R^{\gamma}{}_{\nu\rho\beta} R^{\beta}{}_{\nu'\rho'}{}^{\tau} q^\nu q^\rho q^{\nu'} q^{\rho'} \qquad (20)$$

and $\tilde{B}^\gamma = \tilde{g}^{\star\gamma\tau}\tilde{D}_\tau - \dot{x}^\gamma$ is then given by

$$\begin{aligned}\tilde{B}^\gamma(x(t),q(t)) &= \frac{1}{3} R^{\gamma}{}_{\nu\rho\sigma} q^\nu q^\rho \dot{x}^\sigma + \frac{1}{12} R'^{\gamma}{}_{\nu\rho\sigma} \dot{x}^\sigma q^\nu q^\rho + \frac{1}{60} R''^{\gamma}{}_{\nu\rho\sigma} q^\nu q^\rho \dot{x}^\sigma - \frac{1}{45} R^{\gamma}{}_{\nu\rho\mu} R^{\mu}{}_{\alpha\beta\delta} q^\nu q^\rho q^\alpha q^\beta \dot{x}^\delta \\ &= \frac{1}{3} R^\gamma(\dot{x},q)q + \frac{1}{12} R'^\gamma(\dot{x},q)q + \frac{1}{60} R''^\gamma(\dot{x},q)q - \frac{1}{45} R^\gamma(R(\dot{x},q)q,q)q \end{aligned} \qquad (21)$$

b) $\tilde{K}$ and $\tilde{G}$

In (9.C(27)) we claimed that (for $\eta = 1$), in the midpoint discretization

$$\int_{t_o}^{T} dt(\dot{x}^{\sigma} + \nabla_t q^{\sigma} + \tilde{B}^{\sigma})\tilde{A}^{\star}_{\sigma} = \int_{t_o}^{T} \nabla_t q^{\rho}\tilde{K}_{\rho} + \tilde{G}(q) \tag{22}$$

where $\tilde{K}_{\rho}$ depends only on $F_{\mu\nu}$, and the same is valid for $\tilde{G}(q)$ except for the first term in the development in powers of q which is just $\dot{x}^{\mu}A_{\mu}(x)$. To prove this, and to calculate explicitly $\tilde{K}_{\rho}$ and $\tilde{G}(q)$, we compute the development of $M(u,t) = <\frac{\partial\alpha}{\partial t}, A>$ which for u=1 is $<Q,A>$. One has for $\bar{M}(t) = M(1,t)$

$$\bar{M}(t) = \dot{Q}^{\mu}g_{\mu\nu}A^{\nu} = \dot{Q}^{\mu} \frac{\partial q^{\rho}}{\partial Q^{\mu}} \tilde{g}^{\star}_{\rho\sigma}\frac{\partial q^{\sigma}}{\partial Q^{\nu}} A^{\nu} \tag{23}$$

and using (IV.43), one gets

$$\bar{M}(t) = (\dot{x}^{\rho} + \nabla_t q^{\rho} + \tilde{B}^{\rho})\tilde{A}^{\star}_{\rho} \tag{24}$$

which is precisely what we need. In $\int_{t_o}^{T} dt\ \bar{M}(t)$ which is in the midpoint, we can freely integrate by parts with the ordinary rules of calculus. In order to eliminate boundary terms, we take $\frac{\partial\alpha}{\partial u}(u,t_o) = \frac{\partial\alpha}{\partial u}(u,T) = 0$ and in what follows, we omit the integration $\int dt.M(u,t)$ which should be understood. Computing the developments of M(u,t) we have in first order

$$\frac{\partial M}{\partial u} = <\nabla_t \frac{\partial\alpha}{\partial u}, A> + <\frac{\partial\alpha}{\partial t}, \nabla_u A> \tag{25}$$

and integrating by parts the first term

$$\begin{aligned}\frac{\partial M}{\partial u} &= -<\frac{\partial\alpha}{\partial u}, \nabla_t A> + <\frac{\partial\alpha}{\partial t}, \nabla_u A> \\ &= (\frac{\partial\alpha}{\partial t})^{\mu}(\frac{\partial\alpha}{\partial u})^{\nu}(\nabla_{\nu}A_{\mu} - \nabla_{\mu}A_{\nu}) = (\frac{\partial\alpha}{\partial t})^{\mu}(\frac{\partial\alpha}{\partial u})^{\nu}F_{\nu\mu} = -<\frac{\partial\alpha}{\partial t}, F(\frac{\partial\alpha}{\partial u})>\end{aligned} \tag{26}$$

with the notation $F(d) = F^{\mu}{}_{\nu}d^{\nu}$. From this formula we see that all higher derivatives of M(u,t) will be expressed in terms of $F_{\mu\nu}$ and the curvature tensor. Up to $\frac{\partial^4 M}{\partial u^4}$, we have

$$\frac{\partial^2 M}{\partial u^2} = -<\nabla_t \frac{\partial\alpha}{\partial u}, F(\frac{\partial\alpha}{\partial u})> - <\frac{\partial\alpha}{\partial t}, F'(\frac{\partial\alpha}{\partial u})> \tag{27}$$

$$\frac{\partial^3 M}{\partial u^3} = -<R(\frac{\partial\alpha}{\partial t}, \frac{\partial\alpha}{\partial u})\frac{\partial\alpha}{\partial u}, F(\frac{\partial\alpha}{\partial u})> - 2<\nabla_t \frac{\partial\alpha}{\partial u}, F'(\frac{\partial\alpha}{\partial u})> - <\frac{\partial\alpha}{\partial t}, F''(\frac{\partial\alpha}{\partial u})> \tag{28}$$

$$\frac{\partial^4 M}{\partial u^4} = -<R'(\frac{\partial\alpha}{\partial t},\frac{\partial\alpha}{\partial u})\frac{\partial\alpha}{\partial u}, F(\frac{\partial\alpha}{\partial u})> - <R(\nabla_t\frac{\partial\alpha}{\partial u},\frac{\partial\alpha}{\partial u})\frac{\partial\alpha}{\partial u}, F(\frac{\partial\alpha}{\partial u})>$$

$$- 3<R(\frac{\partial\alpha}{\partial t},\frac{\partial\alpha}{\partial u})\frac{\partial\alpha}{\partial u}, F'(\frac{\partial\alpha}{\partial u})> - 3<\nabla_t\frac{\partial\alpha}{\partial u}, F''(\frac{\partial\alpha}{\partial u})> - <\frac{\partial\alpha}{\partial t}, F'''(\frac{\partial\alpha}{\partial u})> \quad (29)$$

The quantities $\tilde{K}_\rho$ and $\tilde{G}$ can now be read from (26) to (29), using the same rules as before

$$\begin{aligned}
\tilde{K}_\rho^{[o]} &= 0\\
\tilde{K}_\rho^{[1]} &= -\frac{1}{2}F_\rho(q)\\
\tilde{K}_\rho^{[2]} &= -\frac{1}{3}F'_\rho(q)\\
\tilde{K}^{[3]} &= -\frac{1}{24}R_\rho(F(q),q)q - \frac{1}{8}F''_\rho(q)
\end{aligned} \quad (30)$$

$$\begin{aligned}
\tilde{G}^{[o]} &= <\dot{x},A>\\
\tilde{G}^{[1]} &= -<\dot{x},F(q)>\\
G^{[2]} &= -\frac{1}{2}<\dot{x},F'(q)>\\
\tilde{G}^{[3]} &= -\frac{1}{6}<R(\dot{x},q)q,F(q)> - \frac{1}{6}<\dot{x},F''(q)>\\
\tilde{G}^{[4]} &= -\frac{1}{24}<R'(\dot{x},q)q,F(q)> - \frac{1}{8}<R(\dot{x},q)q,F''(q)> - \frac{1}{24}<\dot{x},F'''(q)>
\end{aligned} \quad (31)$$

c) $V^\star$, $\tilde{\Gamma}^{\star\nu}_{\mu\rho}$

The expansion of a scalar is given directly in terms of its covariant derivatives

$$V^\star(x(t),q(t)) = V(x(t)) + V' + \frac{1}{2}V'' + \frac{1}{3}V''' + \frac{1}{4}V'''' + O(q^5) \quad (32)$$

The computation of the extensions of $\tilde{\Gamma}^{\star\nu}_{\rho\mu}$ is done in [V3] . From there we have the expansion

$$\tilde{\Gamma}^{\star\gamma}_{\rho\mu}(x(t),q(t)) = -\frac{1}{3}(R^\gamma_{\rho\mu\nu} + R^\gamma_{\mu\rho\nu})q^\nu - \frac{1}{12}(5\nabla_\nu R^\gamma_{\rho\mu\sigma} + 3\nabla_\sigma R^\gamma_{\nu\rho\mu} + \nabla_\rho R^\gamma_{\sigma\mu\nu})$$
$$q^\nu q^\sigma + O(q^3) \quad (33)$$

This expansion is needed in the form $\tilde{g}^{\star\alpha\gamma}\tilde{\Gamma}^{\star\sigma}_{\alpha\gamma}$, and also in the computation of $(\tilde{\nabla}_\mu\tilde{K}^\mu)$. (33) and (20) yield (using the Biachi identity)

$$\tilde{\Gamma}^{\star\sigma}_{\alpha\gamma}\tilde{g}^{\star\alpha\gamma} = -\frac{2}{3}\mathcal{R}(q) - \frac{1}{2}\mathcal{R}'(q) + \frac{1}{12}g^{\sigma\lambda}(\nabla_\lambda R_{\nu\gamma})q^\nu q^\gamma + O(q^3) \quad (34)$$

and $$\tilde{\nabla}_\mu \tilde{K}^\mu = g^{\star\mu\lambda}\left(\frac{\partial}{\partial q^\mu}\tilde{K}_\lambda - \tilde{\Gamma}^{\star\rho}_{\mu\lambda}\tilde{K}_\rho\right)$$

$$= -\frac{1}{3}\nabla_\rho F^\rho{}_\lambda q^\lambda - \frac{1}{4}\nabla_\rho\nabla_\sigma F^\rho{}_\lambda q^\sigma q^\lambda - \frac{5}{12}<\mathcal{R}(q),F(q)> \tag{35}$$

up to order $q^2$. We shall use the notation

$$= -\frac{1}{3}(\nabla_\rho F^\rho)(q) - \frac{1}{4}(\nabla_\rho F'^\rho)(q) - \frac{5}{12}<\mathcal{R}(q),F(q)> \tag{36}$$

To include the $\eta$ dependence, one should recall

$$g^{\star\mu\nu}(x(t),q(t)) = \frac{\partial q^\mu}{\partial Q^\rho}\frac{\partial q^\nu}{\partial Q^\sigma}g^{\rho\sigma}(Q) = \frac{1}{\eta}g^{\mu\nu}(x) + O(q)$$

since $\left.\frac{\partial q^\mu}{\partial Q^\rho}\right|_{q=0} = \frac{1}{\sqrt{\eta}}\delta^\mu_\rho$ and consequently

$$g^{\star\mu\nu}(x(t),q(t)) = \frac{1}{\eta}\tilde{g}^{\star\mu\nu}(x(t),\sqrt{\eta}\,q(t))$$

For the other quantities one can proceed similarly.

# CHAPTER X : THE SEMICLASSICAL EXPANSION IN CONFIGURATION SPACE

In (I.C), we have seen that perturbation theory in phase space functional integrals is well defined if one uses the notion of discretization. A particular application was the construction of the semiclassical expansion of the propagator I of the Schrodinger and Fokker Planck equations in chapter IX . In the framework of the same application, we shall consider now the perturbation theory in configuration space. The purpose of solving the same problem in configuration space is to show that the notion of discretization also allows us here to give a precise meaning to all the terms in the expansion. Although this was to be expected, the situation is much more complicated due to the occurrence of contractions of the derivatives of q(t), such as $\nabla_t \overbrace{q\; q}(t)$ and $\overbrace{\nabla_t q\; \nabla_{t'} q}$, which at equal time cannot be determined in a formal way, but require the close examination of the discretized expression. The main difference comes from the fact that after splitting the quadratic part defining the free generating functional $Z_o[J]$, one cannot take the continuous limit separately in $Z_o[J]$, and in the interaction terms, but has to study previously the discrete form of the equal time contractions with derivatives. This is in contrast to the phase space situation where the limit could be taken in the free generating functional $Z_o[J,J^*]$, and all the discretization dependence of the functional integral could be taken into account effectively by a prescription on the action of the interaction Hamiltonian on $Z_o[J,J^*]$. Another new feature appearing in the configuration space treatment, is the occurrence of undefined terms in $\delta(0)$ in the expansion, that appear as terms in $\frac{1}{\varepsilon}$ in the discrete, and which we show to cancel in all orders of $\eta$. Similar problems to these have been studied in the literature [C6,G8 ,L29], mostly with operator techniques, or with formal use of the path integral.

We take as starting point the configuration space path integral in $\gamma_1$ (1) discretization, representing the propagator. Then we set up the perturbation expansion, and study in a formal way the contractions. Having seen the difficulties that appear, we study the delicate problem of the equal time contractions in the discrete. Here we find the remarkable result that one has to study $\overbrace{q^\mu(t)\; q^\nu}(t)$ up to terms of order $\varepsilon$, which vanish in the continuous limit, but give a contribution to $\overbrace{\nabla_t q\;\; \nabla_t q}$, and it is this effect that explains in a systematic way the new kind of ambiguities, appearing in configuration space. We also prove in the discrete the cancellation of all singular terms in $\frac{1}{\varepsilon}$ (formally $\delta(0)$) coming from the configuration space measure and the singularities arising from the singular $\overbrace{\nabla_t q^\mu \nabla_t q^\nu}$ contractions. This analysis of the underlying structure in the discrete, will confirm that it is possible (as in phase space) to use a set of simple formal rules, if a suitable discretization is chosen. We shall give these rules for the $\gamma_1(1)$ discretization, and as before we present the interaction terms needed to compute the first order correction and derive from these the short time propagator.

A. Perturbation theory in configuration space

From ( 9.C(29)), one obtains the configuration space functional integral representation of the propagator I in the form

$$I = \eta^{-\frac{d}{2}} \int_{\gamma_1(1)} \mathcal{D}q \left(\frac{\tilde{g}^{\star}(\sqrt{\eta}q,t)}{c^d}\right)^{\frac{1}{2}} \exp\{-\frac{1}{\eta c}\int_{t_o}^{T} dt\, \frac{1}{2}[\dot{x}^{\sigma} + \sqrt{\eta}\nabla_t q^{\sigma} + \tilde{B}^{\sigma}(\sqrt{\eta}\, q)]$$

$$\tilde{g}^{\star}_{\sigma\rho}[\dot{x}^{\rho} + \sqrt{\eta}\nabla_t q^{\rho} + \tilde{B}^{\rho}(\sqrt{\eta}\, q)] + \sqrt{\eta}\, \nabla_t q^{\sigma}\, \tilde{K}_{\sigma}(\sqrt{\eta}\, q) + \tilde{G}(\sqrt{\eta}q) - V(\sqrt{\eta}q)\}$$

$$\exp\{\int_{t_o}^{T} dt\, [\frac{\eta c}{8}\, \tilde{g}^{\star\alpha\gamma}\tilde{\Gamma}^{\star\rho}_{\alpha\gamma}\, \tilde{g}^{\star}_{\rho\sigma}\tilde{g}^{\star\mu\nu}\tilde{\Gamma}^{\star\sigma}_{\mu\nu} + \frac{1}{2}\, \tilde{g}^{\star\alpha\gamma}\partial_{\alpha}\tilde{K}_{\gamma} + \frac{1}{2}(\dot{x}^{\sigma} + \sqrt{\eta}\nabla_t q^{\sigma} + \tilde{B}^{\sigma})$$

$$\tilde{g}^{\star}_{\sigma\rho}\tilde{g}^{\star\alpha\gamma}\tilde{\Gamma}^{\star\rho}_{\alpha\gamma}]_{q\to\sqrt{\eta}q}\}\delta(q(T))\delta(q(t_o)) \quad (1)$$

We repeat that the star indicates the quantities in normal coordinates centered at x(t), while the tilde denotes quantities independent of $\eta$, so that the explicit $\eta$ dependence has to be introduced by $q\to\sqrt{\eta}q$. If we call E the exponential in (1), then we recognize (see App.9.5) in the first part $E_1$ the quantity

$$E_1 = -\frac{1}{\eta c}\, [\frac{1}{2}\, \dot{Q}^{\mu}g_{\mu\nu}\dot{Q}^{\nu} + \dot{Q}^{\mu}A_{\mu}(Q) - V(Q)] \quad (2)$$

evaluated in normal coordinates around $x^{\mu}(t)$. $\tilde{K}$ and $\tilde{G}$ where introduced ( 9.C(27)) to exhibit the gauge invariance of the correction terms to the WKB approximation $E_2$ the remaining part, which is at least of $\mathcal{O}(\sqrt{\eta})$ can be considered as a correction term with this specific form due to the $\gamma_1(1)$ discretization. Developing E we obtain in order $\frac{1}{\eta}$ the classical action, while the terms in $\frac{1}{\sqrt{\eta}}$ vanish owing to the equation of motion. The terms of order $(\eta^{o})$ have the form

$$W_o = -\frac{1}{2c} \int_{t_o}^{T} dt[\nabla_t q^{\mu}g_{\mu\nu}(x)\nabla_t q^{\nu} - \nabla_t q^{\mu}F_{\mu\nu}(x)q^{\nu}$$

$$+ q^{\rho}q^{\nu}(R_{\mu\nu\rho\sigma}\dot{x}^{\mu}\dot{x}^{\sigma} + \nabla_{\rho}F_{\nu\mu}\dot{x}^{\mu} - \nabla_{\rho}\partial_{\nu}V)] \quad (3)$$

These terms also come from $E_1$, and we remark that the expansion of $E_1$ can be read immediately from the expansions of $\bar{L}$, and $\bar{M}$ in App.9.5. The higher order terms from $E_1$ and $E_2$ will be combined in the interaction Lagrangian $\mathcal{L}_I$, as

$$\int dt\mathcal{L}_I(x(t),q(t),\nabla_t q) = \sum_{n\geq 1} (\sqrt{\eta})^n W_n \quad (4)$$

and we see that it is of the form

$$\mathcal{L}_I = -\frac{1}{c}\left[\frac{1}{2}\nabla_t q^\sigma \tilde{g}^{\star(2)}_{\sigma\rho}\nabla_t q^\rho + \nabla_t q^\sigma \tilde{D}_\sigma(t,q) + V_1(t,q)\right] \tag{5}$$

with

$$\tilde{g}^{\star(2)}_{\sigma\rho} = \tilde{g}^{\star}_{\sigma\rho}(\sqrt{\eta}q) - g_{\sigma\rho}(x) = \mathcal{O}(\eta)$$

After taking out the classical action as $I = \eta^{-\frac{d}{2}} \exp\left\{-\frac{1}{\eta c}A_{cl}(X,T,X_\circ,t_\circ)\right\}\tilde{I}$, we obtain a functional integral that should give us the normalization factor and the higher order corrections. Its discretized form is $(x_j = x(t_j))$

$$\tilde{I}_N = \int_{q_o=0}^{q_{N+1}=0} \prod_{j=1}^{N+1} \frac{\sqrt{\tilde{g}^\star(\sqrt{\eta}q_j,\bar{x}_{j-1})}}{(2\pi\varepsilon c)^{d/2}} \prod_{i=1}^{N}(dq_i)\exp\{-\frac{\varepsilon}{2c}\sum_{j=1}^{N+1}[(\nabla_t q^\mu)_j\, g_{\mu\nu}(\bar{x}_{j-1})$$

$$(\nabla_t q^\nu)_j - (\nabla_t q^\mu)_j F_{\mu\nu}(x_j)q^\nu_j + q^\rho_j q^\nu_j (R_{\mu\nu\rho\sigma}(x_j)\dot{x}^\mu_j \dot{x}^\sigma_j + \nabla_\rho F_{\nu\mu}(x_j)\dot{x}^\mu_j -$$

$$\nabla_\rho\partial_\nu V(x_j))] + \varepsilon\sum_{j=1}^{N+1}\mathcal{L}_I(\bar{x}_{j-1},(\nabla_t q)_j,q_j)\} \tag{6}$$

The explicit time dependence of the quantities in (6) comes through the classical solution $x(t)$, and one is free to put $x(t'_j)$ with $t_{j-1} \leqslant t'_j \leqslant t_j$, and even to choose different values of $t'_j$ in different quantities. There is however one exception $\tilde{g}^\star_{\mu\nu}(q,x(t))$ and $\tilde{g}^\star = \det \tilde{g}^\star_{\mu\nu}(q,x(t))$ have to be discretized with the same $t'_j$ that we choose to be $\bar{t}_{j-1} = (t_j + t_{j-1})/2$, and we note $x(\bar{t}_{j-1})$ by $\bar{x}_{j-1}$.
Note that $\bar{x}_{j-1}$ can also be taken as $\frac{1}{2}(x_{j-1} + x_j)$ since the difference is $0(\varepsilon^2)$ and can be checked to give no extra contribution. All q dependence in (6) has been discretized in the postpoint $q_j$, as indicated by $\gamma_1(1)$ in (1). In order to set up the perturbation expansion which will give the higher order corrections to the WKB approximation, we introduce sources $J_{\ell\mu}$ and write (6) as

$$\tilde{I}_N = \prod_{j=1}^{N+1}\sqrt{\frac{\tilde{g}^\star(\sqrt{\eta}q_j,\bar{x}_{j-1})}{g(\bar{x}_{j-1})}}\ \exp\{\varepsilon\sum_{j=1}^{N+1}\mathcal{L}_I(\bar{x}_{j-1},(\nabla_t q)_j,q_j)\}\Big|_{q^\mu_j=\frac{1}{i\varepsilon}\frac{\partial}{\partial J_{j\mu}}} \cdot Z^N_o[J]\Big|_{J=0} \tag{7}$$

$$Z_o^N[J] = \int \prod_{j=1}^{N+1} \sqrt{\frac{g(\bar{x}_{j-1})}{(2\pi\varepsilon c)^d}} \prod_{i=1}^{N} dq_i \exp\{-\frac{\varepsilon}{2c}\sum_{j=1}^{N+1}(\nabla_t q^\mu)_j g_{\mu\nu}(\bar{x}_{j-1})(\nabla_t q^\nu)_j -$$

$$(\nabla_t q^\mu)_j F_{\mu\nu}(x_j) q_j^\nu + q_j^\rho q_j^\nu V_{\rho\nu}(x_j) + i\varepsilon \sum_{\ell=1}^{N} J_{\ell\mu} q_\ell^\mu \} \tag{8}$$

where $V_{\rho\nu}$ can be read from (6) and $q_o = q_{N+1} = 0$. If we formally take the limit $\varepsilon \to 0$ $(N \to \infty)$ separately in $Z_o^N[J]$ and in the exponential acting on it, we expect difficulties in the form of undefined expressions. What we want now is to recognize these difficulties and go back to (8) to show how these can be solved. In the continuous limit

$$\prod_{j=1}^{N+1} \sqrt{\frac{\tilde{g}^\star(\sqrt{\eta}q,\bar{x}_{j-1})}{g(\bar{x}_{j-1})}} = \exp\{\frac{\varepsilon}{2}\sum_{j=1}^{N+1}\frac{\delta_{jj}}{\varepsilon}\ln\frac{(\tilde{g}^\star(\sqrt{\eta}q_j,\bar{t}_{j-1})}{g(\bar{x}_{j-1})}\} \to$$

$$\exp\{\frac{\delta(0)}{2}\int_{t_o}^{T} dt \ln (\frac{\tilde{g}^\star(\sqrt{\eta}q(t),x(t)}{g(x(t))}) \} \tag{9}$$

and then $(\bar{\mathcal{L}}_I = \mathcal{L}_I + \frac{1}{2}\delta(0)\ln(\frac{\tilde{g}^\star(t)}{g(t)}))$

$$\tilde{I} = \exp\{\int_{t_o}^{T} dt\, \bar{\mathcal{L}}_I(x(t),q(t),\nabla_t q)\}\Big|_{q(t)=\frac{1}{i}\frac{\delta}{\delta J(t)}} Z_o[J]\Big|_{J=0} \tag{10}$$

where $Z_o[J] = \lim Z_o^N[J]$, $N\to\infty$ has been computed in (9 .C.52-55,83), where one has to take $J^\star = 0$

$$Z_o[J] = \bar{N}^{(1)} \exp\{-\frac{1}{2}\int_{t_o}^{T} dt dt' J_\mu(t)\bar{G}^{\mu\nu}(t,t')J_\nu(t')\} \tag{11}$$

with

$$\bar{N}^{(1)} = [\frac{g(X_o)}{g(X)}]^{\frac{1}{4}} (2\pi c)^{-\frac{d}{2}} (\det \bar{M}_{\alpha\beta})^{\frac{1}{2}} \tag{12}$$

where $\bar{M}_{\alpha\beta} = -\dfrac{\partial^2 A_{cl}}{\partial X_o^\alpha \partial X^\beta}$ is the Van-Vleck matrix. $\bar{G}^{\mu\nu}(t,t')$ is given in (9 .C.75)

$$\bar{G}^{\mu\nu}(t,t') = c[\theta(t-t')\bar{J}^{\mu\alpha}(t,T)M_{\alpha\beta}(T,t_o)\bar{J}^{\beta\nu}(t_o,t') - \theta(t'-t)\bar{J}^{\mu\alpha}(t,t_o)M_{\alpha\beta}(t_o,T)$$

$$\bar{J}^{\beta\nu}(T,t')] \tag{13}$$

The identity $F(\frac{1}{a}\frac{\partial}{\partial J}) Z_o(J)\Big|_{J=0} = Z_o[\frac{1}{a}\frac{\partial}{\partial q}]F(q)\Big|_{q=0}$ leads to the Wick theorem form of the expansion in (12)

$$\tilde{I} = \bar{N}^{(1)} \exp\{\frac{1}{2}\int_{t_o}^{T} dt dt' \frac{\delta}{\delta q^{\mu}(t)} \bar{G}^{\mu\nu}(t,t') \frac{\delta}{\delta q^{\nu}(t')}\}$$

$$\exp\{\int_{t_o}^{T} dt \bar{\mathcal{L}}_I(x(t),q(t),\nabla_t q)\}\Big|_{q=0} \quad (14)$$

and from this formula we can read the contractions that will appear in the expansion. They are

$$\overbrace{q^{\mu}(t)q^{\nu}}(t') = \exp\{\frac{1}{2}\int_{t_o}^{T} dt_1 dt_2 \frac{\delta}{\delta q^{\rho}(t_1)} \bar{G}^{\rho\sigma}(t_1,t_2) \frac{\delta}{\delta q^{\sigma}(t_2)}\} q^{\mu}(t)q^{\nu}(t')\Big|_{q=0}$$

$$= \bar{G}^{\mu\nu}(t,t') \quad (15)$$

$$\nabla_t \overbrace{q^{\mu}(t)\; q^{\nu}}(t') = \nabla_t \bar{G}^{\mu\nu}(t,t') = c[\theta(t-t')\nabla_t \bar{J}(t,T)M(T,t_o)\bar{J}(t_o,t') -$$

$$- \theta(t'-t)\nabla_t \bar{J}(t,t_o)M(t_o,T)\bar{J}(T,t')]^{\mu\nu} \quad (16)$$

$$\nabla_t \overbrace{q^{\mu}(t)\nabla_{t'} q^{\nu}}(t') = \nabla_t \nabla_{t'} \bar{G}^{\mu\nu}(t,t') = c[\theta(t-t')\nabla_t \bar{J}(t,T)M(T,t_o)\nabla_{t'}\bar{J}(t_o,t')$$

$$-\theta(t'-t)\nabla_t \bar{J}(t,t_o)M(t_o,T)\nabla_{t'}\bar{J}(T,t')] + cg^{\mu\nu}(t)\delta(t-t') \quad (17)$$

where in (16) and (17) the notation means the product of the matrices, and $\nabla_t$ acts on the index associated with t. We can see that at equal times (16) and (17) are not defined. In fact the jump of $\nabla_t \bar{G}^{\mu\nu}(t,t') = S^{\mu\nu}(t,t')$ is

$$S^{\mu\nu}(t,t+\delta) - S^{\mu\nu}(t,t-\delta)\Big|_{\delta\to 0^+} = cg^{\mu\nu}(x(t)) \quad (18)$$

while that of $\nabla_t \nabla_{t'} \bar{G}^{\mu\nu}(t,t')$ is

$$[\nabla_t \nabla_{t'} \bar{G}^{\mu\nu}(t,t')\Big|_{t'>t} - \nabla_t \nabla_{t'} \bar{G}^{\mu\nu}(t,t')\Big|_{t'<t}]_{t=t'} = c\, F^{\mu\nu}(x(t)) \quad (19)$$

The discontinuity in (19) should play no role, since from (5,7) we see that only the symmetric part of the equal time contractions survives, due to the multiplication with $\tilde{g}^{\star(2)}_{\mu\nu}$, but (19) als gives a $\delta(0)$ contribution at t=t' which is undefined. We shall prove that in fact no term containing $\delta(0)$ survives in the expansion of (14) (as is clear from the calculation in phase space) due to cancellations between the singular contributions of (17) and the ones coming from the term $\frac{1}{2}\delta(0)$ ln $[\tilde{g}^{\star}(\sqrt{\hbar}q(t), x(t))/g(x(t))]$ in $\bar{\mathcal{L}}_I$. Moreover, due to the $\gamma_1(1)$ discretization (16) has to be taken at equal times as $S^{\mu\nu}(t,t+\varepsilon)$ $\varepsilon \to 0^+$. In summary we shall see that $\tilde{I}$ is given by (7) with $\bar{\mathcal{L}}_I$ replaced by $\mathcal{L}_I$ with the contractions (15-17) and the equal time contractions

$$\nabla_t q^{\mu}(t)\, q^{\nu}(t) = S^{\mu\nu}(t,t+\varepsilon) = -c[\nabla_t \bar{J}(t,t_o)M(t_o,T)\bar{J}(T,t)] \tag{20}$$

$$\nabla_t q^{\mu}(t)\, \nabla_t q^{\nu}(t) = -c[\nabla_t \bar{J}(t,t_o)M(t_o,T)\nabla_t \bar{J}(T,t)]^{\mu\nu} \tag{21}$$

In (21) we can use as well $c[\nabla_t \bar{J}(t,T)M(T,t_o)\nabla_t \bar{J}(t_o,t)]^{\mu\nu}$ since the difference is c $F^{\mu\nu}(x(t))$. Moreover each time $\delta(0)$ appears when computing $\tilde{I}$ using these rules the term should be dropped. The proof of all this relies on the calculation of $Z^N_o[J]$.

B. The free generating functional $Z_o^N[J]$

In order to compute the equal time contractions (A.20, 21) we go back to the discrete formula (A.8) which can be computed as

$$Z_o^N[J] = \bar{N}_\varepsilon^{(1)} \exp\{ -\frac{\varepsilon^2}{2} \sum_{\ell,k=1}^{N} J_{\ell\mu} G^{\mu\nu}(\ell,k;\varepsilon) J_{k\nu} \} \tag{1}$$

where $\ell$ stands for $t_\ell = t_o + \ell\varepsilon$, $J_{\ell\mu} = J_\mu(t_\ell)$, etc., and $G^{\mu\nu}(\ell, k) = \bar{G}^{\mu\nu}(\ell,k) + \delta G^{\mu\nu}(\ell,k,\varepsilon)$ with $\delta G \to 0$ when $\varepsilon \to 0$. So the indicated $\varepsilon$ dependence is not the one due to the arguments $\ell$, k. In (A.11) we have computed the limiting value of (1) when $\varepsilon \to 0$, and there appears only $\bar{G}^{\mu\nu}(t,t')$. It turns out however that $\delta G$ contributes to the equal time contraction, and consequently we have to determine this correction. We shall treat a more general situation than (A.8) which is in postpoint discretization. By considering the same expression in $\gamma_1(u)$ discretization, i.e. $q_j^\nu$ replaced by $q_{j-1}^{\nu(u)} = q_{j-1}^\nu + u\Delta q_j^\nu$. $T_N^{(u)}$ is the corresponding integral

$$T_N^{(u)} = \int_{q_o=0}^{q_{N+1}=0} \prod_{j=1}^{N+1} \sqrt{\frac{g(\bar{x}_{j-1})}{(2\pi\varepsilon c)^d}} \prod_{i=1}^{N} dq_i \exp\{-\frac{\varepsilon}{2c} \sum_{j=1}^{N+1} \frac{\Delta q_j^\mu}{\varepsilon} g_{\mu\nu}(\bar{x}_{j-1}) \frac{\Delta q_j^\nu}{\varepsilon} +$$

$$+ q_{j-1}^{\mu(u)} B_{\mu\nu}(j) \frac{\Delta q_j^\nu}{\varepsilon} + q_j^\rho q_j^\nu \tilde{V}_{\rho\nu}(x_j) + i\varepsilon J_{j\mu} q_j^\mu \} \tag{2}$$

with

$$B_{\mu\nu}(j) = 2 g_{\rho\nu}(\bar{x}_{j-1}) \dot{x}_j^\gamma \Gamma^\rho_{\gamma\mu}(x_j) + F_{\mu\nu}(x_j) \tag{3}$$

We write the exponential of (2) as

$$E^{(u)} = -\frac{1}{2} \sum_{j,k=1}^{N} q_j^\mu W_{\mu\nu}^{(u)}(j,k;\varepsilon) q_k^\nu \tag{4}$$

If $G_{(u)}$ is then the inverse matrix of $W_{\mu\nu}^{(u)}$ in the sense,

$$\sum_{k=1}^{N} \sum_{\rho=1}^{d} W_{\mu\rho}^{(u)}(j,k;\varepsilon) G_{(u)}^{\rho\nu}(k,\ell;\varepsilon) = \delta_\mu^\nu \delta_{j\ell} \tag{5}$$

then the solution of (2) can be written as

$$T_N^{(u)} = (2\pi)^{-\frac{d}{2}} \prod_{j=1}^{N+1} \left(\frac{g(\bar{x}_{j-1})}{(\varepsilon c)^d}\right)^{\frac{1}{2}} (\det G_{(u)}^{\mu\nu}(\ell,k;\varepsilon))^{\frac{1}{2}} \exp\{ -\frac{\varepsilon^2}{2} \sum_{\ell,k} J_{\ell\mu} G_{(u)}^{\mu\nu}(\ell,k;\varepsilon) J_{k\nu} \} \tag{6}$$

giving in the limit $N \to \infty$

$$T^{(u)} = \bar{N}^{(u)} \exp\{ -\frac{1}{2} \int dt dt' J_\mu(t) \bar{G}^{\mu\nu}(t,t') J_\nu(t') \} \tag{7}$$

and

$$\bar{N}^{(u)} = \left(\frac{g(X_o)}{g(X)}\right)^{\frac{2u-1}{4}} \bar{N}^{(1/2)}, \quad \bar{N}^{(1/2)} = (2\pi c)^{-\frac{d}{2}} (\det \bar{M}_{\alpha\beta})^{1/2} \tag{8}$$

This can easily be checked by shifting in (2) to the postpoint, which only generates an extra contribution to the normalization factor. Putting as before $G = \bar{G} + \delta G_{(u)}$, we want to compute $\delta G_{(u)}$ given by (5)

$$W^{(u)}(\bar{G} + \delta G_u) = 1$$

Here we can use a result of C. Dewitt in [17], giving the exact inverse of $\bar{G}$. Let $A(Q'',t'', Q',t') = \int_{t'}^{t''} dt L_{cl}(Q(t),\dot{Q}(t))$ be the classical action calculated for the classical trajectory $Q(t)$ with boundary conditions $Q(t') = Q'$, $Q(t'') = Q''$. We consider

$$\bar{A} = -\frac{1}{\eta c} \sum_{j=1}^{N+1} A(Q_j,t_j,Q_{j-1},t_{j-1}) \qquad Q_o = X_o,\ Q_{N+1} = X \tag{9}$$

and taking $Q_j = x_j + \sqrt{\eta} q_j$, $q_o = q_{N+1} = 0$, we develop (9) in powers of $q_j$ as $\bar{A} = \sum_{n\geq 0} A^{[n]}$, where $A^{[n]}$ is $\mathcal{O}(q_j^n)$. Writing the quadratic part $A^{[2]}$ in the form

$$A^{[2]} = -\frac{1}{2} \sum_{j,k=1}^{N} q_j^\mu \bar{W}_{\mu\nu}(j,k)\, q_k^\nu \tag{10}$$

one can show that the matrix $\bar{G}$ defined by (A.13) is the exact inverse of $\bar{W}$, i.e. $\bar{W}\bar{G} = 1$ (See app.10.1)

$W^{(u)} = \bar{W} + \delta W$ then gives for $\delta G$ the expansion

$$\delta G = -\bar{G}(\delta W \bar{G} - (\delta W \bar{G})^2 + (\delta W \bar{G})^3 - \ldots) \tag{11}$$

In order to determine $\delta W$ we write (see (4)) $E^{(u)} = A^{[2]} + \delta E$, with $\delta E = \delta E_1 + \delta E_2$, $\delta E_1 = E^{(u)} - E^{(1/2)}$, $\delta E_2 = E^{(1/2)} - A^{[2]}$. In app.10.1 we show that

$$\delta E_2 = -\frac{1}{2} \sum_{j=1}^{N} q_j^\mu K_{\mu\nu}(j) q_j^\nu + \sum_{j=2}^{N+1} q_{j-1}^\mu H_{\mu\nu}(j) \frac{\Delta q_j^\nu}{\varepsilon} + \sum_{j=1}^{N+1} \Delta q_j^\mu \bar{L}_{\mu\nu}(j) \frac{\Delta q_j^\nu}{\varepsilon} \tag{12}$$

with ($R$ any of the three matrices $K, H, \bar{L}$)

$$R = \varepsilon^2 (R^{(1)} + \varepsilon R^{(2)} + \varepsilon^2 R^{(3)} + \ldots) \tag{13}$$

On the other hand we see from (12) that $(y = u - \frac{1}{2})$

$$\delta E_1 = - \frac{\varepsilon}{2} \Sigma \Delta q_j^\mu L^o_{\mu\nu}(x_j) \frac{\Delta q_j^\nu}{\varepsilon} \tag{14}$$

with $L^o_{\mu\nu}(j) = \frac{y}{c} B^{(s)}_{\mu\nu}(j) = \frac{y}{c} \dot{x}^\gamma_j \partial_\gamma g_{\mu\nu}(x_j)$, where $B^{(s)}_{\mu\nu} = \frac{1}{2}(B_{\mu\nu} + B_{\nu\mu})$

We note that in (12) K and $\bar{L}$ are symmetric in $(\mu\nu)$ while $H_{\mu\nu}$ can be decomposed in a symmetric and an antisymmetric part as $H_{\mu\nu} = H^{(s)}_{\mu\nu} + H^{(a)}_{\mu\nu}$. Putting $L = \varepsilon(L^{(0)} + \varepsilon L^{(1)} + \varepsilon^2 L^{(2)} + \ldots)$ one has that $\delta E = -\frac{1}{2} \sum_{j,k} q^\mu_j \delta W_{\mu\nu}(j,k) q^\nu_k$ is given by (12) but with $\bar{L}_{\mu\nu}$ replaced by $L_{\mu\nu}$. It is important to remark here that the discretizations $\gamma_1(u)$ are defined as in (25), but that one can change the time dependence of $B_{\mu\nu}(j)$ and $V_{\rho\nu}(x(t_j))$ to $x(t'_j)(t_{j-1} \leqslant t'_j \leqslant t_j)$, and also change $q^\rho_j q^\nu_j \to q^{(v)\rho}_{j-1} q^{(w)\nu}_{j-1}$. These changes give an equivalent discretization in the sense that they do not alter the limit of $T^{(u)}_N$, $N \to \infty$, but they change $\delta G$, since they change the matrices K, H, $\bar{L}$ in (12), which remain however of the form (13). This leads to a certain arbitrariness in (12) (13) and (14), which could be removed by fixing the explicit time dependence everywhere, but since we shall see that the precise nature of the expansion in (12), and the time dependence in (14) is unimportant, we shall not seek to do this. From (12)(with $\bar{L} \to L$) we determine $\delta W$

$$\begin{aligned}\delta W_{\mu\nu}(k,\ell) &= \delta_{k\ell}[K_{\mu\nu}(k) - \frac{1}{\varepsilon}(H^{(s)}_{\mu\nu}(k+1) - L_{\mu\nu}(k) - L_{\mu\nu}(k+1))] \\ &+ \delta_{k,\ell-1} \frac{1}{\varepsilon}[\frac{1}{2} H_{\mu\nu}(k+1) - L_{\mu\nu}(k+1)] \\ &+ \delta_{k,\ell+1} \frac{1}{\varepsilon}[\frac{1}{2} H_{\nu\mu}(k) - L_{\mu\nu}(k)]\end{aligned} \tag{15}$$

We can now calculate $Y = \delta W \bar{G}$ which we need in (11). Using (15) one obtains

$$\begin{aligned}Y^\rho_\mu(k,j) &= Y^\rho_{(1)\mu}(k,j) - \frac{1}{\varepsilon}[L_{\mu\nu}(k+1)(\bar{G}^{\nu\rho}(k+1,j) - \bar{G}^{\nu\rho}(k,j)) - L_{\mu\nu}(k) \\ &(\bar{G}^{\nu\rho}(k,j) - \bar{G}^{\nu\rho}(k-1,j))]\end{aligned} \tag{16}$$

where in $Y_{(1)}$ we have included all the contributions containing K and H. Studying each of the terms in (16) one can see that Y is of the form

$$Y^\rho_\mu(k,j) = \varepsilon\delta_{kj} C^\rho_\mu(j) + \varepsilon^2 D^\rho_\mu(k,j) \tag{17}$$

where $D = \sum_{n\geqslant 0} \varepsilon^n D_{(n)}$ and we say that these last terms are of $\mathcal{O}(\varepsilon^2)$. All the terms in $Y_{(1)}$ are of $\mathcal{O}(\varepsilon^2)$ while the last part of (16) determines $C^\rho_\mu(j)$. This is clear because the dominant contribution there is $-\varepsilon \frac{\partial}{\partial t} L_{\mu\nu}(t) \frac{\partial}{\partial t} \bar{G}^{\nu\rho}(t,t')$ and since

$$\frac{d^2}{dt^2}\bar{G}^{\nu\rho}(t,t') \sim -cg^{\nu\rho}(t) \quad \delta(t-t').$$

To show this in more detail, we follow the exact procedure for these terms. We have

$$-\frac{1}{\varepsilon}(L_{\mu\nu}(k+1)[\bar{G}^{\nu\rho}(k+1,j) - \bar{G}^{\nu\rho}(k,j)] - L_{\mu\nu}(k)[\bar{G}^{\nu\rho}(k,j) - \bar{G}^{\nu\rho}(k-1,j)])$$

$$= -(\dot{L}_{\mu\nu}(k) + \frac{\varepsilon}{2}\ddot{L}_{\mu\nu}(k) + \ldots)[\bar{G}^{\nu\rho}(k+1,j) - \bar{G}^{\nu\rho}(k,j)]$$

$$-\frac{1}{\varepsilon}L_{\mu\nu}(k)[\bar{G}^{\nu\rho}(k+1,j) - 2\bar{G}^{\nu\rho}(k,j) + \bar{G}^{\nu\rho}(k-1,j)]$$

The first part is $\mathcal{O}(\varepsilon^2)$ since $L = \mathcal{O}(\varepsilon)$ and $\bar{G}^{\nu\rho}(k+1,j) - \bar{G}^{\nu\rho}(k,j) = \mathcal{O}(\varepsilon)$. In the second part we distinguish between $k = j$ and $k \neq j$. If $k \neq j$, we develop $\bar{G}^{\nu\rho}(k+1,j)$ and $\bar{G}^{\nu\rho}(k-1,j)$ around $\bar{G}^{\nu\rho}(k,j)$, giving

$$-\frac{2}{\varepsilon}L_{\mu\nu}(k)[\frac{\varepsilon^2}{2}\ddot{G}^{\nu\rho}(k,j) + \frac{\varepsilon^4}{4}\overset{....}{G}(k,j) + \ldots]$$

which is clearly of $\mathcal{O}(\varepsilon^2)$, while for $k = j$, we use the form (A.13), and [App 9.1, (15, 26)],

$$-\frac{c}{\varepsilon}L_{\mu\nu}(j)\{[\bar{J}(j+1,T)M(T,t_o)\bar{J}(t_o,j) - \bar{J}(j,T)M(T,t_o)\bar{J}(t_o,j)]^{\nu\rho}$$

$$+ [\bar{J}(j,t_o)M(t_o,T)\bar{J}(T,j) - \bar{J}(j-1,t_o)M(t_o,T)\bar{J}(T,j)]^{\nu\rho}\}$$

$$= -c\, L_{\mu\nu}(j)\{\frac{\partial}{\partial t}\bar{J}(t,j)^{\nu\rho}\Big|_{t=j} + \mathcal{O}(\varepsilon)\}$$

$$= c\, L_{\mu\nu}(j)g^{\nu\rho}(j) + \mathcal{O}(\varepsilon^2)$$

This gives $C^{\rho}_{\mu}(j)$ in (17) as $c\, L^{(0)}_{\mu\nu}(j)\, g^{\nu\rho}(j)$, or

$$C^{\rho}_{\mu}(j) = y\dot{x}^{\gamma}_{j}\partial_{\gamma}g_{\mu\nu}(j)g^{\nu\rho}(j) \tag{18}$$

while all other terms are of order $\varepsilon^2$. In the same way one easily shows that $Y^n$ for $n\geqslant 2$ contains only terms of order $\varepsilon^2$, and we obtain

$$\delta G^{\mu\nu}(i,j) = -\varepsilon\bar{G}^{\mu\alpha}(i,j)\, C^{\nu}_{\alpha}(j) - \varepsilon^2\sum_{k=1}^{N}\bar{G}^{\mu\alpha}(i,k)\,\bar{D}^{\nu}_{\alpha}(k,j) \tag{19}$$

with $\bar{D} = \bar{D}_{(0)} + \varepsilon\bar{D}_1 + \varepsilon^2\bar{D}_2 + \ldots$ . The last term has also a part of order $\varepsilon$, since when $N \to \infty$, tends to $\varepsilon\int dt'\bar{G}^{\mu\alpha}(t,t')\,\bar{D}^{\alpha\nu}(t',t'')$, but we shall see that it does not contribute to the equal time contractions. $\nabla_t q\, \nabla_t q$.

We only need the limiting value (for $N \to \infty$) of the normalization factor in (6), and its value was given in (8), obtained by the methods used in (IV.C). Another method can be used however, based on the form given in (6), and this is a good opportunity to present it. Using $\det G_u = \det[\bar{G}(1+\bar{Y})] = \det(\bar{G})\exp\{\mathrm{tr}\ln(1+\bar{Y})\}$ where $\bar{Y}$ can be read from (19), we see that only the first term contributes when $N\to\infty$, and one obtains, replacing $C^{\nu}_{\alpha}(j)$ by its value (18)

$$\lim_{N\to\infty}\det[\bar{G}^{-1}G_u] = \lim_{N\to\infty}\exp\{\mathrm{tr}(-y\,\varepsilon\sum_{j=1}^{N}\dot{x}^{\gamma}_{j}\partial_{\gamma}g_{\mu\nu}g^{\nu\rho})\}$$

$$= \exp\{-y\int_{t_o}^{T} dt\,\frac{d}{dt}\ln(g(t))\} = \left(\frac{g(X_o)}{g(X)}\right)^{y} \tag{20}$$

For the main part of the proof, where it is shown that

$$\lim_{N\to\infty}(2\pi)^{-\frac{d}{2}}\prod_{J=1}^{N+1}\left(\frac{g(x_{j-1})}{(\varepsilon c)^{d}}\right)^{1/2}(\det\bar{G}^{\mu\nu}(\ell,k))^{\frac{1}{2}} = \bar{N}^{(\frac{1}{2})} \tag{21}$$

we refer to [App.10.], (21) and (20) lead again to (8).

More important is the computation of the contractions in the discrete. We define

$$\mathcal{D}_{(u)} = \sum_{\ell,k}\frac{\partial}{\partial q^{\mu}_{\ell}}(\bar{G}^{\mu\nu}(\ell,k) + \delta G^{\mu\nu}(\ell,k))\frac{\partial}{\partial q^{\nu}_{k}} \tag{22}$$

which we write as $\mathcal{D}_{(u)} = \bar{\mathcal{D}} + \mathcal{D}$ where $\bar{\mathcal{D}}$ is the term with $\bar{G}$ and $\mathcal{D}$ the one with $\delta G$. We first compute the $\overbrace{\dot{q}^{\mu}(j)\;q^{\nu}(j)}$ contraction. Since $\delta G$ plays no role here

$$\overbrace{\dot{q}^{\mu}(j)\;q^{\nu}(j)}\Big|_{\gamma_1(1)} = \bar{\mathcal{D}}\,\frac{\Delta q^{\mu}_{j}}{\varepsilon}\,q^{\nu}_{j} = \frac{1}{\varepsilon}(\bar{G}^{\mu\nu}(j.j) - \bar{G}^{\mu\nu}(j-1,j)) \tag{23}$$

and putting $\bar{S}^{\mu\nu}(t,t') = \frac{\partial}{\partial t}\bar{G}^{\mu\nu}(t,t')$ we have from (A.13), up to order $\varepsilon$

$$\overbrace{\dot{q}^{\mu}(j)\;q^{\nu}(j)} = -c\,[\dot{\bar{J}}(j,t_o)M(t_o,T)\bar{J}(T,j)]^{\mu\nu} = \bar{S}^{\mu\nu}(t,t+\delta)\Big|_{\delta\to 0^{+}} \tag{24}$$

In $\gamma_1(u)$ discretization, the equal time contraction is $\bar{\mathcal{D}}\,\frac{\Delta q^{\mu}_{j}}{\varepsilon}\,q^{\nu(u)}_{j-1}$ and since

$$\bar{\mathcal{D}}\,\frac{\Delta q^{\mu}_{j}}{\varepsilon}\,\Delta q^{\nu}_{j} = \bar{S}^{\mu\nu}(t,t+\delta) - \bar{S}^{\mu\nu}(t,t-\delta)\Big|_{\delta\to 0^{+}} = c\,g^{\mu\nu}(t) \tag{25}$$

one obtains for $\overbrace{(\nabla_t q^{\mu})_j\;q^{\nu}(j)}\Big|_{\gamma_1(u)} = \bar{\mathcal{D}}\left(\frac{\Delta q^{\mu}_{j}}{\varepsilon} + \dot{x}^{\gamma}_{j}\,\Gamma^{\mu}_{\gamma\alpha}(j)q^{(u)\alpha}_{j-1}\right)q^{\nu(u)}_{j-1}$ the expression

$$\overbrace{\nabla_t q^{\mu}q^{\nu}}(t)\Big|_{\gamma_1(u)} = S^{\mu\nu}(t,t+\varepsilon) + (u-1)c\,g^{\mu\nu}(x(t)) \tag{26}$$

which proves (A.20) putting $u = 1$.

We study now the contraction $\overbrace{\dot{q}^{\mu}(j)\ \dot{q}^{\nu}}(j) = (\bar{\mathcal{D}}+\mathcal{D})\frac{\Delta q^{\mu}_j}{\varepsilon}\ \frac{\Delta q^{\nu}_j}{\varepsilon}$, where we have to use both $\bar{\mathcal{D}}$ and $\mathcal{D}$. Using [app.9.1(26)] we find

$$\bar{\mathcal{D}}\ \frac{\Delta q^{\mu}_j}{\varepsilon}\ \frac{\Delta q^{\nu}_j}{\varepsilon} = \frac{c}{\varepsilon^2}\ [(\bar{J}(j,T) - \bar{J}(j-1,T))M(T,t_o)(\bar{J}(t_o,j) - \bar{J}(t_o,j-1))]^{\mu\nu}$$

$$+ \frac{c}{\varepsilon^2}\ \bar{J}^{\mu\nu}(j-1,j) \tag{27}$$

Developing the last part

$$\frac{d}{dt}\ \bar{J}^{\mu\nu}(t,j)\Big|_{t=j} = -g^{\mu\nu}(x_j)$$

and

$$\ddot{J}^{\mu\nu}(t,j)\Big|_{t=j} = g^{\mu\alpha}B_{\beta\alpha}g^{\beta\nu}$$

where the last equality is read from the Jacobi equation (IV.C.56) and B is given in (3). We only need the symmetric part (see A.19), and since $\partial_{\sigma}g^{\mu\rho} = -\Gamma^{\rho}_{\sigma\lambda}g^{\mu\lambda} - \Gamma^{\mu}_{\sigma\lambda}\ g^{\rho\lambda}$ we obtain

$$\frac{c}{\varepsilon^2}\ [\bar{J}^{\mu\nu}(j-1,j)]^s = \frac{1}{\varepsilon}\ g^{\mu\nu}(x_j) - \frac{1}{2}\ \dot{x}^{\gamma}_j\partial_{\gamma}g^{\mu\nu}(x_j) = \frac{1}{\varepsilon}g^{\mu\nu}(\bar{x}_{j-1}) \tag{28}$$

giving

$$\bar{\mathcal{D}}\ \frac{\Delta q^{\mu}_j}{\varepsilon}\ \frac{\Delta q^{\nu}_j}{\varepsilon} = c[\dot{\bar{J}}\ (j,T)M(T,t_o)\dot{\bar{J}}(t_o,j)]^{\mu\nu} + \frac{c}{\varepsilon}\ g^{\mu\nu}(\bar{x}_{j-1}) \tag{29}$$

Using the previous result (26) we then obtain

$$\bar{\mathcal{D}}\ (\nabla_t q^{\mu})_j(\nabla_t q^{\nu})_j\Big|_{\gamma_1(u)} = c[\nabla_t\bar{J}(j,T)M(T,t_o)\nabla_t\bar{J}(t_o,j)]^{\mu\nu} + \frac{c}{\varepsilon}\ g^{\mu\nu}(\bar{x}_{j-1}) -$$

$$- y\ \dot{x}^{\gamma}\partial_{\gamma}g^{\mu\nu}(x_j) \tag{30}$$

We have to calculate now the contribution of $\delta G$ which is $\mathcal{D}(\nabla_t q^{\mu})_j(\nabla_t q^{\nu})_j = \mathcal{D}\ \frac{\Delta q^{\mu}_j}{\varepsilon}\ \frac{\Delta q^{\nu}_j}{\varepsilon}$

The first part of (19) contributes as (see (25),(18))

$$\mathcal{D}\ \frac{\Delta q^{\mu}_j}{\varepsilon}\ \frac{\Delta q^{\nu}_j}{\varepsilon} = -\ C^{\nu}_{\alpha}(j)\ [\bar{S}^{\mu\alpha}(j,j+\varepsilon) - \bar{S}^{\mu\alpha}(j,j-\varepsilon)] = -\ c\ C^{\nu}_{\alpha}(j)g^{\mu\alpha}(j)$$

$$= yc\ \dot{x}^{\gamma}_j\partial_{\gamma}g^{\mu\nu}(j) \tag{31}$$

and cancels the last noncovariant term in (30). The other part of (19) leads to

$$\sum_k \varepsilon \frac{[G^{\mu\alpha}(i,k) - G^{\mu\alpha}(i-1,k)]}{\varepsilon} [\bar{D}^{\nu}_{\alpha}(k,j) - \bar{D}^{\nu}_{\alpha}(k,j-1)]$$

and is of order $\varepsilon$, as we have mentioned before. The final result is then that, neglecting terms of $\mathcal{O}(\varepsilon)$,

$$(\nabla_t q^{\mu})_j (\nabla_t q^{\nu})_j = c[\nabla_t \bar{J}(j,T)M(T,t_o)\nabla_t \bar{J}(t_o,j)]^{\mu\nu} + \frac{c}{\varepsilon} g^{\mu\nu}(\bar{x}_{j-1}) \tag{32}$$

irrespective of u in the $\gamma_1(u)$ discretization.

This proves (A.21) since we shall see in the next section that the term $\frac{c}{\varepsilon} g^{\mu\nu}(\bar{x}_{j-1})$ (if it appears as $c\delta(0)g^{\mu\nu}(t)$ in the continuum, i.e. without sum) can be dropped. The fact that (32) is the naive result is accidental, and comes from the fact that $B^{(s)}_{\mu\nu}$ is here determined by the first term in (3). If $B^{(s)}_{\mu\nu}$ has an extra contribution, it will also appear in (32). We study first the effect of using $\tilde{g}^{\star}_{\mu\nu}(\sqrt{\eta}q_j, x^{(v)}_{j-1})$ in the dominating exponential of (A.1), in order to see the most general discretization effects that can occur in a perturbation expansion. This effect is simple: the free generating functional $Z^N_o[J]$ (A.8) has now $g_{\mu\nu}(x^{(v)}_{j-1})$ in the exponential and $g(x^{(v)}_{j-1})$ in the measure. Since $(y_1 = v - \frac{1}{2})$

$$\frac{1}{\varepsilon} (\Delta^{\mu}_j g_{\mu\nu}(x^{(v)}_{j-1})\Delta^{\nu}_j) = \frac{1}{\varepsilon} (\Delta^{\mu}_j g_{\mu\nu}(\bar{x}_{j-1})\Delta^{\nu}_j + \Delta^{\mu}_j\, y_1 \dot{x}^{\rho}_j \partial_{\rho} g_{\mu\nu}(j)\Delta^{\nu}_j)$$

we obtain an extra term in $L^{(o)}_{\mu\nu}$ and hence in $C^{\rho}_{\mu}(j)$. The correction term is now (see 31) $(y + y_1)c\dot{x}^{\rho}_j \partial_{\rho} g^{\mu\nu}(j)$ which can be incorporated in (32) writing $\frac{c}{\varepsilon} g_{\mu\nu}(x^{(v)}_{j-1})$, and we see that this has to be taken into account when studying the cancellation of the terms proportional to $\delta(0)$. That the normalization factor remains unchanged can also be checked; doing a calculation similar to the one in (20),(21).

The exponential in (A.1) can come in several ways. To illustrate this we add there a vanishing term ($Z_{\mu\nu}(t)$ is symmetric in $\mu\nu$)

$$-\frac{1}{2c} \int_{t_o}^{T} dt^{\gamma_1(w)} q^{\mu} Z_{\mu\nu}(t)\dot{q}^{\nu}(t) - \frac{1}{4c} \int_{t_o}^{T} dt q^{\mu} q^{\nu} \dot{Z}_{\mu\nu}(t) +$$

$$\frac{y_2}{2} \int dt \tilde{g}^{\star\mu\nu}[\sqrt{\eta}q, x(t)] Z_{\mu\nu}(t) \tag{33}$$

$\gamma_1(w)$ stands as usual for the discretization $q^{\gamma_1(w)}_{j-1} = \bar{q}_{j-1} + y_2 \Delta q_j$ $(y_2 = w - \frac{1}{2})$, and (33) is proved by partial integration of the first term [app.4.3] and use of the replacements $\Delta q^{\mu}_j \Delta q^{\nu}_j \doteq \varepsilon c \tilde{g}^{\star\mu\nu}(\sqrt{\eta}q, x_j)$. We include $\frac{y_2}{2} \int dt\ \tilde{g}^{\star(2)\mu\nu}(\sqrt{\eta}q, x(t))Z_{\mu\nu}(t)$ in the interaction Lagrangian, while the rest will be added to the free generating functional $T^{(1)}_N$ (2). These terms

$$- \frac{1}{c}[\int dt^{\gamma}{}_{1}{}^{(w)} \frac{1}{2} q^{\mu}\dot{q}^{\nu} Z_{\mu\nu}(t) + \frac{1}{4}\int dt q^{\mu}q^{\nu}\dot{Z}_{\mu\nu}(t) + \int dt \frac{y_2}{2} g^{\mu\nu}(x(t))Z_{\mu\nu}(t)] \quad (34)$$

do not change the propagator $\bar{G}^{\mu\nu}(t,t')$ which is obtained for $N \to \infty$, and we expect a modification of the equal time contractions $\nabla_t q^{\mu} \nabla_t q^{\nu}$ which should cancel the extra interaction terms.

$$\varepsilon \sum_{j=1}^{N+1} (\frac{1}{2} q_j^{\mu(1/2)} \frac{\Delta q_j^{\nu}}{\varepsilon} Z_{\mu\nu}(j) + \frac{1}{4} q_j^{\mu} q_j^{\nu} \dot{Z}_{\nu j}) = \varepsilon \sum_{1}^{N} \frac{1}{4} q_j^{\mu} q_j^{\nu} (\dot{Z}_{\mu\nu}(j) - \frac{[Z_{\mu\nu}(j+1) - Z_{\mu\nu}(j)]}{\varepsilon}) \quad (35)$$

does not change $W^{(1)}$ in (4) effectively (as far as $\nabla_t q \nabla_t q$ contractions are concerned) so that the only effect changing $L^{o}_{\mu\nu}$ (and $C^{\nu}_{\mu}$) is due to

$$- \frac{1}{c} \sum_{j=1}^{N+1} (\frac{y_2}{2} \Delta q_j^{\mu} Z_{\mu\nu}(j) \Delta q_j^{\nu}) \quad (36)$$

giving a modified $\nabla_t q^{\mu} \nabla_t q^{\nu}$ contraction

$$\delta(\nabla_t q_j^{\mu} \nabla_t q_j^{\nu}) = - c y_2 g^{\mu\alpha} Z_{\alpha\beta}(j) g^{\beta\nu}(j) \quad (37)$$

It is clear that the normalization factor remains the same. This can be shown by combining the last term in (34) and the correction to $G^{(1)}$ (compare 20) responsible for (37). The conclusion is then that the modification of equal time contraction given in (37) is equivalent to the addition of a new potential term. $\Delta\mathcal{L}_I = - \frac{y_2}{2} \int dt \tilde{g}^{\star\mu\nu(2)}(\sqrt{\eta} q(t), x(t)) Z_{\mu\nu}(t)$, the interaction. This can also be shown explicitly in perturbation theory, using the same methods as in appendix 10.2. From (33) it is clear that the modified rules (37) can always be avoided, by bringing any term of the form $- \frac{1}{2c} \int_{t_o}^{T} dt^{\gamma_1(w)} q^{\mu} Z_{\mu\nu}(t) \dot{q}^{\nu}(t)$ to the midpoint. To finish this section we repeat the essential point in a formal way. In a free generating functional of the form

$$\int_{o}^{o} \mathcal{D}q (\frac{g(t)}{c^d})^{1/2} \exp\{- \frac{1}{2c} \int dt\, \dot{q}^{\mu} g_{\mu\nu}(t) \dot{q}^{\nu} + \dot{q}^{\mu} B_{\mu\nu} q^{\nu} + C_{\mu\nu} q^{\mu} q^{\nu} + i \int dt J_{\mu} q^{\mu}\} \quad (38)$$

one must specify the discretization of the crossterm, while the discretization of the explicit time dependence in $g_{\mu\nu}(t)$ is arbitrary, provided one uses the same discretization in $g(t)$. Partially integrating and symmetrizing, we obtain the exponential in the form

$$-\frac{1}{2c}\int dt q^{\mu}(t)\ \{-\frac{d}{dt}g_{\mu\nu}\frac{d}{dt}-\frac{1}{2}(B_{\mu\nu}-B_{\nu\mu})\frac{d}{dt}+C_{\mu\nu}-\frac{1}{2}\frac{d}{dt}(B_{\mu\nu})\}q^{\nu}(t) \quad (39)$$

where the self-adjoint Jacobi operator (with respect to zero boundary conditions) corresponds to the symmetrized matrix in the discrete. $\bar{G}^{\mu\nu}(t,t')$ is the inverse, or Green's function of the operator in (39). If one realizes that $\bar{G}^{\mu\nu}(t,t')$ corresponds in some sense [app.10.1] to the midpoint discretization, we can study corrections to the midpoint by taking $g_{\mu\nu}(t^{w}_{j-1})$ and $q^{v}_{j-1}$. These corrections have the form $\varepsilon\Sigma\frac{\Delta^{\mu}}{\varepsilon}U_{\mu\nu}\Delta^{\nu}$, or formally $\varepsilon\int\dot{q}^{\mu}U_{\mu\nu}(t)\dot{q}^{\nu}(t)$, or more precisely

$$\frac{\varepsilon}{2c}\int\dot{q}^{\mu}\{(w-\frac{1}{2})\frac{d}{dt}g_{\mu\nu}+(v-\frac{1}{2})\frac{d}{dt}B^{(s)}_{\mu\nu}\}\dot{q}^{\nu}=\varepsilon\int\dot{q}^{\mu}U_{\mu\nu}(t)\dot{q}^{\nu} \quad (40)$$

giving the correction $-\varepsilon\frac{d}{dt}U_{\mu\nu}\frac{d}{dt}$ to the operator in (39). The correction to the Green's function is then in order $\varepsilon$

$$\delta G(t,\ t')=-\bar{G}\delta W\bar{G}=\varepsilon\int d\tau\bar{G}^{\mu\nu}(t,\tau)\frac{d}{d\tau}U_{\nu\lambda}(\tau)\frac{d}{d\tau}\bar{G}^{\lambda\rho}(\tau,t') \quad (41)$$

Only the part $U_{\nu\lambda}(\tau)\frac{d^2}{d\tau^2}\bar{G}^{\lambda\rho}(\tau,t')$ is relevant, since it yields a $\delta$-function $(-cg^{\lambda\rho}(\tau)\delta(\tau-t'))$ and leads to

$$-c\varepsilon\bar{G}^{\mu\nu}(t,t')\ U_{\nu\lambda}(t')g^{\lambda\rho}(t') \quad (42)$$

In a contraction $\dot{q}(t)\ \dot{q}(t')$ only such a term can give the $\delta(0)=\frac{1}{\varepsilon}$ at equal times to make the result finite. Indeed, $\frac{d}{dt}\frac{d}{dt'}\bar{G}^{\mu\nu}(t,t')\rightarrow cg^{\mu\nu}(t)\delta(t-t')$ leads to

$$\delta G(t,t')=-\varepsilon c\delta(t-t')g^{\mu\nu}(t)U_{\nu\lambda}(t)g^{\lambda\rho}(t) \quad (43)$$

while all other terms remain of order $\varepsilon$. This summarizes in a formal way one of the essential points that we explained in this section. The terms of the form $\varepsilon\Delta^{\mu}\Delta^{\nu}$, of which we knew already that they contributed to the normalization factor of the free generating functional, also have to be taken into account, if equal time contractions $\dot{q}^{\mu}\dot{q}^{\nu}$ have to be evaluated, although these modify the Green's function $\bar{G}^{\mu\nu}(t,t')$ only in terms of order $\varepsilon$.

C. Cancellation of the singularities

The perturbation expansion generated by (A.7) is of the form

$$\tilde{I}_N = \prod_{j=1}^{N+1} \left(\frac{\tilde{g}^\star(\sqrt{\eta}q_j,\bar{x}_{j-1})}{g(\bar{x}_{j-1})}\right)^{\frac{1}{2}} \exp\{-\frac{\varepsilon}{c}\sum_{j=1}^{N+1}[\tfrac{1}{2}(\nabla_t q^\sigma)_j \tilde{g}^{\star(2)}_{\sigma\rho}(q_j,\bar{x}_{j-1})(\nabla_t q^\rho)_j +$$

$$(\nabla_t q^\sigma)_j \tilde{D}_\sigma(q_j,x_j) + V_1(q_j,x_j)]\Big|_{q^\mu_j=\frac{1}{i\varepsilon}\frac{\partial}{\partial J_{j\mu}}}\} Z^N_o[J]\Big|_{J=0} \qquad (1)$$

Using the results for the contractions of the previous section, we can write this effectively as

$$\tilde{I}_N = \bar{N}^{(1)}_\varepsilon \prod_{j=1}^{N+1} \left(\frac{\tilde{g}^\star(\sqrt{\eta}q_j,\bar{x}_{j-1})}{g(\bar{x}_{j-1})}\right)^{\frac{1}{2}} \exp\{-\frac{\varepsilon}{c}\sum_{j=1}^{N+1}\frac{1}{2}(Q^\sigma_j+p^\sigma_j)\tilde{g}^{\star(2)}_{\sigma\rho}(Q^\rho_j+p^\rho_j) +$$

$$+ (Q^\sigma_j+p^\sigma_j)\tilde{D}_\sigma(q_j,x_j) + V_1(q_j,x_j)\}\Big|_{\substack{Q^\sigma_j=\frac{1}{i\varepsilon}\frac{\partial}{\partial K_{j\sigma}}\\ p^\sigma_j=\frac{1}{i\varepsilon}\frac{\partial}{\partial J^\star_{j\sigma}}\\ q^\sigma_j=\frac{1}{i\varepsilon}\frac{\partial}{\partial J_{j\sigma}}}} \bar{Z}_o[J^\star,J,K]\Big|_{J^\star=J=K=0} \qquad (2)$$

$$\text{with } \bar{Z}_o[J^\star,J,K] = \exp\{-\frac{\varepsilon}{2}\sum_{j=1}^{N+1} K_{j\mu}cg^{\mu\nu}(\bar{x}_{j-1})K_{j\nu}\}\bar{Z}_o[J,J^\star] \qquad (3)$$

$$\text{and } \bar{Z}_o[J,J^\star] = \exp\{-\varepsilon^2\sum_{j,k}\frac{1}{2}J_{j\mu}(\overbrace{q^\mu_j q^\nu_k})J_{k\nu} + J^\star_{j\mu}(\overbrace{\nabla_t q^\mu_j q^\nu_k})J_{k\nu} + \frac{1}{2}J^\star_{j\mu}[\overbrace{\nabla_t q^\mu_j \nabla_t q^\nu_k} -$$

$$cg^{\mu\nu}(\bar{x}_{j-1})\frac{\delta_{jk}}{\varepsilon}]J^\star_{k\nu}\} \qquad (4)$$

where the contractions in (3) have the meaning of the exact contractions, derived from the discretized $Z^N_o[J]$, and which were evaluated in (B) for $\varepsilon \to 0$. Introducing an integration over the $Q^\sigma_j$, formula (2) can be written as

$$\tilde{I}_N = \bar{N}^{(1)}_\varepsilon \prod_{j=1}^{N+1}\left(\frac{\tilde{g}^\star(\sqrt{\eta}q_j,\bar{x}_{j-1})}{c^d}\right)^{1/2}\int \prod_1^{N+1}(\frac{\varepsilon}{2\pi})^d dQ_j \exp\{-\frac{\varepsilon}{c}\sum_1^{N+1}[\tfrac{1}{2}Q^\sigma_j\tilde{g}^\star_{\sigma\rho}$$

$$(\sqrt{\eta}q_j,\bar{x}_{j-1})Q^\rho_j + Q^\sigma_j B_\sigma(q_j,p_j) + V_2(q_j,p_j)]\}\Big|_{\substack{p^\mu_j=\frac{1}{i\varepsilon}\frac{\partial}{\partial J^\star_{j\mu}}\\ q^\mu_j=\frac{1}{i\varepsilon}\frac{\partial}{\partial J_{j\mu}}}} \bar{Z}_o[J,J^\star]\Big|_{J=J^\star=0} \qquad (5)$$

What we did was the reversed procedure of setting up the perturbation expansion, expressing the K dependent part of $\bar{Z}_o[J^\star,J,K]$ as a functional integral, and taking the interaction Lagrangian in the functional integral again, putting $K = 0$. B and $V_2$ can be read from (2). Doing the Q-integration in (5) we finally obtain

$$I_N = \bar{N}_\varepsilon^{(1)} \exp\{\frac{\varepsilon}{c}\sum_{j=1}^{N+1}[\frac{1}{2}B_\sigma(q_j,p_j)\tilde{g}^{\star\sigma\rho}B_\rho(q_j,p_j) - V_2(q_j,p_j)]\}\Bigg|_{\substack{q_j^\mu=\frac{1}{i\varepsilon}\frac{\partial}{\partial J_{j\mu}}\\ p_j^\mu=\frac{1}{i\varepsilon}\frac{\partial}{\partial J^\star_{j\mu}}}} \bar{Z}_o[J,J^\star]\Big|_{J=J^\star=0} \tag{6}$$

We have now two expressions for $\tilde{I}_N$, (2) and (6), that we want to compare. If in (2), we write the preexponential factor in the form

$$\prod_{j=1}^{N+1}\left(\frac{\tilde{g}^\star(\sqrt{\eta}q_j,\bar{x}_{j-1})}{g(\bar{x}_{j-1})}\right)^{\frac{1}{2}} = \exp\{-\frac{1}{2\varepsilon}\sum_{j=1}^{N+1}\varepsilon\ln\left(\frac{\tilde{g}^\star(\sqrt{\eta}q_j,\bar{x}_{j-1})}{g(\bar{x}_{j-1})}\right) \tag{7}$$

we can see that any term in the perturbation expansion will be of the form

$$\sum_{n\geqslant 1}\left(\frac{1}{\varepsilon}\right)^n A_n + B_n \tag{8}$$

where $A_n$ and $B_n$ are discretized expressions, functions of the contractions given in (4), and where $A_n$, $B_n$ have a finite limit for $N \to \infty$. On the other hand, if we compute $\tilde{I}_N$ from (6), we only find terms of the type $B_n$, i.e. all singular terms have cancelled (or equivalently each $A_n$ vanishes). Practically, this means that we can calculate $I_N$ from (2) with the prescription that each time we find a term of the singular form $\varepsilon^{-n}A_n$, it should be dropped, because finally they cancel, and since all terms coming from (7) are of this form, these can be ignored. Moreover the cancellation of these singular terms allows us to drop in the remaining terms, of the form B, the corrections of order $\varepsilon$, since they will play no role in the limit $N \to \infty$.

The reasoning we followed was general, and was based on the fact that one can write the perturbation expansion in a different way, without singularities. In fact, we should have been more careful in the specification of the terms $A_n$ and $B_n$, since the $A_n$ (being discretized expressions) contain parts of $\mathcal{O}(\varepsilon^n)$ making that part finite. That is why we defined the $A_n$, $B_n$ as functions of the complete discretized contractions (which is a series in $\varepsilon$). Moreover one should specify that in going from (2) to (6), one must not touch the explicit time discretization, which can be taken rather arbitrarily in (2), since one could think of expanding

$x_{j-1}$ around $x_j$ in some term of the form $A_n$ leading again to $\varepsilon^n$. With this specification it is clear that this proof is then valid for any time discretization we chose in $\tilde{g}^{\star}_{\mu\nu}(x^{\nu}_{j-1})$ and $\tilde{g}^{\star}(x^{\nu}_{j-1})$. In app.10.2 we shall give another proof of this cancellation, where it is seen how the singular parts of the $\nabla_t q \nabla_t q$ contractions sum up to cancel (7).

We conclude then by repeating the statements of (A), i.e. $\bar{I}$ can be computed from (1) working in the continuum, using $\mathcal{L}_I$(A.5) as interaction Lagrangian, and ignoring the measure, using the contractions imposed by the $\gamma_1$(1) discretization.

$$\overbrace{q^{\mu}(t)\ q}^{}{}^{\nu}(t) = \bar{G}^{\mu\nu}(t,t') = c\{\theta(t-t')\bar{J}(t,T)M(T,t_o)\bar{J}(t_o,t') - \theta(t'-t) \bar{J}(t,t_o)M(t_o,T)\bar{J}(T,t')\}^{\mu\nu} \tag{9}$$

$$\nabla_t \overbrace{q^{\mu}(t)\ q}^{}{}^{\nu}(t') = \nabla_t\bar{G}^{\mu\nu}(t,t') = c\{\theta(t-t')\nabla_t\bar{J}(t,T)M(T,t_o)\bar{J}(t_o,t') - \theta(t'-t) \nabla_t\bar{J}(t,t_o)M(t_o,T)\bar{J}(T,t')\}^{\mu\nu} \tag{10}$$

$$\nabla_t \overbrace{q^{\mu}(t)\ q}^{}{}^{\nu}(t) = \nabla_t\bar{G}^{\mu\nu}(t,t+\delta)\Big|_{\delta\to 0^+} = -c[\nabla_t\bar{J}(t,t_o)M(t_o,T)\bar{J}(T,t)]^{\mu\nu} \tag{11}$$

$$\nabla_t \overbrace{q^{\mu}(t)\ \nabla_{t'} q}^{}{}^{\nu}(t') = c\{\theta(t-t')\nabla_t\bar{J}(t,T)M(T,t_o)\nabla_{t'}\bar{J}(t_o,t') - \theta(t'-t)\nabla_t\bar{J}(t,t_o) M(t_o,T)\nabla_{t'}\bar{J}(T,t')\}^{\mu\nu} + cg^{\mu\nu}(t)\delta(t-t') \tag{12}$$

and dropping all terms in $\delta(0)$ coming from (12).

D. The first order correction and the short time propoagator

As in (9.C.4) we give here the explicit form of the interaction Lagrangian up to $\mathcal{O}(\eta)$ which we need to compute the first order correction to the WKB-approximation. If $\tilde{I} = \bar{N}^{(1)}(1 + \eta I^{(1)} + \mathcal{O}(\eta^2))$ then

$$I^{(1)} = [\int_{t_o}^{T} dt\, W_2(t) + \frac{1}{2}\int_{t_o}^{T}\int_{t_o}^{T} dtdt' W_1(t)W_1(t')]\,\bar{Z}_o[J]\Big|_{J=0,\; q^\mu = \frac{1}{i}\frac{\delta}{\delta J_\mu}} \quad (1)$$

where $W_1$ and $W_2$ can be computed from (A.1). The rules to be used have been given in (C).

Using the notations and the results of app.9.5 we have

$$W_1 = -\frac{1}{6c}[<\dot{x},R'(\dot{x},q)q> + 4<R(\dot{x},q)q,\nabla_t q> - <R(\dot{x},q)q,F(q)> -2<\nabla_t q,F'(q)> - <\dot{x}, F''(q)> - V''(q)] - \frac{1}{3}[\frac{1}{2}\nabla_\alpha F^\alpha(q) + <\dot{x},\mathcal{R}(q)>] \quad (2)$$

$$W_2 = -\frac{1}{24c}[<R''(x,q)q,\dot{x}> + 6<R'(\dot{x},q)q,\nabla_t q> + 4<R(\nabla_t q,q)q,\nabla_t q> + 4<R(\dot{x},q)q, ,R(\dot{x},q)q> -<R'(\dot{x},q)q,F(q)> - <R(\nabla_t q,q)q,F(q)> - 3<R(\dot{x},q)q, F'(q)> - - 3<\nabla_t q,F''(q)> - <\dot{x},F'''(q)> - V''''(q)] - [\frac{1}{24}<\mathcal{R}(q),F(q)> + \frac{1}{4}<\dot{x},\mathcal{R}'(q)> - \frac{1}{24}<q,\nabla_t\mathcal{R}(q)> + \frac{1}{3}<\nabla_t q,\mathcal{R}(q)> + \frac{1}{8}(\nabla_\alpha F^\alpha(q))''] \quad (3)$$

In (9.E.1) we computed the contribution of $I^{(1)}$ to the short time propagator $I(X,\varepsilon,X_o,0)$ developed up to $\mathcal{O}(\varepsilon)$. The same procedure can be followed here. It consists of a scaling of the time, the computation of the dominant contribution to the propagator, ($t = \varepsilon u$)

$$\bar{G}^{\mu\nu}(u,u') = c\varepsilon[\theta(u-u')(1-u)u' + \theta(u'-u)(1-u')u]g^{\mu\nu}(X_o) \quad (4)$$

and an effective power counting. Here we have $q = \mathcal{O}(\varepsilon^{1/2})$ and $\nabla_t q = \mathcal{O}(\varepsilon^{-1/2})$. One can see then that only $W_2$ will contribute, and this with only two terms, which give

$$I^{(1)} \approx \varepsilon\int_o^1 du[-\frac{1}{6c}R_{\mu\nu\rho\sigma}(\dot{q}^\sigma q^\rho q^\nu \dot{q}^\mu + \dot{q}^\sigma q^\rho q^\nu \dot{q}^\mu) - \frac{1}{3}R_{\mu\nu}\dot{q}^\mu q^\nu] \quad (5)$$

This gives the same integrals as in (9.E.1.10) and the result

$$\eta I^{(1)} \approx -\frac{1}{12}\varepsilon\eta c\, R(X_o) \quad (6)$$

Appendix 10.1

In chapter 10 we used a special form of discretization for the argument in the exponential of the free generating functional. Some of its remarkable properties were given in [ D14 ], and we want to comment here on some aspects which were not given there.

Let $A(Q'',t'';Q',t') = \int_{t'}^{t''} d\tau L_{cl}(Q(t),\dot{Q}(t))$ be the classical action calculated for the classical trajectory Q(t) with boundary conditions $Q(t') = Q'$, $Q(t'') = Q''$. We shall take the classical Lagrangian, used in (V)

$$L_{cl} = \frac{1}{2} g_{\mu\nu}(Q)\, \dot{Q}^{\mu}\dot{Q}^{\nu} + \dot{Q}^{\mu}A_{\mu}(Q) - V(Q) \tag{1}$$

and we develop

$$\bar{A} = -\frac{1}{\eta c} \sum_{j=1}^{N+1} A(Q_j,t_j,Q_{j-1},t_{j-1}) \qquad Q_o = X_o,\ Q_{N+1} = X \tag{2}$$

which is to be considered as a function of the variables $Q_j$, $j = 1, \ldots N$ using $Q^{\mu}_j = x^{\mu}_j + \sqrt{\eta}\, q^{\mu}_j$, with x(t) the classical path having boundary conditions $x(t_o) = X_o$, $x(T) = X$. The terms quadratic in $q_j$ can be written as

$$A^{[2]} = -\frac{1}{2} \sum_{j,k=1}^{N} q^{\mu}_j \bar{W}_{\mu\nu}(j,k) q^{\nu}_k \tag{3}$$

Using the functions $\bar{J}(t,t')$, $M(t,t')$ [app. 9.1] corresponding to the Jacobi equation ( 9.C.58), it is possible to write $A^{[2]}$ as

$$A^{[2]} = -\frac{1}{2c} [\sum_{2}^{N} q^{\mu}_{k-1} q^{\nu}_k M_{\mu\nu}(k-1,k) - q^{\mu}_{k-1} q^{\nu}_k M_{\nu\mu}(k,k-1) + \sum_{1}^{N} q^{\mu}_k q^{\nu}_k M_{\mu\rho}(k,k-1)$$
$$\bar{J}^{\rho\sigma}(k-1,k) M_{\sigma\nu}(k+1,k)\,] \tag{4}$$

We used the shorthand notation $\bar{J}(j,k) = \bar{J}(t_j,t_k)$ and $t_j = t_o + j\varepsilon$. Formula (4) was given in [D14 ] and we refer to there for its proof. Remark however, that for the off-diagonal terms

$$(-\eta c) \left. \frac{\partial^2 A(x_i+q_i,t_i;x_{i-1}+q_{i-1},t_{i-1})}{\partial q^{\mu}_{i-1}\, \partial q^{\nu}_i} \right|_{q_i = q_{i-1} = o} = M_{\mu\nu}(i-1,i)$$

is precisely [App. 9.1(22)]

(4) gives the matrix elements of $\bar{W}_{\mu\nu}(j,k)$ as

$$\bar{W}_{\mu\nu}(k,k) = \frac{1}{c}\,(M_{\mu\rho}(k,k-1)\ \bar{J}^{\rho\sigma}(k-1,k+1)M_{\sigma\nu}(k+1,k))$$

$$\bar{W}_{\mu\nu}(k,k+1) = \frac{1}{c}\,(M_{\mu\nu}(k,k+1)) \tag{5}$$

$$\bar{W}_{\mu\nu}(k+1,k) = -\frac{1}{c}\,(M_{\mu\nu}(k+1,k))$$

Using this explicit form for $\bar{W}_{\mu\nu}(k,\ell)$, it is not so difficult to prove that we have exactly

$$\sum_{j=1}^{N}\ \sum_{\mu=1}^{d} \bar{G}\,(i,j)^{\alpha\mu}\bar{W}(j,k)_{\mu\lambda} = \delta_{i,k}\delta^{\alpha}_{\lambda} \tag{6}$$

where $\bar{G}$ is the Green's function of the Jacobi equation ( 9 .C.72) with zero boundary conditions

$$\bar{G}^{\mu\lambda}(t,t') = c\{\theta(t-t')\bar{J}^{\mu\sigma}(t,T)M_{\sigma\rho}(T,t_o)\bar{J}^{\rho\lambda}(t_o,t') - \theta(t'-t)\bar{J}^{\mu\sigma}(t,t_o)$$

$$M_{\sigma\rho}(t_o,T)\bar{J}^{\rho\lambda}(T,t')\} \tag{7}$$

We want to study now the connection between (4) and the discretized expressions that we use in our free generating functional. To see which discretization corresponds to (4), we can expand the matrix elements of $\bar{W}$ in powers of $\varepsilon$, using the boundary conditions of $\bar{J}^{\mu\nu}(t,t')$ and the Jacobi equation. But we shall use here the known expansion of $A(Q_k,Q_{k-1}) = \text{St.} \int_{t_{k-1}}^{t_k} L_{cl}dt$, which we have given in normal coordinates in (5.104) and which in general coordinates reads [L13]

$$A(Q_k,Q_{k-1}) = \frac{1}{\varepsilon}\,[\frac{1}{2}\,g_{\mu\nu}\Delta^{\mu}\Delta^{\nu} + \frac{1}{2}\,g_{\mu\nu}\Gamma^{\mu}_{\lambda\rho}\Delta^{\nu}\Delta^{\lambda}\Delta^{\rho} + (\frac{1}{12}\partial_{\mu}\partial_{\lambda}g_{\nu\rho} - \frac{1}{24}\,\Gamma^{\sigma}_{\mu\lambda}g_{\sigma\tau}\Gamma^{\tau}_{\nu\rho})$$

$$\Delta^{\mu}\Delta^{\lambda}\Delta^{\nu}\Delta^{\rho} + [A^{\mu}\Delta^{\mu} + \frac{1}{2}\,\partial_{\mu}A_{\nu}\Delta^{\mu}\Delta^{\nu}] - \varepsilon V + O(\frac{\Delta^5}{\varepsilon},\ \Delta^3,\ \varepsilon\Delta,\ \varepsilon^2) \tag{8}$$

where the coefficients are evaluated in $Q_{k-1}$ and $\Delta = Q_k - Q_{k-1}$. Expanding this expression in powers of $\bar{\Delta}_j = q_j - q_{j-1}$ and $q_{j-1}$, using $Q_{j-1} = x_{j-1} + q_{j-1}$ and $\Delta_j = \bar{\Delta}_j + \varepsilon\dot{x}_j + O(\varepsilon^2)$, we obtain the quadratic term

$$\frac{1}{2}\,\frac{\bar{\Delta}^{\mu}\bar{\Delta}^{\nu}}{\varepsilon}\ \ g_{\mu\nu}[\bar{x}_{j-1}] + (\frac{q^{\mu}_{j} + q^{\mu}_{j-1}}{2})\bar{\Delta}^{\nu}(\partial_{\mu}g_{\nu\lambda}\dot{x}^{\lambda} + \partial_{\mu}A_{\nu}) + \frac{\varepsilon}{2}\,q^{\mu}_{j-1}q^{\nu}_{j-1}$$

$$(\frac{1}{2}\partial_{\mu}\partial_{\nu}g_{\rho\sigma}\dot{x}^{\rho}\dot{x}^{\sigma} + \partial_{\mu}\partial_{\nu}A_{\sigma}\dot{x}^{\sigma} - \partial_{\mu}\partial_{\nu}V) \tag{9}$$

if we keep only the terms up to order $\varepsilon$, using that $\bar{\Delta} = \mathcal{O}(\sqrt{\varepsilon})$. We neglect in fact terms of the form $\bar{\Delta}\bar{\Delta}\,\varepsilon$, $\bar{\Delta}q\varepsilon, qq\varepsilon^2$, and higher order terms. It can easily be seen that the terms neglected in (8) also give no contribution to (9). It can also be rewritten in the more familiar covariant form (10.A(3))

$$\frac{1}{2}\nabla_t q^\mu g_{\mu\nu}(x)\nabla_t q^\nu - \nabla_t q^\mu F_{\mu\nu} q^\nu + q^\rho q^\nu (R_{\mu\nu\rho\sigma}\dot{x}^\mu \dot{x}^\sigma + \nabla_\rho F_{\nu\mu}\dot{x}^\mu - \nabla_\rho \partial_\nu V) \quad (10)$$

adding that this expression is to be discretized in the midpoint $\bar{\Delta}^\nu(\frac{q_j^\mu + q_{j-1}^\mu}{2})$ and that the explicit time dependence of $g_{\mu\nu}$ in the dominating term has to be taken in the midpoint $\frac{x_j + x_{j-1}}{2}$, or $x\,\frac{(t_j + t_{j-1})}{2}$ which is equivalent. This shows that this midpoint discretization differs from the one used in (3) (where one has the exact inverse of $\bar{G}^{\mu\lambda}(t_i, t_j)$ in terms of the form (10.B.(12)), a fact that was used there.

The structure of $\bar{W}$ and $\bar{G}$ can be used to obtain the determinant of $\bar{G}^{\mu\nu}(\ell,k)$. It is also evaluated in [D14] where one finds

$$\det \bar{G}^{\mu\nu}(\ell,k) = \frac{\det M(T,t_o)}{\prod_1^{N+1} \det M(j,j-1)}\, c^{dN} \quad (11)$$

and this can be used in the computation of the normalization of the free generating functional (10.B.(21)), if one takes into account the discretization. We have

$$\bar{N}^{(1/2)} = (2\pi)^{-d/2} \lim_{N\to\infty} \prod_1^{N+1} \left(\frac{g(\bar{x}_{j-1})}{(\varepsilon c)^d}\right)^{1/2} (\det \bar{G}^{\mu\nu}(\ell,k))^{1/2}$$

$$= (2\pi c)^{-d/2} (\det M(T,t_o))^{1/2} \lim_{N\to\infty} \prod_{j=1}^{N+1} \left(\frac{g(\bar{x}_{j-1})}{\varepsilon^d M(j,j-1)}\right)^{1/2} \quad (12)$$

and a small calculation gives us

$$\det M(j,j-1) = \varepsilon^{-d} \det[g_{\mu\nu}(\bar{x}_{j-1}) + \frac{\varepsilon}{2}(F_{\mu\nu} + g_{\lambda\nu}\Gamma^\lambda_{\mu\rho}\dot{x}^\rho - g_{\lambda\mu}\Gamma^\lambda_{\nu\rho}\dot{x}^\rho)]$$

$$= \varepsilon^{-d} \det(g_{\mu\nu}(\bar{x}_{j-1}))\,[1 + \mathcal{O}(\varepsilon^2)] \quad (13)$$

since the term of order $\varepsilon$ is antisymmetric. We finally obtain in the limit

$$\bar{N}^{(1/2)} = (2\pi c)^{-d/2} \det(\bar{M}(t_o,T)) \quad (14)$$

which has been shown before by other means (9.C.(78))

Appendix 10.2

We show in perturbation theory the cancellation of all singular terms arising in the perturbation expansion of (X.C.1)

$$I_N = \prod_{j=1}^{N+1} \left(\frac{\tilde{g}^{\star}(\sqrt{\eta}\, q_j, \bar{x}_{j-1})}{g(\bar{x}_{j-1})}\right)^{1/2} \exp\{-\frac{\varepsilon}{c} \sum_{j=1}^{N+1} [\tfrac{1}{2}(\nabla_t q^\sigma)_j\, \tilde{g}^{\star(2)}_{\sigma\rho}(\nabla_t q^\rho)_j$$

$$+ (\nabla_t q^\sigma)_j\, \tilde{D}_\sigma(q_j, x_j) + V_1(q_j, x_j)]\Big|_{q_j^\mu = \frac{1}{i\varepsilon}\frac{\partial}{\partial J_{j\mu}}}\} Z_o^N[J]\Big|_{J=0} \tag{1}$$

Such singular terms come from the preexponential factor, which is written in the form

$$\prod_{j=1}^{N+1} \left(\frac{\tilde{g}^{\star}(\sqrt{\eta}\, q_j, \bar{x}_{j-1})}{g(\bar{x}_{j-1})}\right)^{1/2} = \exp\{\frac{1}{2\varepsilon} \sum_1^N \varepsilon \ln\left(\frac{\tilde{g}^{\star}(\sqrt{\eta}\, q_{j-1}, \bar{x}_{j-1})}{g(\bar{x}_{j-1})}\right)\} \to$$

$$\exp\{\frac{1}{2}\delta(o) \int_{t_o}^{T} dt\, \ln\left(\frac{\tilde{g}^{\star}(\sqrt{\eta} q, x)}{g(x)}\right)\} \tag{2}$$

and from contractions of the form $\overline{\nabla_t q\ \nabla_t q}$ , since (X.B.32), neglecting terms of order $\varepsilon$

$$\overline{(\nabla_t q^\mu)_j\ (\nabla_t q^\nu)_j} = c(\nabla_t \bar{J}(j,T) M(T,t_o) \nabla_t \bar{J}(t_o,j)) + \frac{c}{\varepsilon} g^{\mu\nu}(\bar{x}_{j-1}) \tag{3}$$

In (X.B) we have studied the other contractions, and these do not give any contribution of order $\frac{1}{\varepsilon}$ . What we have in mind is to show that one can use the rule

$$\overline{\nabla_t q^\mu(t)\ \nabla_{t'} q^\nu(t')} = c(\nabla_t \bar{J}(t,T) M(T,t_o) \nabla_{t'} \bar{J}(t_o,t')) + c g^{\mu\nu}(x(t))\delta(t-t') \tag{4}$$

in the continuum, dropping any term with a $\delta(o)$, in particular the preexponential factor (2). The method we use is similar to the one used in [L29 ], and consists in a summation of graphs in which we contract only the part giving singularities. Factorizing the terms that can give such singularities, and keeping in mind their discretization, we have (we use now $g^{(2)}_{\sigma\rho}$ instead of $\tilde{g}^{\star(2)}_{\sigma\rho}$)

$$\exp\{\frac{1}{2}\delta(o)\int_{t_o}^{T} dt\ \ln(\frac{\tilde{g}^{\star}(\sqrt{\eta}q,x)}{g(x)}) - \frac{1}{2c}\int_{t_o}^{T} dt\nabla_t q^{\sigma} g^{(2)}_{\sigma\rho}\nabla_t q^{\rho}$$

$$= \sum_{0}^{\infty}\frac{1}{n!}\sum_{k=0}^{n}\binom{n}{k}\Big(\frac{\delta(o)}{2}\int_{t_o}^{T} dt\ \ln(\frac{\tilde{g}^{\star}(\sqrt{\eta}q,x)}{g(x)})\Big)^k\Big(-\frac{1}{2c}\int_{t_o}^{T}\nabla_t q^{\sigma} g^{(2)}_{\sigma\rho}\nabla_t q^{\rho}\Big)^{n-k} \tag{5}$$

One must realize first that, as a generalisation of (3), contractions of the form

$$\int dt_1 \ldots dt_m\ (\nabla_{t_1} q^{\alpha_1}\nabla_{t_1} q^{\lambda_1})(\nabla_{t_2} q^{\alpha_2}\nabla_{t_2} q^{\lambda_2})\ldots(\nabla_{t_m} q^{\alpha_m}\nabla_{t_m} q^{\lambda_m}) \tag{6}$$

also give $\delta(o)$ (taking the last term in (4) for each contraction), and that only such loop wise contractions of a set of pairs $(\nabla_t q\nabla_t q)$ are singular. The singular contribution of such a loop of m pairs is

$$\delta(o)(-\frac{1}{2c})^m\int_{t_o}^{T} dt g^{(2)}_{\alpha_1\lambda_1} g^{\lambda_1\alpha_2} g^{(2)}_{\alpha_2\lambda_2}\ \ldots\ g^{(2)}_{\alpha_m\lambda_m} g^{\lambda_m\alpha_1}\ (c)^m$$

$$= \delta(o)\ \frac{1}{2m}\int_{t_o}^{T} dt\ \mathrm{tr}(-Y)^m \tag{7}$$

which is obtained by writing $g^{\lambda_i\alpha_{i+1}} = \sigma^{\lambda_i}_{\ k}\sigma^{\alpha_{i+1}}_{\ k}$ and introducing the matrix

$$Y_{ij} = \sigma^{\lambda}_{\ i} g^{(2)}_{\lambda\mu}\sigma^{\mu}_{\ j} \tag{8}$$

Of course there is also a finite contribution of such a loop, but we shall not need that part. Using this result, we want to take out of (5) the terms giving $\delta(o)^{\ell}$. These come from the terms

$$\sum_{n=2}^{\infty}\frac{1}{n!}\sum_{k=0}^{\ell}\binom{n}{k}\Big[\frac{\delta(o)}{2}\int_{t_o}^{T} dt[\ln(\frac{\tilde{g}^{\star}(\sqrt{\eta}\ q,x)}{g(x)})]^k\ \Big[-\frac{1}{2c}\int_{t_o}^{T} dt\nabla_t q^{\sigma} g^{(2)}_{\sigma\rho}\nabla_t q^{\rho}\Big]^{n-k} \tag{9}$$

where $(\nabla_t q\nabla_t q)^{n-k}$ is contracted into $\ell$-k loops of the form (6), while the remaining $\nabla_t q\nabla_t q$ terms are contracted in such a way that they do not give $\delta(o)$, either, because of the topology of the contractions, or because they give the finite part of a contraction like (6). To be complete we recall that we have expanded only part of (1), so that contractions with that second part are also possible.

Consideration of (9) leads to the combinatorial problem in how many ways one can take from $(\nabla_t q \nabla_t q)^{n-k}$ $\ell-k$ groups of $m_1, \dots m_{\ell-k}$ pairs such that $\Sigma m_i \leq n-k$. This is solved by writing all possibilities in the form

$$\frac{1}{(\ell-k)!} \underset{\Sigma m_i \leq n-k}{\sum_{m_1} \cdots \sum_{m_{\ell-k}}} \frac{(n-k)!}{m_1! \cdots m_{\ell-k}!(n-k-\Sigma m_i)!} \tag{10}$$

The last combinatorial gives the number of ways one can put $m_1$ elements in box 1, ..., $m_\ell$ elements in box $\ell$. Summing $\sum_{m_1} \dots \sum_{m_{\ell-k}}$ and dividing by $(\ell-k)!$ then eliminates the distinction made between boxes, containing the same number of elements. (10) has to be combined with a second factor giving the number of ways $(\nabla_t q \nabla_t q)^{m_i}$ can be contracted in a loop like (6). This factor is $2^{m_i-1}$ $(m_{i-1})!$ since, fixing one pair, one remains with the permutation symmetry of the $(m_i-1)$ other pairs, each giving an extra factor 2. From (9) we then find as coefficient of $\delta(o)^\ell [\nabla_t q^\sigma g^{(2)}_{\sigma\rho} \nabla_t q^\rho]^r$, if these do not give any supplementary $\delta(o)$.

$$\sum_{n=\ell+r}^{\infty} \frac{1}{n!} \sum_{k=0}^{\ell} \binom{n}{k} \frac{1}{(\ell-k)!} \underset{\Sigma m_i = n-k-r}{\sum_{m_1} \cdots \sum_{m_{\ell-k}}} \frac{(n-k)!}{m_1! \cdots m_{\ell-k}! r!} \Big[\prod_{i=1}^{\ell-k} (\int dt \frac{tr}{2}(-Y)^{m_i}$$

$$(m_i - 1)!\Big]\Big[\int dt \frac{1}{2} \ln \Big(\frac{\tilde{g}^\star}{g}\Big)\Big]^k$$

$$= \frac{1}{r!} \sum_{n=\ell+r} \sum_{k=0}^{\ell} \frac{1}{k!} \frac{1}{(\ell-k)!} \underset{\Sigma m_i = n-k-r}{\sum_{m_1} \cdots \sum_{m_{\ell-k}}} \Big[\prod_{i=1}^{\ell-k} \Big(\int dt \frac{tr}{2m_i} (-Y)^{m_i}\Big)\Big]$$

$$\Big[\int dt \frac{1}{2} \ln\Big(\frac{\tilde{g}^\star}{g}\Big)\Big]^k \tag{11}$$

changing $n = \ell + r + n'$, and using

$$\sum_1^\infty \int dt \frac{1}{2m} \operatorname{tr} (-Y)^m = -\int dt \frac{1}{2} \ln \det(1+Y) = -\frac{1}{2} \int dt \ln\Big(\frac{\tilde{g}^\star}{g}\Big)$$

we see the cancellation in the form

$$\frac{1}{r!} \frac{1}{\ell!} \sum_{k=0}^{\ell} \binom{\ell}{k} \Big[-\frac{1}{2} \int dt \ln \Big(\frac{\tilde{g}^\star}{g}\Big)\Big]^{\ell-k} \Big[\frac{1}{2}\int dt \ln \Big(\frac{g^\star}{g}\Big)\Big]^k \tag{13}$$

To use this proof for the discrete perturbation series in (1), one must make remarks similar to the ones in (V.C). In particular we mean that the remaining contractions, which are understood, or the remaining part of the singular contractions, have to be kept together, and that one must fix the time discretization. It is also easy to convince oneself that the cancellation is valid for any discretization of $g^{\star}_{\mu\nu}(t^{v}_{j-\frac{1}{2}})$, if $g^{\star}$ has the same discretization (or differs from it in terms of order $\varepsilon^{2}$).

# CHAPTER XI : OTHER APPROACHES

For the construction of the semiclassical expansion of the propagator I, we have benefited from the work of C. Dewitt [D14,D17] on some technical points as the study of the Jacobi equation [App.9.1] and the lattice calculation of the inverse of the propagator $\bar{G}^{\mu\nu}(t_i,t_j)$ [App.10.1]. In fact, one of the aims of these works was precisely to calculate the correction terms to the WKB-approximation in curved spaces. The functional integral methods that were used [D12,D13] do not seem to give an interpretation however to the undefined quantities that appear in the expansion, and that were recognized in [D14,D15,]. In D17,D18 a method is followed based on the work of Truman and Elworthy [E2,E3] who use stochastic differential geometry to arrive at the WKB-approximation. At this point, we can agree with the procedure, since we indicated in section 7.6 that the manipulations we make have a stochastic origin, and that, like for the Ito and Stratonovich integrals, the lattice offers the opportunity to control the stochastic nature. In this sense, we do not see any fundamental opposition between the two methods, and cannot agree with the assertion that a "kinetic energy term $g_{\alpha\beta}(Q)\dot{Q}^{\alpha}\dot{Q}^{\beta}$ cannot be treated by the lattice approximation and must be handled by methods suitable for stochastic variables on curved spaces"[D17]. We seem to have shown here the opposite.

Phase space path integrals for the semiclassical expansion were discussed formally in [M15]. In [M16] it is conjectured that the expansion can be obtained, using correspondence rules and a related prescription for the $\overline{pq}$ tadpoles. An application to the anharmonic oscillator is considered in [M17]

In an other series of papers [T7-T10] Truman uses a limiting procedure based on polygonal paths to define his functional integrals. Apart from the mathematical question on the kind of potentials that can be integrated, and a rigorous formulation of the Cameron-Martin transformation that is used, he also studies the WKB-approximation (for the cases without ordering problems), specifying the conditions on the classical problem, and including the Maslov index. In [E2,E3] the WKB-approximation on curved spaces is studied, using stochastic methods.

We give a brief survey of the method used in [E2], but, at the risk of oversimplifying their mathematics, we shall introduce discretized expressions that give us the possibility to make the link to the formulas and methods that we use. In fact they treat the solution of the equation ($\mu^2$ corresponds to our $\eta$)

$$\frac{\partial}{\partial t}\psi^\mu = \frac{\mu^2}{2}\Delta\psi^\mu + \frac{V}{\mu^2}\psi^\mu \tag{1}$$

with initial condition $\psi^\mu(t=t_o) = \psi^\mu_o(x) = T_o(x)\exp - \frac{S_o(x)}{\mu^2}$, and they start with the construction of the quasiclassical representation (in the words of Truman) of the solution, starting from the Feynman-Kac formula

$$\psi^\mu(X,t) = \int_\Omega \exp\{\frac{1}{\mu^2}\int_{t_o}^{t} V[x^\mu(s,\omega)]ds\}\,\psi^\mu_o(x^\mu(t,\omega)d_\gamma(\omega) \tag{2}$$

Here $\int_\Omega d_\gamma(\omega)$ indicates the Wiener integral, and $x^\mu(t,\omega)$ is the solution of a stochastic differential equation. In their work it is made mathematically precise, how one can define a stochastic process on a manifold starting from a Wiener process, by means of the development mapping, which is the stochastic version (up to the factor $\mu$) of the Cartan equations

$$\dot{x}^\lambda = \mu\sigma^\lambda_i(x)\frac{d\omega^i}{dt} \tag{3}$$

$$\dot{\sigma}^\lambda_i = -\Gamma^\lambda_{\mu\kappa}\dot{x}^\mu\sigma^\kappa_i$$

In the work of Cartan [C3], these equations allow to "project" a curve $x^\lambda(t)$ on the manifold onto Euclidean space (the tangent space at the origin $x(t_o)$) by taking $\dot{\omega}(t)$ as the parallel transport of $\dot{x}(t)$ to this origin. As is well-known the integrability condition for (3) is the vanishing of the Riemann Christoffel tensor. The stochastic process $x^\mu(t,\omega)$ in the Feynman-Kac formula (2) is the stochastic version of (3), which is interpreted as a Stratonovich stochastic differential equation, and $d\omega^i$ is the Wiener process in $\mathbb{R}^d$. The initial conditions are $\omega(t_o) = 0$, $x^\mu(t_o,\omega) = X^\mu$

The relation of (2) to our representation [App. 9.4(4)] of the propagator can be seen, by writing it in discretized form as

$$\psi^\mu(X,t) = \int \prod_1^{N+1} d\omega^i \exp\{-\frac{1}{2}\sum_{j=1}^{N+1}\frac{(\Delta\omega^i_j)^2}{\varepsilon} + \frac{\varepsilon}{\mu^2}V(x^\mu_j)\}\psi_o[x^\mu_{N+1}(\omega)] \tag{4}$$

Writing the first of the equations (3) in Ito form, using the second one:

$$dx^\lambda = \mu\sigma_i^\lambda d\omega^i + \frac{1}{2}\mu d\sigma_i^\lambda d\omega^i = \mu\sigma_i^\lambda d\omega^i - \frac{\mu}{2}\Gamma^\lambda_{\rho\kappa} dx^\rho \sigma_i^\kappa d\omega^i$$

$$= \mu\sigma_i^\lambda d\omega^i - \frac{\mu^2}{2}\Gamma^\lambda_{\rho\kappa}\sigma_i^\kappa\sigma_j^\rho\, d\omega^i\omega^j$$

$$= \mu\sigma_i^\lambda d\omega^i - \frac{\mu^2}{2}\Gamma^\lambda_{\rho\kappa} g^{\rho\kappa} dt \quad (I) \tag{5}$$

and changing variables in (4) from $\omega^i$ to $x^i$ (see section 7.6), we see that the Jacobian of the transformation is $\prod_1^{N+1} \det[\sigma_i^\lambda(x_{i-1})]^{-1/2} = \prod_1^{N+1} (g(x_{i-1}))^{1/2}$ (we took the coefficient of dt in the prepoint) and we arrive at the representation ($x_o = X$)

$$\psi^\mu(X,t) = \int \prod_1^{N+1} \frac{dx_i}{(2\pi\varepsilon\mu^2)^{d/2}} (g(x_{i-1}))^{\frac{1}{2}} \exp\{-\frac{\varepsilon}{2\mu^2}\sum_1^{N+1}[\frac{\Delta x_j^\lambda}{\varepsilon} + \frac{\mu^2}{2}(\Gamma^\lambda_{\rho\kappa}g^{\rho\kappa})_{j-1}]$$

$$g_{\lambda\sigma}(x_{j-1})[\frac{\Delta x_j^\sigma}{\varepsilon} + \frac{\mu^2}{2}\Gamma^\sigma_{\alpha\beta}g^{\alpha\beta}_{j-1}] + \frac{\varepsilon}{\mu^2}\sum_1^{N+1} V(x_j)\}\psi_o[x^\mu_{N+1}] \tag{6}$$

Reversing the paths, by taking $q^i = x^{N+1-i}$, we obtain ($q^{N+1} = x_o = X$) .

$$\psi^\mu(X,t) = \int \prod_o^N dq_i \left(\frac{g(q_{i+1})}{(2\pi\varepsilon\mu^2)^d}\right)^{\frac{1}{2}} \exp\{-\frac{\varepsilon}{2\mu^2}\sum_1^{N+1}[\frac{\Delta q_j^\lambda}{\varepsilon} - \frac{\mu^2}{2}(\Gamma^\lambda_{\rho\kappa}g^{\rho\kappa})_j]g_{\lambda\sigma}(q_j)$$

$$[\frac{\Delta q_j^\sigma}{\varepsilon} - \frac{\mu^2}{2}\Gamma^\sigma_{\alpha\beta}g_j^{\alpha\beta}] + \frac{\varepsilon}{\mu^2}\sum_1^{N+1} V(q_j)\}\ \psi_o(q_o) \tag{7}$$

$$= \int dX_o g(X_o) I(X,t,X_o,t_o) \tag{8}$$

and this is the representation that we use for the propagator $I(X,t,X_o t_o)$
Instead of treating the initial value problem (2), we shall discuss the propagator, by taking $S_o = 0$, $T_o = \delta(x - X_o)$ (the consequence is that one has to use different boundary conditions for the classical path T10 ]).For simplicity, we also take $V = 0$. Let $Z(t)$ then be the geodesic with boundary conditions $Z(t_o) = X$, $Z(T) = X_o$. We introduce the orthonormal frame $\sigma_i^\lambda(Z(t))$, parallel along $Z(t)$. As a consequence $\dot{Z}^\lambda(t) = \sigma_i^\lambda \dot{\gamma}^i(t)$ implies $\ddot{\gamma}^i(t) = 0$.
A first step towards the WKB-approximation of

$$I^\mu(X,T;\ X_o,t_o) = \int_\Omega \delta(x^\mu(T,\omega) - X_o)d_\gamma(\omega) \tag{9}$$

is a transformation in the variable $\omega$, which makes it possible to take out the classical action. This transformation is written as

$$d\omega'^{i} = d\omega^{i} - \mu^{-1}\dot{\gamma}^{i} ds \qquad \omega'(t_o) = 0 \tag{10}$$

Reexpressing (9) in terms of $\omega'$, we have, since the Jacobian $\omega \to \omega'$ is 1.

$$\int \prod_{1}^{N+1} d\omega_i \exp\{-\frac{1}{2}\sum_{1}^{N+1}\frac{(d\omega^i)^2}{\varepsilon}\} = \int \prod_{1}^{N+1} d\omega_i' \exp\{-\frac{1}{2}\sum_{1}^{N+1}(d\omega'^i + \dot{\gamma}^i\mu^{-1}\varepsilon)^2\} \tag{11}$$

and $\int (\dot{\omega}^i + \dot{\gamma}^i\mu^{-1})^2 dt = \mu^{-2}\int\dot{\gamma}^{i2} + \int\dot{\omega}'^{i2} + 2\mu^{-1}[(\omega'(T)\,\dot{\gamma}\,(T) - \omega'(t_o)\,\dot{\gamma}\,(t_o)]$

The new representation of $I^{\mu}(X,T;\, X_o,t_o)$ is then

$$I^{\mu} = \exp\{-\frac{1}{2\mu^2}\int\dot{\gamma}^{i2}\}\int d_{\gamma'}(\omega')\}\exp\{-\omega'(T)\,\dot{\gamma}\,(T)\mu^{-1}\}\delta(y^{\mu}(T,\omega)-X_o) \tag{12}$$

where $y^{\mu}(t,\omega)$ is the solution of the stochastic differential equation

$$\text{a) } dy^{\lambda} = \mu\sigma^{\lambda}_{\ i}(y)\ (d\omega^i + \mu^{-1}\,\dot{\gamma}^i dt)$$

$$\text{(S)}$$

$$\text{b) } d\sigma^{\lambda}_{i} = -\Gamma^{\lambda}_{\mu\kappa} dy^{\mu}\sigma^{\kappa}_{i} \tag{13}$$

In the theory of stochastic processes the transformation rule from (9) to (12) is known as Girsanovs formula. The factorization of the classical action $\int\dot{\gamma}^{i2} = \int\dot{Z}^{\mu}g_{\mu\lambda}\dot{Z}^{\nu}$ in (12) is only the first step, and a second transformation should lead to a Wiener integral representing the normalization factor. In the work of Truman and Elworthy this is done by considering the stochastic differential equation satisfied by $u^i(\omega)$

$$u^i(\omega) = \sigma_{\lambda i}(Z(t))\ .\ \delta y^{\lambda}(t,\omega) \tag{14}$$

$$\text{where}\quad \delta y^{\lambda}(\omega) = \frac{\partial}{\partial\mu}\,y^{\lambda}(t,\omega)\Big|_{\mu=0} \tag{15}$$

Still following their notation, we write

$$\delta y^{\lambda}(\mu,t) = \frac{\partial}{\partial\mu}\,y^{\lambda}(\mu,t,\omega) \qquad \delta y^{\lambda}(0,t) = \delta y^{\lambda}(t)$$

$$\delta^2 y^{\lambda}(\mu,t) = \nabla_{\mu}\frac{\partial}{\partial\mu}\,y^{\lambda}(\mu,t,\omega) \qquad \delta^2 y^{\lambda}(0,t) = \delta^2 y^{\lambda}(t) \tag{16}$$

We take now the covariant derivative along the $\mu$-curves in (13), and use

$$\nabla_{\mu}\frac{\partial y^{\lambda}}{\partial s} = \nabla_s\partial_{\mu}y^{\lambda} = \nabla_s\delta y^{\lambda}(\mu,t)$$

Then

$$\nabla_s\delta y^{\lambda}(\mu,s) = \sigma^{\lambda}_{i}\frac{d\omega^i}{ds} + (\nabla_{\mu}\,\sigma^{\lambda}_{i})\mu(\frac{d\omega^i}{ds} + \mu^{-1}\,\dot{\gamma}^i)$$

and for $\mu \to 0$

$$\nabla_s \delta y^\lambda(s) = \sigma_i^\lambda(Z(s)) \frac{d\omega^i}{ds} + (\nabla_\mu \sigma_i^\lambda)_{\mu=0} \dot{\gamma}^i \tag{17}$$

where we used the fact that $y^\lambda(\mu,t) \overset{\mu\to 0}{\to} Z(t)$, since for $\mu = 0$ (13) reduces to a set of deterministic equations.
Multiplying at the left with $\sigma_i^\lambda[Z(s)]$ we obtain the stochastic differential equation for u

$$\frac{d}{ds} u^i(s) = \frac{d}{ds} \omega^i + \sigma_{\lambda i}(Z(s)) \dot{\gamma}^j (\nabla_\mu \sigma_j^\lambda)_{\mu=0} \tag{18}$$

The last term is evaluated, using

$$\nabla_s (\nabla_\mu \sigma_j^\lambda)_{\mu=0} = [(\nabla_s \nabla_\mu - \nabla_\mu \nabla_s)\sigma_j^\lambda]_{\mu=0} = R^\lambda_{\mu\nu\rho}(Z(t))\delta y^\rho \dot{Z}^\nu \sigma_j^\mu(Z(t)) \tag{19}$$

which allows us to check immediately the relation

$$\sigma_{\lambda i}(Z(s))[\nabla_\mu \sigma_j^\lambda(y)]_{\mu=0} = \int_{t_o}^s \sigma_{\lambda i}(r) R^\lambda_{\mu\nu\rho}(Z(r))\sigma_j^\mu(Z(r))\dot{Z}^\nu(r)\delta y^\rho(r)dr$$

by taking the derivative with respect to s. Hence, with $u_o = 0$,

$$\frac{du^i}{ds} = \frac{d\omega^i}{ds} + \int_{t_o}^s R_{\lambda\mu\nu\rho}(Z(r))\sigma_i^\lambda(Z(r))\sigma_j^\mu(Z(r))\sigma_\ell^\rho(Z(r))u^\ell Z^\nu \dot{\gamma}^j(s)dr \tag{20}$$

Writing this in the discrete, we see that the Jacobian of the transformation $\omega^i \to u^i$ is 1, because the transformation matrix is triangular, while the second term will contribute in $\mathcal{O}(\varepsilon^2)$ on the diagonal. To compute $\int (\frac{d\omega}{dt})^2$, we do not use (20), but consider

$$\langle \frac{dy}{ds}, \frac{\bar{dy}}{ds} \rangle = (\mu \frac{d\omega^i}{ds} + \dot{\gamma}^i)^2$$

and therefore

$$\frac{\partial^2}{\partial\mu^2} \langle \frac{dy}{ds}, \frac{dy}{ds} \rangle = 2(\frac{d\omega^i}{ds})^2$$

while on the other hand

$$\frac{\partial}{\partial\mu} \langle \frac{dy}{ds}, \frac{dy}{ds} \rangle = 2 \langle \nabla_\mu \frac{\partial y}{\partial s}, \frac{\partial y}{\partial s} \rangle = 2 \langle \nabla_s \delta y(\mu), \frac{dy}{ds} \rangle$$

$$\frac{\partial^2}{\partial\mu^2} < \frac{dy}{ds}, \frac{dy}{ds} > = 2 < \nabla_\mu \nabla_s \delta y(\mu), \frac{dy}{ds} > + 2 < \nabla_s \delta y(\mu), \nabla_s \delta y(\mu) >$$

$$= 2 < \nabla_s \nabla_\mu \delta y, \frac{dy}{ds} + < R(\frac{\partial y}{\partial s}, \frac{\partial y}{\partial \mu}) \frac{\partial y}{\partial \mu}, \frac{dy}{ds} > + < \nabla_s \delta y, \nabla_s \delta y >$$

and for $\mu \to 0$

$$(\frac{d\omega'^i}{ds})^2 = <\nabla_s \delta y, \nabla_s \delta y> + <\nabla_s \delta^2 y, \dot{Z}> + <R(\dot{Z}, \delta y)\delta y, \dot{Z}>$$

$$= (\frac{du^i}{ds})^2 + R_{\lambda\mu\nu\rho}\dot{Z}^\rho \delta y^\nu \delta y^\mu \dot{Z}^\lambda + <\nabla_s \delta^2 y, \dot{Z}> \tag{21}$$

In (11) the exponential can be rewritten, by means of (20)

$$(\omega'^i(T)\dot{\gamma}^i(T)) = \int_{t_o}^T d\omega^i \dot{\gamma}^i(t) = \int_{t_o}^T du^i(t)\dot{\gamma}^i(t) - \int_{t_o}^T ds \dot{\gamma}^j(s)\dot{\gamma}^i(s)$$

$$[\int_{t_o}^s R_{\lambda\mu\nu\rho}(Z(r))\sigma_i^\lambda(r)\sigma_j^\mu(r)\sigma_\ell^\rho(r)\dot{z}^\nu(r)u^\ell(r)dr]$$

The second part vanishes owing to the antisymmetry of the Riemann-Christoffel tensor, and therefore $\omega'^i(T)\dot{\gamma}^i(T)$ can be replaced by $u^i(T)\dot{\gamma}^i(T)$
Since $\delta^2 y(0) = 0$, we obtain finally, in our notation

$$I(X,T;X_o,t_o) = \exp\{-\frac{1}{\mu^2}A_{cl}\}\int \mathcal{D}u \quad \exp\{-\frac{1}{2}\int \dot{u}^{i2} + R_{\lambda\mu\nu\rho}\dot{z}^\lambda \dot{z}^\rho \sigma^\mu{}_k \sigma^\nu_\ell u^k u^\ell\}$$

$$\exp\{-\frac{1}{\mu}(u^i(T)\dot{\gamma}^i(T)) + \frac{1}{2}\delta^2 y^\lambda(T)\dot{Z}_\lambda(T)\}\delta[y^\mu(T,\omega(u)) - X_o] \tag{23}$$

We shall obtain the normalization factor in the form [App.9.2(1)] if we develop the argument of the $\delta$-function as

$$\delta(y^o(T,\omega(u)) + \mu\delta y(T,\omega(u)) + \frac{\mu^2}{2}\delta^2 y(T,\omega'(u)) + \ldots - X_o)$$

$$= \delta(\mu\sigma_i^\lambda(Z(T))[u^i(T) + \frac{1}{2}\mu\sigma_{\rho i}(Z(T))\delta^2 y^\rho(T,\omega(u)) + \mathcal{O}(\mu^2)]) \tag{23}$$

It is clear now that the boundary terms in (22) vanish, owing to (23), while for the other terms, one can take $\delta(u^i(T))$, neglecting $\mathcal{O}(\mu)$. Finally

$$I_{WKB} = \exp\{-\frac{1}{\mu^2}A_{cl}\ (\frac{g(X_o)}{\mu^{2d}})^{\frac{1}{2}} \int \mathcal{D}u \exp\{-\frac{1}{2}\int \dot{u}^{i2} + R_{\lambda\mu\nu\rho}\dot{z}^\lambda \dot{z}^\rho \sigma^\mu{}_\kappa \sigma^\nu_\ell u^\kappa u^\ell\}$$

$$\delta(u^i(T))$$

$$= \exp\{-\frac{1}{\mu^2}A_{cl}\}(\frac{g(X_o)}{\mu^{2d}})^{\frac{1}{2}}[g(X_o)^{-\frac{1}{2}}\det(-\frac{\partial^2 A}{\partial X_o^\alpha \partial X^\beta})\ g(X)^{-\frac{1}{2}}]^{1/2} \tag{24}$$

Since this last functional integral has been computed in app.9.2(30).

The method that is used is rigorous for the case of the diffusion equation, and formal for the case of the Schrodinger equation.

It is clear that the validity of the semiclassical expansion that we give is strongly restricted by the structure of the classical problem. We refer to [T10,M8] for the precise conditions.

They correspond to the existence of a unique classical path connecting $(X_o,t_o)$ and $(X,T)$, without points conjugate to $(X_o,t_o)$ on that trajectory. The WKB-approximation for the Schrodinger equation has been given for more general cases [M7,M8] and contains in fact a sum over several classical paths. It is well known that there is a slight modification in the expression of the WKB-approximation in the case of the Schrodinger equation when there are conjugate points on the trajectory. These can be defined as the points where $\det[\bar{J}^{\mu\nu}(t_o,t)]$[App.9.1] vanishes, and they are assigned a multiplicity (d-rank of $\bar{J}^{\mu\nu}(t_o,t)$) [D17]. It is a consequence of Morse theory [M13] that the number of conjugate points, each counted with its multiplicity, also called the Morse index, is precisely the number of negative eigenvalues of the Jacobi operator (with zero boundary conditions in $t_o,T$). Computation of the free generating functional $Z_o(0,0)$ (9.C.(78)) in the form $\int \pi\, dx_i \exp\{i\, x_i A_{ij} x_j\}$ then gives the modified phase factor $\exp\{-\frac{i\pi}{2}(\frac{d}{2} - m(X,T,X_o,t_o))$ with $m(X,T,X_o,t_o)$ the morse index of the trajectory, since each eigenvalue $\lambda$ gives

$$F(\lambda) = \int_{-\infty}^{+\infty} \exp\{i\ \frac{\lambda}{2}\, x^2\}dx = \sqrt{\frac{2\pi}{\lambda}}\ \exp\ \frac{i\pi}{4}\qquad \lambda > 0$$

$$= \sqrt{\frac{2\pi}{-\lambda}}\ \exp\ \frac{-i\pi}{4}\qquad \lambda < 0$$

In one dimension a conjugate point is characterized by $\frac{\partial x_{cl}(t)}{\partial p_o} = 0$ and the appearance of the phase factor can easily be seen in a discrete calculation of the normalization factor [T10,G6]

A particular application is the harmonic oscillator which is computed in [C10,H8] Further discussion and references are given in [B7].

# CHAPTER XII : COMPUTATION OF THE PROPAGATOR ON THE SPHERE $S^3$

We want to show here how the result (9.E.52) can be obtained globally, without resorting to a perturbation expansion. In that formula one can see that the WKB-approximation, including the normalization factor with the Van Vleck determinant is exact up to the factor $\exp\{\frac{\eta c K}{2}(T-t_o)\} = \exp\{-\frac{\eta c R}{12}(T-t_o)\}$ .
It has been shown ( [D22] and ref.) that for free motion on the manifold of a semi-simple Lie group ($S^3$ is an example of this, being the manifold of SU(2)) one has the same factor $\exp\{-\frac{\eta c R}{12}(T-t_o)\}$. This result is derived, [D22,D23] using the WKB-approximation in the equation for the propagator, or comparing to the eigenfunction expansion [M2,M3,S3] using the transformation properties of the $\theta$ - function. In the Feynman definition (sect.6.2) the short-time propagator for the sphere $S_3$, includes the factor $\exp\{\frac{\eta c}{2} K\varepsilon\}$ , and is also valid for finite times. Rewriting this short time propagator with the help of the eigenfunctions, one has in fact a method to do the iteration from the short time propagator to the global one [S3] .
Starting from the path integral representation of the propagator in normal coordinates, and postpoint discretization we show that it is possible to calculate the functional integral explicitly, making use of polar coordinates on the group. This procedure seems to be an answer (in the case of SU(2)) to the remark of Dowker [D26] that "it should be possible to derive this result directly from the functional integral, rather like the calculations for the free particle and harmonic oscillator". At the same time this will allow us to give some comments on functional integrals in polar coordinates.
The problems arising from the fact that SU(2) is compact, can be circumvented by treating the manifold as the quotient of the Lie algebra and the characteristic lattice [M2] , or following the discussion of Dowker [D22] on the boundary conditions of the propagator. This implies that the solution on the sphere $S^3$ is obtained by summing (9.E.52) over all equivalent points ($x + 2\pi n$), or stated in another way, by taking into account all classical paths connecting two points. The eigenfunction expansion yields this complete propagator (with the sum) , and hence, the iteration method of Schulman that we referred to, is for the complete propagator, while our iteration method is for the partial propagator on the Lie algebra of SU(2).
Now we start immediately from the metric (9.E.19) in normal coordinates q ($\sum_{\mu=1}^{3} (q^\mu)^2 = \phi^2$)

$$g_{\mu\nu}(q) = \frac{\sin^2 \sqrt{K}\,\phi}{K\phi^2}\left(\delta_{\mu\nu} - \frac{q^\mu q^\nu}{\phi^2}\right) + \frac{q^\mu q^\nu}{\phi^2} \quad . \tag{1}$$

It is our aim to construct the fundamental solution $I(x,T;0,t_o)$ satisfying

$$\frac{\partial I}{\partial T} = \frac{\eta c}{2}\,\Delta I \tag{2}$$

and that reduces to a delta function at $T=t_o$. To get rid of the K dependence in (1) we make the time independent transformation $q'^\mu = \sqrt{K}q^\mu$ which implies $g'_{\mu\nu}(q') = \frac{1}{K}\, g_{\mu\nu}(\frac{q'}{\sqrt{K}})$. Rescaling time to include the conformal factor $\frac{1}{K}$, and maintaining the $\delta$-normalization at $T=t_o$, we see that we can express the propagator in terms of the one corresponding to K=1. Indeed, $g'_{\mu\nu}(q') = \frac{1}{K} g^{(K=1)}_{\mu\nu}(q')$ which implies $\Delta'^K = K\Delta'^{(K=1)}$, and the $\delta$-function gives $K^{3/2}$. Indicating explicitly the dependence on $x^\mu, \eta c(T-t_o)$ and K, we have

$$I(x^\mu,\ \eta c(T-t_o), K) = I\,(\sqrt{K}x^\mu,\ K\eta c(T-t_o), 1)K^{3/2} \tag{3}$$

and that's why we compute the case K=1. By (3) we shall be able to include the K dependence afterwards.
The path integral representation for the propagator I can be written in postpoint discretization as [App.9.4(4)]

$$I = \int_{\gamma_1(1)} \mathcal{D}q(t)\ \sqrt{g(q(t))}\ \exp\left\{-\frac{1}{2\eta c}\int_{t_o}^{T} (\dot q^\mu - \frac{\eta c}{2}\, g^{\lambda\rho}\Gamma^\mu_{\lambda\rho})g_{\mu\sigma}(\dot q^\sigma - \frac{\eta c}{2}\, g^{\tau\nu}\Gamma^\sigma_{\tau\nu}\ )\right\}$$

$$\delta(q(T)-x)\,\delta(q(t_o)\ )$$

$$= \lim_{N\to\infty} \frac{\sqrt{g(x)}}{(2\pi\varepsilon\eta c)^{3/2}} \int \prod_1^N dq_i\ \frac{\sqrt{g(q_i)}}{(2\pi\varepsilon\eta c)^{3/2}}\ \exp\Big\{-\frac{\varepsilon}{2\eta c}\sum_1^{N+1}\Big[\frac{(q_i^\mu - q_{i-1}^\mu)}{\varepsilon} - \frac{\eta c}{2} g^{\lambda\rho}(q_i)$$

$$\Gamma^\mu_{\lambda\rho}(q_i)\Big]\ g_{\mu\sigma}(q_i)\ \Big[\frac{(q_i^\sigma - q_{i-1}^\sigma)}{\varepsilon} - \frac{\eta c}{2}\, g^{\tau\nu}(q_i)\Gamma^\sigma_{\tau\nu}(q_i)\Big]\Big\}$$

$$q_o = 0,\ q_{N+1} = x_{N+1} = x. \tag{4}$$

In this form, the calculation of $I(x,T;0,t_o)$ seems to be hopeless, and that's why we shall make a transformation to polar coordinates. The normal coordinates, used in (1) coincide with the canonical coordinates used in the description of SU(2). Using the Pauli-matrices $\sigma_i$, these parametrize an element U of SU(2) by

$$\begin{aligned} U &= \exp\,\{i\, q^\alpha \sigma^\alpha\} \qquad 0 \leqslant |q| = \phi \leqslant \pi \\ &= \cos\phi\ + i\,\vec\sigma\,.\,\hat\omega\ \sin\phi \end{aligned} \tag{5}$$

The last form arises if one describes the vector q by specifying its length and its direction $\hat{\omega}$: $q^\alpha = \phi\omega^\alpha$. In these coordinates, we have $U(\phi,\hat{\omega}) = U(2\pi-\phi,-\hat{\omega})$, and $U(\phi+2\pi,\hat{\omega}) = U(\phi)$.
We shall describe $\hat{\omega}$ by means of spherical coordinates $\lambda,\psi$ . The explicit transformation is then

$$\begin{aligned} q^1 &= \phi\cos\lambda & 0 \leqslant \phi \leqslant \pi \\ q^2 &= \phi\sin\lambda\sin\psi & 0 \leqslant \lambda \leqslant \pi \\ q^3 &= \phi\sin\lambda\cos\psi & 0 \leqslant \psi \leqslant 2\pi \end{aligned} \tag{6}$$

The variable $\phi$ in (6) is in fact the radial variable arising in the description of SU(2) by means of polar coordinates. Here it coincides with the radial variable in $\mathbb{R}^3$, but the notion can be generalized to any semisimple Lie group, where there are as many radial variables as the rank of the group [M2] . The connection of (6) with the **parametrization** of the unit sphere $S^3$ is found by writing U as

$$U = y_o\, 1 + i\,\vec{\sigma}\,.\,\vec{y}$$

or
$$\begin{aligned} y_o &= \cos\phi \\ y_1 &= \sin\phi\cos\lambda \\ y_2 &= \sin\phi\sin\lambda\sin\psi \\ y_3 &= \sin\psi\sin\lambda\cos\psi \end{aligned}$$

showing that $(\phi,\psi,\lambda)$ are spherical coordinates on $S^3$.

In (4) we leave the variables $q_i$ unrestricted $-\infty<q^i< +\infty$, and under the transformation (6) we then have $0 \leqslant \phi \leqslant \infty$ , $0 \leqslant \lambda \leqslant \pi$, $0 \leqslant \psi \leqslant 2\pi$ .
The actual range of the variables $0 \leqslant \phi \leqslant \pi$ will have to be considered afterwards. In the new coordinates, the metric is now diagonal, and of the form

$$(ds)^2 = (d\phi)^2 + \sin^2\phi(d\lambda)^2 + \sin^2\phi\sin^2\lambda(d\psi)^2 . \tag{7}$$

Instead of using this metric to construct the new short time propagator in these variables, we want to make the explicit change of variables in the discrete version of the propagator, and we shall not use (4), in postpoint discretization, but another form which is related to it by a change of discretization.
We think that this new change of discretization is unessential, that we could start as well from (4) and do the change of variables there immediately.

The reason to do it, is to avoid complications in the integral that we obtain after the change of variables. The nature of these integrals has not been studied well enough yet, and we shall explain why we refrain from doing there certain manipulations, that have been used by other authors.
The metric to be used in (4) is in postpoint discretization

$$g_{\mu\nu}(q_j) = D_j(\delta_{\mu\nu} - \frac{q_j^\mu q_j^\nu}{\phi_j^2}) + \frac{q_j^\mu q_j^\nu}{\phi_j^2} = D_j X_j^{\mu\nu} + Y_j^{\mu\nu} \tag{8}$$

where

$$D_j = \frac{\sin^2\phi_j}{\phi_j^2} \tag{9}$$

and

$$Y_j^{\mu\nu} = \frac{q_j^\mu q_j^\nu}{\phi_j^2}, \quad X_j^{\mu\nu} = \delta_{\mu\nu} - Y_j^{\mu\nu}. \tag{10}$$

In (4) we also need the Christoffel symbols [(compare to (9.E.38)] $(D' = \frac{\partial D}{\partial\phi})$

$$\Gamma^\tau_{\rho\lambda} = \frac{1}{2}\{\frac{D'}{D\phi}(q^\rho X^{\lambda\tau} + q^\lambda X^{\rho\tau}) + (\frac{2(1-D)}{\phi^2} - \frac{D'}{\phi})\, q^\tau X^{\rho\lambda}\} \tag{11}$$

giving

$$g^{\rho\lambda}\Gamma^\tau_{\rho\lambda} = q^\tau\{\frac{2(1-D)}{D\phi^2} - \frac{D'}{\phi D}\}. \tag{12}$$

The first change we do in (4) is to introduce a metric with a discretization $[\bar{q}^\mu_{j-1} = \frac{1}{2}(q_j^\mu + q_{j-1}^\mu)]$

$$\tilde{g}_{\mu\nu}(j) = \delta_{\mu\nu}D_j + (1-D_j)\frac{\bar{q}^\mu_{j-1}\bar{q}^\nu_{j-1}}{\phi_j^2} \tag{13}$$

which will be used in the dominating term.
In what follows we drop the indices j,j-1 and indicate by $q^\mu, D, \phi, g_{\mu\nu}$ the quantities $q_j^\mu, D_j, \phi_j, g_{\mu\nu}(j)$
$\bar{q}^\mu$ means $\bar{q}^\mu_{j-1}$, while $\tilde{g}_{\mu\nu}$ is given by (13). The relation between $g_{\mu\nu}$ and $\tilde{g}_{\mu\nu}$ is given by

$$g_{\mu\nu} = \tilde{g}_{\mu\nu} + \frac{(1-D)}{2\phi^2}(\bar{q}^\mu\Delta^\nu + \bar{q}^\nu\Delta^\mu) + \frac{(1-D)}{4\phi^2}\Delta^\mu\Delta^\nu \tag{14}$$

and the inverse $\tilde{g}^{\mu\nu}$, that we need to do the replacements, is up to order $\Delta$

$$\tilde{g}^{\sigma\lambda} = g^{\sigma\lambda} + q^\sigma\Delta^\nu g^{\nu\lambda} + g^{\sigma\mu}\Delta^\mu q^\lambda + O(\Delta^2). \tag{15}$$

Introducing the new metric in the dominating term of the exponential in (4) we can write it as

$$\exp -\frac{1}{2}\ \{\frac{\Delta^{\mu}\tilde{g}_{\mu\sigma}\Delta^{\sigma}}{\varepsilon\eta c} + [\frac{(1-D)}{\phi^2}\ \frac{\Delta^{\mu}\bar{q}^{\mu}\Delta^{\sigma}\Delta^{\sigma}}{\varepsilon\eta c} - \Delta^{\mu}g_{\mu\sigma}g^{\lambda\rho}\Gamma^{\sigma}_{\lambda\rho}] + \eta c[\frac{\varepsilon}{4}\ g^{\lambda\rho}\Gamma^{\rho}_{\lambda\rho}g_{\mu\sigma}\Gamma^{\sigma}_{\nu\tau}g^{\nu\tau} + \frac{(1-D)}{4\phi^2}\ \frac{\Delta^{\mu}\Delta^{\mu}\Delta^{\nu}\Delta^{\nu}}{\varepsilon}\ ]\} \tag{16}$$

restricting ourself to the j'th term. To do the replacements in the part of order $\sqrt{\varepsilon}$ , we expand it as

$$\exp -\frac{1}{2}\ \{\frac{(1-D)}{2}\ \frac{\Delta^{\mu}\bar{q}^{\mu}\Delta^{\sigma}\Delta^{\sigma}}{\varepsilon\eta c} - \Delta^{\mu}g_{\mu\sigma}g^{\lambda\rho}\Gamma^{\sigma}_{\lambda\rho}\ \}$$

$$= 1 - \frac{1}{2}\ \{\frac{(1-D)}{\phi^2}\ \frac{\Delta^{\mu}\bar{q}^{\mu}\Delta^{\sigma}\Delta^{\sigma}}{\varepsilon\eta c} - \Delta^{\mu}g_{\mu\rho}g^{\lambda\rho}\Gamma^{\sigma}_{\lambda\rho}\} + \frac{1}{8}\ \{\frac{(1-D)}{\phi^2}\ \frac{\Delta^{\mu}\Delta^{\nu}\Delta^{\nu}\bar{q}^{\mu}\Delta^{\sigma}\bar{q}^{\sigma}\Delta^{\rho}\Delta^{\rho}}{(\varepsilon\eta c)^2} + 2\ \frac{(1-D)}{\phi^2}\ \frac{\Delta^{\mu}\Delta^{\nu}\Delta^{\nu}\bar{q}^{\mu}}{\varepsilon\eta c}\ \Delta^{\tau}g_{\tau\sigma}g^{\lambda\rho}\Gamma^{\sigma}_{\lambda\rho} + \Delta^{\kappa}\Delta^{\tau}g_{\kappa\sigma}\ ^{\lambda\rho}\Gamma^{\sigma}_{\lambda\rho}g_{\tau\alpha}\Gamma^{\alpha}_{\mu\nu}g^{\mu\nu}\ \} \tag{17}$$

and use the replacements $\Delta^{\mu}\Delta^{\sigma} \doteq \varepsilon\eta c\ \tilde{g}^{\mu\sigma}$ , etc. (5.35) in the terms with $\Delta^3,\Delta^4$, $\Delta^6$. Bringing this back to the exponential we have

$$\exp -\frac{1}{2}\ \{[\frac{(1-D)}{\phi^2}\ (2\tilde{g}^{\mu\sigma}\bar{q}^{\mu}\Delta^{\sigma} + \tilde{g}^{\sigma\sigma}\Delta^{\mu}\bar{q}^{\mu}) - \Delta^{\mu}\ g_{\mu\sigma}g^{\lambda\rho}\Gamma^{\sigma}_{\lambda\rho}]\ \}$$

$$\cdot\exp\ \{\ \frac{\varepsilon\eta c}{8}\ [\ \frac{(1-D)^2}{\phi^2}\ \bar{q}^{\mu}\bar{q}^{\sigma}(2g^{\nu\rho}g^{\nu\rho}g^{\mu\sigma} + 4g^{\mu\rho}g^{\nu\rho}g^{\nu\sigma}\ ]\ \} \tag{18}$$

and expressing this again in the postpoint, we find

$$\exp -\frac{1}{2}\ \{\frac{(1-D)}{\phi^2}\ (2\tilde{g}^{\mu\sigma}\Delta^{\sigma} + \tilde{g}^{\sigma\sigma}\Delta^{\mu}\ )\bar{q}^{\mu}\ \}$$

$$= \exp -\frac{1}{2}\ \{\frac{(1-D)}{\phi^2}\ (2 + g^{\nu\nu})q^{\mu}\Delta^{\mu}\}\exp\{\varepsilon\eta c[\frac{1}{4}\ \frac{(1-D)}{\phi^2}\ (2g^{\mu\nu}g^{\nu\mu} + g^{\nu\nu}g^{\mu\mu}\ ) - \frac{1}{4}\ (\frac{(1-D)}{\phi^2})^2\ (2\phi^2\ g^{\sigma\nu}g^{\nu\sigma} + 4\phi^2)]\} \tag{19}$$

The last step is to consider the complete crossterm in $q^{\mu}\Delta^{\mu}$ ,using (12), and to bring $q^{\mu}\Delta^{\mu}$ to $\bar{q}^{\mu}\Delta^{\mu}$ again

$$\exp - \frac{1}{2} \{[\frac{(1-D)}{\phi^2} \Delta^\mu q^\mu (3 + \frac{2}{D}) - \Delta^\mu q^\mu (\frac{2(1-D)}{D\phi^2} - \frac{D'}{D\phi})] \}$$

$$= \exp - \frac{1}{2} \{\Delta^\mu q^\mu [\frac{3(1-D)}{\phi^2} + \frac{D'}{D\phi}]\} = \exp - \frac{1}{2}\{\Delta^\mu \bar{q}^\mu [\frac{3(1-D)}{\phi^2} + \frac{D'}{D\phi}]\}$$

$$\cdot \exp \{- \frac{\varepsilon\eta c}{4} (\frac{2}{D} + 1) (\frac{3(1-D)}{\phi^2} + \frac{D'}{D\phi})\}. \tag{20}$$

Here we used the first of the following 2 formulas. The second will be used afterwards.

$$g^{\nu\nu} = \mathrm{tr} [\frac{1}{D} X + Y] = (\frac{2}{D} + 1) \tag{21}$$

$$g^{\mu\nu} g^{\nu\mu} = \mathrm{tr}[\frac{1}{D^2} X + Y] = (\frac{2}{D^2} + 1) . \tag{22}$$

(20) completes the rewriting of (16). Collecting terms from (18),(19),(20), using (21),(22) and doing the remaining replacements in (16) we obtain the expression

$$(16) = \exp \{- \frac{1}{2\eta c} [\frac{\Delta^\mu \tilde{g}_{\mu\nu} \Delta^\nu}{\varepsilon} + \eta c \Delta^\mu \bar{q}^\mu (\frac{3(1-D)}{\phi^2} + \frac{D'}{D\phi})]\}$$

$$\exp \{\varepsilon\eta c[- \frac{1}{4} (\frac{2}{D} + 1)(\frac{3(1-D)}{\phi^2} + \frac{D'}{D\phi}) - \frac{1}{8} \phi^2 (\frac{2(1-D)}{D\phi^2} - \frac{D'}{D\phi})^2 + \frac{1}{8} \frac{(1-D)}{\phi^2} .$$

$$\cdot [2(\frac{2}{D^2} + 1) + (\frac{2}{D} + 1)^2] - \frac{1}{8} (\frac{1-D}{\phi^2})^2 (6\phi^2 + \frac{4\phi^2}{D^2})]\} . \tag{23}$$

It will now become clear why we did all these manipulations. The terms of order $\varepsilon$ in (23) are already expressed in the new radial variable $\phi$ . Introduction of the new variables $(\phi,\lambda,\psi)$ in the other terms gives, using the subscripts $j,j-1$ again,

$$\Delta_j^\mu \bar{q}_j^\mu = \frac{1}{2} (\phi_j^2 - \phi_{j-1}^2) \tag{24}$$

and

$$\Delta_j^\mu \tilde{g}_{\mu\sigma j} \Delta_j^\sigma = \Delta_j^\mu \Delta_j^\mu D_j + \frac{(1-D_j)}{\phi_j^2} (\Delta_j^\mu \bar{q}_j^\mu)^2$$

$$= (\phi_j^2 + \phi_{j-1}^2 - 2\phi_j \phi_{j-1} \cos \theta_{j,j-1}) + \frac{(1-D_j)}{\phi_j^2} \frac{1}{4} (\phi_j^2 - \phi_{j-1}^2)^2 \tag{25}$$

where $\cos\theta_{j,j-1} = \cos\lambda_j \cos\lambda_{j-1} + \sin\lambda_j \sin\lambda_{j-1} \cos(\psi_j - \psi_{j-1})$. (26)

In the new variables the measure is

$$dq_j \sqrt{g(q_j)} = d\phi_j d\lambda_j d\psi_j (\sin^2\phi_j \sin\lambda_j) \tag{27}$$

giving us finally I as an integral over the variables $\phi_j, \lambda_j, \psi_j$

$$I = \lim_{N\to\infty} \frac{\sqrt{g(x)}}{(2\pi\varepsilon\eta c)^{3/2}} \int \prod_1^N \frac{d\phi_i d\lambda_i d\psi_i}{(2\pi\varepsilon\eta c)^{3/2}} \sin^2\phi_i \sin\lambda_i$$

$$\exp\{-\frac{1}{2\eta c} \sum_1^{N+1} (\phi_j^2 + \phi_{j-1}^2 - 2\phi_j\phi_{j-1} \cos\theta_{j,j-1})\frac{D_j}{\varepsilon} + \frac{(1-D_j)}{4\phi_j^2} \frac{(\phi_j^2 - \phi_{j-1}^2)^2}{\varepsilon}$$

$$\exp\{-\frac{1}{4} \sum_1^{N+1} (\frac{3(1-D_j)}{\phi_j^2} + \frac{D_j'}{D_j\phi_j})(\phi_j^2 - \phi_{j-1}^2) + \varepsilon\eta c \sum_1^{N+1} V_j \tag{28}$$

where $V_j$ is given by the last part of (23).
The integration over the angular variables can be done explicitly. Taking into account that $\phi_o = 0$, we have to compute

$$\prod_{i=1}^N \int_o^\pi d\lambda_i \int_o^{2\pi} d\psi_i \sin\lambda_i \exp \frac{\phi_i\phi_{i+1}D_{i+1}}{\varepsilon\eta c} \cos\theta_{i+1,i} \tag{29}$$

$\theta_{i+1,i}$ is the angle between the vector $q_i$ and $q_{i+1}$. Exploiting the rotational invariance of the measure, we introduce spherical coordinates such that $\theta_{i+1,i}$ is the new angle with respect to the Z-axis along $q_{i+1}$. Then (29) becomes

$$\prod_{i=1}^N \int_o^\pi d\lambda_i \int_o^{2\pi} d\psi_i \sin\lambda_i \exp\{\frac{\phi_i\phi_{i+1}D_{i+1}}{\varepsilon\eta c} \cos\lambda_i\}$$

$$= \prod_{i=1}^N (2\pi) \int_{-1}^{+1} du_i \exp \frac{\phi_i\phi_{i+1}D_{i+1}}{\varepsilon\eta c} u_i \}$$

$$= \prod_{i=1}^N \frac{2\pi\varepsilon\eta c}{\phi_i\phi_{i+1}D_{i+1}} (\exp\{\frac{\phi_i\phi_{i+1}D_{i+1}}{\varepsilon\eta c}\} - \exp - \{\frac{\phi_i\phi_{i+1}D_{i+1}}{\varepsilon\eta c}\}) \tag{30}$$

Inserting this into (28) we obtain a multidimensional integral over the variables $\phi_i$ (with $0 \leq \phi_i \leq \infty$ ) of the form

$$\prod_1^N \int_o^\infty d\phi_i\ F(\phi_j^2)\ \frac{1}{\phi_1} \prod_2^{N+1} (\exp\{\frac{\phi_i\phi_{i-1}D_i}{\varepsilon\eta c}\} - \exp\{\frac{-\phi_i\phi_{i-1}D_i}{\varepsilon\eta c}\}). \tag{31}$$

Here we used the fact that the integrand in (28), except for the part leading to (30) is expressed in terms of the radial variables $\phi_j$, and that these terms are invariant under substitution $\phi_j \to -\phi_j$. Indeed, only

$D_j = \frac{\sin^2\phi_j}{\phi_j^2}$ , $\phi_j^2$ and $\frac{D_j'}{\phi_j} = 2\frac{\sin\phi_j}{\phi_j}\frac{\cos\phi_j}{\phi_j^2} - 2\frac{\sin^2\phi_j}{\phi_j^4}$ occur, which are

symmetric functions of $\phi_j$ , while

$$\prod_2^{N+1} (\phi_j\phi_{j-1}D_j)^{-1} = \frac{1}{\phi_{N+1}} \prod_2^{N+1} (\sin^2\phi_j)^{-1}\ \frac{1}{\phi_1}. \tag{32}$$

In (31) $F(\phi_j^2)$ then stands for all remaining factors. It is important to see that (31) can be brought to the form

$$\prod_1^N \int_{-\infty}^{+\infty} d\phi_i F(\phi_j^2)\ \frac{1}{\phi_1}\ \exp\{\sum_2 \frac{D_j\phi_j\phi_{j-1}}{\varepsilon\eta c}\} \tag{33}$$

where all the integrations are now from $-\infty$ to $+\infty$ . That this is true can be seen by writing (33) as

$$\int_{-\infty}^{+\infty} \frac{d\phi_1}{\phi_1} \prod_2^N \int_o^\infty d\phi_j\ (\sum_+ \exp\{\sum_2^{N+1} \pm \frac{D_j\phi_j\phi_{j-1}}{\varepsilon\eta c}\})F(\phi_j^2) \tag{34}$$

$\sum_+$ means that we sum over the $2^N$ terms which have a sequence (+,-,-,+,...,+,-) in the exponential with an even number of minus signs. This sum arises when we split $\int_{-\infty}^{+\infty} d\phi_i\ f(\phi_i)$ as $\int_o^\infty d\phi_i(f(\phi_i) + f(-\phi_i))$ and remark that each time two terms, $\frac{D_i}{\varepsilon\eta c}\phi_i\phi_{i-1}$ and $\frac{D_{i+1}}{\varepsilon\eta c}\phi_{i+1}\phi_i$ , change sign. Splitting also the integration over $\phi_1$ , we obtain

$$\prod_1^N \int_o^\infty d\phi_i\ \frac{1}{\phi_1} F(\phi_j^2)\ (\sum_+ \exp\{\sum_2^{N+1} \pm \frac{D_i\phi_i\phi_{i-1}}{\varepsilon\eta c} - \sum_- \exp\{\sum_2^{N+1} \pm \frac{D_i\phi_i\phi_{i-1}}{\varepsilon\eta c}\}) \tag{35}$$

where $\sum_-$ indicates that there is an odd number of minus signs in the exponential. Expansion of the product in (31) leads to precisely the same form. (33) gives then a representation of I by means of an "ordinary" functional integral in one variable, which is integrated from $-\infty$ to $+\infty$

$$I = \lim_{N\to\infty} \frac{\sqrt{g(x)}}{(2\pi\varepsilon\eta c)^{3/2}} \int \prod_{1}^{N} \frac{d\phi_i}{\sqrt{2\pi\varepsilon\eta c}} \frac{\sin^2\phi_i}{\phi_i\phi_{i+1}D_{i+1}} \exp\{ -\frac{1}{2\eta c} \sum_{1}^{N+1}$$

$$(\phi_j^2 + \phi_{j-1}^2 - 2\phi_j\phi_{j-1}) \frac{D_j}{\varepsilon} + \frac{(1-D_j)}{\varepsilon\,\phi_j^2} \frac{(\phi_j^2 - \phi_{j-1}^2)^2}{4\varepsilon} \}$$

$$\cdot\exp\{ -\frac{1}{4} \sum_{2}^{N+1} [ \frac{3(1-D_j)}{\phi_j^2} + \frac{D_j'}{D_j\phi_j} ] (\phi_j^2 - \phi_{j-1}^2) + \varepsilon\eta c \sum_{1}^{N+1} V_j \}. \quad (36)$$

In the measure we have

$$\prod_{1}^{N} \frac{\sin^2\phi_i\phi_{i+1}^2}{\phi_i\phi_{i+1}\sin^2\phi_{i+1}} = \frac{\sin^2\phi_1}{\phi_1} \frac{\phi_{N+1}}{\sin^2\phi_{N+1}} \quad (37)$$

while in the exponential the dominating part $\frac{\Delta_j^2}{\varepsilon}$ $(\Delta_j = \phi_j - \phi_{j-1})$ appears

$$\exp\{ -\frac{1}{2\eta c} \sum_{1}^{N+1} \Delta_j^2 \frac{D_j}{\varepsilon} + \frac{(1-D_j)}{\varepsilon\phi_j^2} \Delta_j^2 \bar{\phi}_j^2 = \exp\{ -\frac{1}{2\eta c} \sum_{1}^{N+1} \frac{\Delta_j^2}{\varepsilon} - \frac{(1-D_j)}{\phi_j} \frac{\Delta_j^3}{\varepsilon} +$$

$$+ \frac{(1-D_j)}{4\varepsilon\phi_j^2} \Delta_j^4 \} \quad (38)$$

and after doing the replacements we obtain

$$\exp\{ -\frac{1}{2\eta c} \sum_{j=1}^{N+1} \frac{\Delta_j^2}{\varepsilon} \} \exp\{ \sum_{1}^{N+1} \frac{3}{2} \frac{(1-D_j)}{\phi_j} \Delta_j + \varepsilon\eta c \left(\frac{3}{4} \frac{(1-D_j)}{\phi_j^2} - \frac{3}{8} \frac{(1-D_j)}{\phi_j^2} \right) \}. \quad (39)$$

The term of order $\sqrt{\varepsilon}$ in (36) is then

$$\exp\{ -\frac{1}{2} \sum_{1}^{N+1} [\frac{3(1-D_j)}{\phi_j^2} + \frac{D_j'}{D_j\phi_j} ]\Delta_j\bar{\phi}_j - \frac{3(1-D_j)}{\phi_j} \Delta_j \}$$

$$\doteq \exp\{ \sum_{1}^{N+1} -\frac{1}{2} \Delta_j \left(\frac{D_j'}{D_j}\right) + \varepsilon\eta c \left(\frac{3}{4} \frac{(1-D_j)}{\phi_j^2} + \frac{1}{4} \frac{D_j'}{D_j\phi_j} \right) \}$$

$$\doteq \exp\{ \sum_{1}^{N+1} -\frac{1}{2}\Delta_j \left(\frac{D_j'}{D_j}\right)^{[1/2]} + \left(\frac{3}{4} \frac{(1-D_j)}{\phi_j^2} + \frac{1}{4} \left( \frac{D_j'}{D_j\phi_j} - \frac{1}{4} \left( \frac{D_j'}{D_j}\right)' \right)\varepsilon\eta c \right\} \quad (40)$$

Here $\{\sum_1^{N+1} \Delta_j \, (\frac{D'_j}{D_j})^{[1/2]}$ means that $\frac{D'}{D}$ is discretized in the midpoint so that integration is allowed to:

$$\int_{t_o}^{T} dt \, \frac{D'}{D} \, \dot{\phi} = \ln(D) \Big|_{t_o}^{T} = \ln \left( \frac{D(\phi)}{D(0)} \right). \tag{41}$$

(36) is then finally rewritten with the help of (37)(39)(40) and (41) as

$$I = \lim_{N\to\infty} \frac{\sqrt{g(x)}}{(2\pi\varepsilon\eta c)^{3/2}} \; \frac{\Phi}{\sin^2\Phi} \left(\frac{D(o)}{D(\phi)}\right)^{\frac{1}{2}} \int \prod_1^N \frac{d\phi_i}{\sqrt{2\mu\varepsilon\eta c}} \; \frac{\sin^2\phi_1}{\phi_1}$$

$$\cdot \exp\left\{ -\frac{1}{2} \frac{\Sigma \Delta_j^2}{\varepsilon\eta c} + \varepsilon\eta c \sum_1^{N+1} V_j + \frac{3}{4} \frac{(1-D_j)^2}{\phi_j^2} + \frac{3}{8} \frac{(1-D_j)}{\phi_j^2} + \frac{1}{4} \frac{D'_j}{D_j \phi_j} - \frac{1}{4} \left(\frac{D'_j}{D_j}\right)' \right\} \tag{42}$$

$$\phi_o = 0, \; \phi_{N+1} = \Phi .$$

Calculating the terms of order $\varepsilon$ in (42) with the help of (23), we find that they sum up to give

$$\sum_{j=1}^{N+1} \frac{\varepsilon\eta c}{2} = \frac{\eta c}{2} (T-t_o) \tag{43}$$

On the other hand we still have $\frac{\sin^2\phi_1}{\phi_1} = \phi_1(1 - \frac{1}{3}\phi_1^2 + \ldots) = \phi_1(1 - \frac{1}{3}\Delta_1^2) \doteq \phi_1 \, (1 - \frac{1}{3}\varepsilon)$. But we only have to keep $\phi_1$ (since the term in $\varepsilon$ is not summed) giving

$$I = \lim_{N\to\infty} \frac{1}{(2\pi\eta \; c)^{3/2}} \; \frac{\Phi}{\sin\Phi} \; \frac{1}{\sqrt{2\pi\varepsilon\eta c}} \int \prod_1^N \frac{d\phi_i}{\sqrt{2\pi\varepsilon\eta c}} \left(\frac{\phi_1}{\varepsilon\phi}\right) \exp\left\{-\frac{1}{2} \sum_{j=1}^{N+1} \frac{\Delta_j^2}{\varepsilon\eta c}\right\}$$

$$\exp\left\{ \frac{\eta c}{2} (T-t_o) \right\}. \tag{44}$$

Up to the factor $\frac{\Phi}{\sin\Phi}$ this is precisely the functional integral representation of the propagator $P(X,T;0,t_o)$ of a free particle in $\mathbb{R}^3$, when we transform it to spherical coordinates. Indeed

$$P(X,T;0,t_o) = \frac{1}{(2\pi\varepsilon\eta c)^{3/2}} \int \prod_1^N \frac{dq_i}{(2\pi\varepsilon\eta c)^{3/2}} \exp\left\{ \sum_1^{N+1} \frac{-1}{2\varepsilon\eta c} \right.$$

$$\left. \sum_{\mu=1}^{3} (q_i^\mu - q_{i-1}^\mu)^2 \right\} \delta(q_o) \; \delta(q_{N+1} - X) \tag{45}$$

leads under the transformation (6) to

$$P(X,T;0,t_o) = \frac{1}{(2\pi\varepsilon\eta c)^{3/2}} \int \Pi d\phi_i \frac{d\lambda_i d\psi_i}{(2\pi\varepsilon\eta c)^{3/2}} \phi_i^2 \sin\lambda_i$$

$$\exp\{-\frac{1}{2\varepsilon\eta c} \sum_1^{N+1} (\phi_i^2 + \phi_{i-1}^2 - 2\phi_i\phi_{i-1}\cos\theta_{i,i-1})\} \tag{46}$$

doing then the integration over the angular variables, and extending the $\phi_i$ integration again to $-\infty < \phi_i < +\infty$ we obtain (44) up to the factor $\frac{\Phi}{\sin\Phi} \exp\{\frac{\eta c}{2}(T-t_o)\}$ and since

$$P(X,T;0,t_o) = \frac{1}{(2\pi(T-t_o)\eta c)^{3/2}} \exp\{-\frac{1}{2\eta c}\frac{(X^\mu)^2}{(T-t_o)}\} =$$

$$\frac{1}{(2\pi(T-t_o)\eta c)^{3/2}} \exp\{-\frac{1}{2\eta c}\frac{\Phi^2}{(T-t_o)}\} \tag{47}$$

we have for I

$$I = \frac{I}{[2\pi\eta c(T-t_o)]^{3/2}} \frac{\Phi}{\sin\Phi} \exp\{-\frac{\Phi^2}{2\eta c(T-t_o)}\} \exp\{\frac{\eta c}{2}(T-t_o)\} \tag{48}$$

(44) can also be computed directly as

$$\int \frac{d\phi_1}{\sqrt{2\pi\varepsilon\eta c}} \frac{\phi_1}{\varepsilon} \exp\{-\frac{1}{2}\frac{\phi_1^2}{\varepsilon\eta c}\} \exp\{-\frac{1}{2\varepsilon\eta c}\frac{(\Phi-\phi_1)^2}{(T-t_o-\varepsilon)}\}$$

now using the fact that it is related to free motion in 1 dimension up to the factor $(\frac{\phi_1}{\varepsilon})$, and doing first the integrations $\int d\phi_1 \ldots d\phi_N$.

Coming back now to the change of discretization we made in (13), it is clear that we did it to obtain in (29) an integral over the angular variables which could be calculated exactly. Use of (4) would have led us to an integral with $\cos^2\theta_{i+1,i}$ in the exponential apart from the term in $\cos\theta_{i+1,i}$, which leads to a set of error functions, and the problem is then to justify a procedure which should lead again to the integral representation (36). An essential point is also that we keep in (31) the two terms

$$\exp\{\frac{\phi_i\phi_{i-1}D_i}{\varepsilon\eta c}\} - \exp\{-\frac{\phi_i\phi_{i-1}D_i}{\varepsilon\eta c}\}$$

which allows us (33-35) to extend the integrals to $-\infty<\phi_i< +\infty$. In the literature on polar coordinates [A9,E1,L27,P4] we think that this point has not been adequately discussed up to now, and leads there to certain formulas which cannot be correct. We first give a simple example. The path integral representation for a free particle in one dimension is

$$P(Q,t;Q_o,t_o) = \frac{1}{\sqrt{2\pi(T-t_o)}} \exp\{-\frac{(Q-Q_o)^2}{2(T-t_o)} = \frac{1}{\sqrt{2\pi\varepsilon}} \int_{-\infty}^{+\infty} \prod_i^N \frac{dq_i}{\sqrt{2\pi\varepsilon}}$$

$$\exp\{-\frac{1}{2}\sum_1^{N+1} \frac{(q_i - q_{i-1})^2}{\varepsilon}\} \quad \begin{matrix} q_o = Q_o \\ q_{N+1} = Q \end{matrix} \tag{49}$$

We want to write this as a multiple integral where the integrations range from 0 to $\infty$, by splitting (as in (34)) $\int_{-\infty}^{+\infty} dq_i\ f(q_i)$ as $\int_0^\infty dq_i(f(q_i)+f(-q_i))$, and remarking that two terms in the sum $\frac{1}{\varepsilon}\Sigma q_i q_{i-1}$ change sign. This leads to the representation

$$P(Q,T;Q_o,t_o) = \frac{1}{\sqrt{2\pi\varepsilon}} \int_0^\infty \prod_1^N \frac{dq_i}{\sqrt{2\pi\varepsilon}} \exp\{-\frac{1}{2}\sum_1^{N+1}\frac{(q_1^2 + q_{i-1}^2)}{\varepsilon}\}$$

$$\sum_+ \exp\{\pm\frac{q_o q_1}{\varepsilon} \pm \ldots \pm \frac{q_N q_{N+1}}{\varepsilon}\} \tag{50}$$

where the sum is over all sequences $\frac{1}{\varepsilon}\sum_1^{N+1} \pm q_i q_{i-1}$ with an even number of $-$ signs. We further remark that we obtain all terms with an odd number of $-$ signs. by replacing $Q_o$ by $(-Q_o)$, i.e. if $\sum_- \exp \sum_1^{N+1} \pm \frac{q_i q_{i-1}}{\varepsilon}$ indicates summing over

the $2^N$ terms each containing an odd number of minus signs in $\sum \pm q_i q_{i-1}$ then

$$\sum_- \exp\{\sum_1^{N+1} \frac{q_i q_{i-1}}{\varepsilon} = \sum_+ \exp\{\pm\frac{q_i q_{i-1}}{\varepsilon}\}\Big|_{q_o \to -q_o} . \tag{51}$$

This is the basis for the following representations of $P(Q,T;Q_o,t_o) \pm P(Q,T;-Q_o,t_o)$

$$P(Q,T;Q_o,t_o) \pm P(Q,T;-Q_i,t_o) = \frac{1}{\sqrt{2\pi\varepsilon}} \int_0^\infty \prod_1^N \frac{dq_i}{\sqrt{2\pi\varepsilon}} \exp\{-\frac{1}{2}\frac{(q_i^2 + q_{i-1}^2)}{\varepsilon}\}$$

$$\left(\sum_+ \exp\{\sum \pm\frac{q_i q_{i-1}}{\varepsilon}\} \pm \sum_- \exp\{\sum \pm \frac{q_i q_{i-1}}{\varepsilon}\}\right)$$

$$= \frac{1}{\sqrt{2\pi\varepsilon}} \int_0^\infty \prod_1^N \frac{dq_i}{\sqrt{2\pi\varepsilon}} \exp\{-\frac{1}{2}\sum\frac{(q_i^2 + q_{i-1}^2)}{\varepsilon}\}$$

$$\prod_{1}^{N+1} (\exp \{\frac{q_i q_{i-1}}{\varepsilon}\} \pm \exp \{- \frac{q_i q_{i-1}}{\varepsilon}\}) \tag{52}$$

We see here clearly that the second term of each factor in the product is important and cannot be neglected, arguing that $\exp\{- \frac{q_i q_{i-1}}{\varepsilon}\}$ is "exponentially small" with respect to $\exp\{\frac{q_i q_{i-1}}{\varepsilon}\}$. Proceeding backwards, the sign in front of $\exp\{- \frac{q_i q_{i-1}}{\varepsilon}\}$ determines how the integral can be rewritten as the sum or difference of two propagators on $(-\infty, +\infty)$. In fact the 2 expressions

$$P_1 = \frac{1}{\sqrt{2\pi(T-t_o)}} \exp \{- \frac{1}{2} \frac{(Q^2 + Q_o^2)}{T-T_o}\} \, 2 \cosh (\frac{QQ_o}{T-t_o}) \tag{53}$$

and

$$P_2 = \frac{1}{\sqrt{2\pi(T-t_o)}} \exp \{- \frac{1}{2} \frac{(Q^2 + Q_o^2)}{T-t_o}\} \, 2 \sinh (\frac{QQ_o}{T-t_o}) \tag{54}$$

which are iterated in (52) are the propagators for the free particle (or here rather free diffusion) on the half axis. Just like $P(Q,T,Q_o,t_o)$ in (49) satisfies the Chapman-Kolmogorov equation on the whole axis (integrating the intermediate variable from $-\infty$ to $+\infty$), these satisfy the same equation, but now on the half axis. They are the simplest illustration of propagators on quotient spaces (here $\mathbb{R}_+ = \mathbb{R}/Z_2$) [D24]. This simple example will show us that certain path integral representations in polar coordinates which have been given, are not correct and that the problem needs further study. In [E1,P4] one represents the propagator of the free particle in 3 dimensions as

$$G(Q,Q_o,t,t_o) = \sum_{\ell=0}^{\infty} \sum_{m=-\ell}^{+\ell} G_\ell^m (R,R_o;t,t_o) \, Y_\ell^m(\theta,\phi) \bar{Y}_\ell^m (\theta_o,\phi_o) \tag{55}$$

Starting from the usual representation of $G_\ell^m$ in cartesian coordinates, and transforming to spherical coordinates by (6), one easily arrives at the representation (which is still exact in the sense that it is even not necessary to take the limit $N \to \infty$).

$$G_\ell^m(R,R_o,T,t_o) = \frac{1}{\varepsilon} \int_0^\infty \prod_1^N dr_i \frac{r_i^2}{\varepsilon} \prod_1^{N+1} \frac{1}{\sqrt{r_i r_{i-1}}} \exp \{- \frac{1}{2\varepsilon} \sum_1^{N+1} (r_i^2 + r_{i-1}^2)\}$$

$$\prod_1^{N+1} I_{\ell+1/2} (\frac{r_i r_{i-1}}{\varepsilon}) \tag{56}$$

for $\ell = 0$ $\quad I_{1/2}(z) = \frac{1}{\sqrt{2\pi z}} (e^z - e^{-z})$, and

$$G_o^o(R,R_o;T,t_o) = \lim_{N\to\infty} \frac{1}{\sqrt{2\pi\varepsilon}} \frac{1}{RR_o} \int_0^\infty \prod_1^N \frac{dr_i}{\sqrt{2\pi\varepsilon}}$$

$$\exp\{-\frac{1}{2} \sum_1^{N+1} (r_i^2 + r_{i-1}^2)\} \prod_1^{N+1} [\exp\{\frac{r_i r_{i-1}}{\varepsilon}\} - \exp\{-\frac{r_i r_{i-1}}{\varepsilon}\}] \quad (57)$$

which is of the form (52). We obtain the result

$$G_o^o(R,R_o,T,t_o) = \lim_{N\to\infty} \frac{1}{\sqrt{2\pi\varepsilon}} \frac{1}{RR_o} \int_{-\infty}^{+\infty} \prod_1^N \frac{dr_i}{\sqrt{2\pi\varepsilon}} \exp\{-\frac{1}{2} \sum_1^{N+1} \frac{(r_i - r_{i-1})^2}{\varepsilon}\}$$

$$\{[r_o = R_o] - [r_o = -R_o]\}$$

$$= \frac{1}{RR_o} \frac{1}{\sqrt{2\pi(T-t_o)}} \exp\{-\frac{(R^2+R_o^2)}{2(T-t_o)}\} 2 \sinh(\frac{RR_o}{T-t_o}) \quad (58)$$

which is different from the representations given in [E1,P4]. The difference is due to the fact that they neglect $\exp\{-\frac{r_i r_{i-1}}{\varepsilon}\}$ compared to $\exp\{\frac{r_i r_{i-1}}{\varepsilon}\}$. The point is discussed in [L27], in the case of the Schrödinger equation, where one has $\exp\{\frac{i r_i r_{i-1}}{\varepsilon}\}$ and $\exp\{\frac{-i r_i r_{i-1}}{\varepsilon}\}$. Using results of Nelson [N2] they show how to reduce this case to that of the diffusion equation by means of analytic continuation in the mass parameter. They do not realize however, that even in that case (which corresponds to (57)), one cannot neglect the second part.

The expression given in (48) for the propagator $I(x,T,0,t_o)$ is not yet the final result, since it was computed as a propagator for the range of variables $q^i \in \mathbb{R}^3$. In spherical coordinates the variable $\phi$ is given by (6) and restricted to $0 \leqslant \phi \leqslant \pi$, while we used $0 \leqslant \phi \leqslant \infty$. The general situation, for semisimple Lie groups, has been treated by Dowker [D23] and Marinov [M2]. We only apply their results to the case of SU(2). There the propagator, which is in principle a function of 6 variables (3 of the final, and 3 of the initial point) can be expressed in terms of 1 variable. This is a consequence of the underlying group structure and the invariance of the associated Laplacian under right and left shifts, which leaves only the radial variable. The construction of the propagator on the group, then starts with taking this variable first in the algebra $-\infty < \phi < +\infty$, and to restrict it further, using the invariance under the Weyl group (here $\phi \to -\phi$), and the mapping from the algebra onto the group, which implies $I^{SU(2)}(\phi+2\pi) = I^{SU(2)}(\phi)$. Looking at the propagator (48) that we constructed on the half axis, we see that indeed $I(\phi) = I(-\phi)$. Extending it to the whole axis, we must take $\frac{1}{2} I(\phi)$, where the factor $\frac{1}{2}$ comes from the fact that for $(T-t_o) \to 0$, $I$ is normalized to a delta function on the half axis in the point $\phi = 0$. Proceeding then as explained, $U(\phi+2\pi,\hat{\omega}) = U(\phi,\hat{\omega})$ (5) implies that the propagator is defined on the torus, the quotient space $\mathbb{R}/2\pi Z$ , and is obtained by summing over the equivalent points [D17,D24,M2,M3,S3] as

$$\frac{1}{2} \sum_{n=-\infty}^{+\infty} I(\phi+ 2\pi n)$$

We can take here as fundamental domain $[-\pi,\pi]$ . The final step is to reduce this domain to $[0,\pi]$ by invoking invariance under the Weyl group (this is another quotient, by the discrete group $Z_2$). Summing again, (but using $I(\phi) = I(-\phi)$) we obtain

$$I^{SU(2)} = \sum_{n=-\infty}^{+\infty} I(\phi+ 2\pi n) \tag{59}$$

and we see immediately that the factor 2 will give us again, for $T-t_o \to 0$ the correct delta function on $[0,\pi]$ for $\phi = 0$.

This result can also be checked by comparing it to the eigenfunction expansion of the propagator.

$$I^{SU(2)} = \frac{1}{[2\pi\eta c(T-t_o)]^{3/2}} \sum_{n=-\infty}^{+\infty} \frac{\phi+2n\pi}{\sin\phi} \exp\{-\frac{(\phi+2n\pi)^2}{2\eta c(T-t_o)}\} \exp\{\frac{\eta c}{2}(T-t_o)\} \tag{60}$$

can be written as

$$\frac{1}{\sin\Phi} \frac{1}{2\pi} \frac{1}{(2\pi\eta c(T-t_o))^{1/2}} \left(-\frac{\partial}{\partial\Phi}\right) \sum_{-\infty}^{+\infty} \exp\{-\frac{(\Phi+2n\pi)^2}{2\eta c(T-t_o)}\} \exp\{\frac{\eta c}{2}(T-t_o)\}$$

Using the Poisson transformation formula for this $\theta$-function, we obtain

$$\frac{1}{4\pi^2} \frac{1}{\sin\Phi} \left(-\frac{\partial}{\partial\Phi}\right) \sum_{-\infty}^{+\infty} \exp\{in\Phi - \frac{n^2\eta c(T-t_o)}{2}\} \exp\{\frac{\eta c}{2}(T-t_o)\} \tag{61}$$

or

$$I^{SU(2)} = \frac{1}{2\pi^2} \sum_{1}^{\infty} n \frac{\sin n\Phi}{\sin\Phi} \exp\{-(n^2-1)\frac{\eta c}{2}(T-t_o)\} \tag{62}$$

and this coincides with the expansion of I in terms of the characters of the representations of SU(2) [D11,D23,M2,M4,S3] . Our normalization $\frac{1}{2\pi^2}$ corresponds to the volume of the group

$$\int_0^{\pi} d\phi \sin^2\phi \int_o^{\pi} d\lambda \sin\lambda \int_o^{2\pi} d\phi = 2\pi^2 \tag{63}$$

(61) allows a quick evaluation of the propagator for $(T-t_o) = 0$. It gives

$$\frac{1}{2\pi} \frac{1}{\sin\phi} \left( - \frac{\partial}{\partial\phi} \right) \sum_{-\infty}^{+\infty} \delta(\phi-2n\pi) = \frac{1}{2\pi} \frac{1}{\sin^2\phi} \sum_{-\infty}^{+\infty} \delta(\phi-2n\pi) \tag{64}$$

giving the correctly normalized $\delta$-function with respect to the measure in (63).

REFERENCES


A1. Abbott, L.F.,Wise, M.B.,Am.J.Phys.49,37(1981)

A2. Abers,E.S.,Lee,B.W.,Phys.Rep.9,1(1973)

A3. Agarwal,B.S.,Wolf,E.,Phys.Rev.D2,2161,2187,2206(1970)

A4. Albeverio,S.,Combe,Ph.,Hoegh-Krohn,R.,Rideau,G.,Sirugue-Collin,M., Sirugue,M.,Stora,R.(Eds.), Feynman Path Integrals, Lecture Notes in Physics 106,Springer-Verlag, Berlin Heidelberg New York, 1979

A5. Amit,D.J.,Field Theory, the Renormalization Group and Critical Phenomena, McGraw Hill, New York,1978

A6. Antoine,J.-P.,Tirapegui,E.(Eds.), Functional Integration, Theory and Applications, Plenum Press, New York and London, 1980

A7. Arnold,L., Stochastic Differential Equations, Wiley Interscience, New York 1974

A8. Arthurs, A.M.(Ed.),Functional Integration and its Applications, Clarendon Press, Oxford, 1975

A9. Arthurs,A.M.,Proc.Roy.Soc.(London) A313,445(1962)

B1. Bach,A.,Dürr,D.,Phys.Lett.70A,279(1979)

B2. Bausch,R.,Jansen,H.K.,Wagner,H.,Z.Phys.B24,113(1976)

B3. Bender,C.M.,Coopers,F.,Guralnik,G.,Rose,H.A.,Sharp,D.H.,J.Stat.Phys.22, 647(1980)

B4. Berezin,F.A.,Theor.Math.Phys.6,194(1971)

B5. Berezin,F.A., The Method of Second Quantization, Academic Press

B6. Berezin,F.A.,Marinov,M.S.,Ann.Phys.104,336(1977)

B7. Berry,M.V.,Mount,K.E.,Rep.Progr.Phys.35,315(1972)

B8. Bertrand,J.,Irac,M.,Lett.Math.Phys.3,97(1979)

B9. Brachet,M.-E.,Thèse troisième cycle,Paris(1979)

B10. Brachet,M.-E.,Tirapegui,E.,Phys.Lett.71A,174(1979)

C1. Cahill,K.E.,Glauber,R.J.,Phys.Rev.177,1857(1969)

C2. Cameron,R.H.,Martin,W.T., Trans.Am.Math.Soc.58,184(1945)

C3. Cartan,E., Leçons sur la Géometrie de Riemann, Gauthier Villars, Paris,1963

C4. Chang,S.S., Inconsistency in Feynman and Hibbs'evaluation of path integrals by Fourier expansions, preprint 4210-91,Syracuse University, New York 1977

C5. Cheng, K.S.,J. Math.Phys.13,1723(1972)

C6. Christ,N.H.,Lee,T.D.,Phys.Rev.D22,939(1980)

C7. Cohen,L.,J.Math.Phys.7,781(1966)

C8. Cohen,L.,J.Math.Phys.11,3296(1970)

C9. Cohen,L.,J.Math.Phys.17,597(1976)

C10. Combe,Ph.,Rodrigues,R.,Rideau,G.,Sirugue-Collin,M.,Rep.Math.Phys.13,279(1978)

D1. De Dominicis,C.,Peliti,L.,Phys.Rev.B18,353(1978)
D2. Deininghaus,U.,Graham,R.,Z.Phys.B34,211(1979)
D3. Dekker H.,Phys.Lett.A65,388(1978)
D4. --,Phys.Lett.A69,241(1978)
D5. --,Phys.Rev.A19,2102(1979)
D6. --, in A6
D7., --,Physica A103,586(1980)
D8. De Witt,B.S.,Phys.Rev.85,653(1952)
D9. De Witt,B.S.,Rev.Mod.Phys.29,377(1957)
D10. De Witt,B.S., Dynamical Theory of Groups and Fields, Gordon and Breach, New York (1965)
D11. De Witt-Morette,C., Ann.Inst.H. Poincaré 21,153(1969)
D12. --,Comm.Math.Phys.28,47(1972)
D13. --,Comm.Math.Phys.37,63(1974)
D14. --,Ann.Phys.97,367(1976)
D15. De Witt-Morette,C. Maheshwari,A.,in Proc.XIIth Winter School of Theor.Phys. Karpacz,Vol.I,241(1976)
D16. De Witt-Morette,C., Maheshwari,A.,Nelson,B.,Gen.Rel.& Grav.8,581(1977)
D17. --, Phys.Rep.50,257(1979)
D18. De Witt-Morette,C.,Elworth,K.D., Nelson,B.L.,Sammelman,G.S., Ann.Inst.H. Poincaré,32,327(1980)
D19. Dirac,P.A.M.,Phys.Z.Sowj.3,band 3 Heft1(1933),reprinted in S5
D20. Doob,S.L.,Stochastic Processes, Wiley,New York,1953
D21. Dowker,J.S.,Mayes,I.W.,Nucl.Phys.B29,259(1971)
D22. Dowker,J.S.,J.Phys.A3,451(1970)
D23. --,Ann.Phys.62,361(1971)
D24. --,J.Phys.A5,936(1972)
D25. --,J.Phys.A7,1256(1974)
D26. --,in A8
D27. --,J.Math.Phys.17,1873(1976)
D28. Dürr,D.,Bach,A.,Comm.Math.Phys.60,153(1978)
D29. Dürr,D.,Bach,A.,Z.Phys.B32,413(1979)
D30. Dekker,H.,Proof of identity of Graham and Dekker covariant lattice propagators preprint TNO Den Haag,March 1981

E1. Edwards,S.F.,Gulyaev,Y.V.,Proc.Roy.Soc.(London)A279,229(1964)
E2. Elworthy,K.D.,Truman,A.J.,Classical Mechanics,the Diffusion (Heat) Equation and the Schrödinger Equation on Riemannian Manifolds, Warwick University and Heriot-Watt University preprint (1979)

E3. Elworthy,K.D.,Truman,A.J., in A6

E4. Enz,C.P.(Ed.),Pauli's Lectures on Physics, Cambridge Mass.MIT Press,1973 Volume 6: Selected Topics in Field Quantization

E5. Enz,C.P.,in Critical Phenomena, Lecture Notes in Physics 54, Brey,J., Jones,R.B.(Eds.), Springer-Verlag,Berlin Heidelberg New York,1976

F1. Faddeev,L.D.,in Methods in Field Theory, Balian,R.,Zinn-Justin,J.(Eds.), North Holland, Amsterdam 1976

F2. Falkoff,D.,Ann.Phys.4,325(1958)

F3. Feynman,R.P., Rev.Mod.Phys.20,367(1948),reprinted in S5

F4. Feynman,R.P.,Statistical Mechanics(Frontiers in Physics),Benjamin 1972

F5. Feynman,R.P.,Hibbs,A.R., Quantum Mechanics and Path Integrals;McGraw Hill, New York,1965

F6. Fujita,T.,Kotani,S., The Onsager-Machlup function for diffusion processes, preprint Kyoto University

G1. Garczynski,W.,Rep.Math.Phys.4,21(1973)

G2. Garczynski,W., in A6

G3. Garrico,L.,San Miguel,M.,Progr.Theor.Phys.59,40,55(1978)

G4. Garrido,L.,Seglar,P.,Shepherd,P.J.(Eds.), Stochastic Processes in Nonequilibrium Systems,Lecture Notes in Physics 132, Springer-Verlag, Berlin Heidelberg New York,1980

G5. Garrido,L.(Ed.), Systems far from Equilibrium,Lecture Notes in Physics 132, Springer-Verlag, Berlin Heidelberg New York,1980

G6. Gel'fand,I.M.,Fomin,S.V.,Calculus of Variations, Prentice Hall, Englewood Cliffs,N.J.,1963

G7. Gel'fand,I.M.,Yaglom,A.M.,J.Math.Phys.1,48,1960

G8. Gerstein,I.S.,Jackiw,R.,Lee,B.W.,Weinberg,S.,Phys.Rev.D3,2486(1971)

G9. Gervais,J.L.,Jevicki,A.,Nucl.Phys.B110,93,113(1976)

G10. Goldstein,H.,Classical Mechanics, Reading Mass.,1950

G11. Grabert,H.,Green,M.S.,Phys.Rev.A19,1747(1979)

G12. Gradstein,I.S.,Ryzhik,I.M.,Tables of Integrals,Series and Products, Academic Press,New York,1965

G13. Graham,R., in Springer Tracts in Modern Physics 66,Springer-Verlag, Berlin Heidelberg New York,1973

G14. --,in Fluctuations , Instabilities and Phase Transitions, Tormard Riste (Ed.), Plenum Press, New York,1975

G15. --, Z. Phys.B26,281(1977)

G16. --, Z. Phys.B26,397(1977)

G17. Graham,R.,Ann.Isr.Phys.Soc.,2,948(1978)

G18. Graham,R.,in G4

G19. Graham,R.,in A6

H1. Haken,H.,Z.Phys.B24,321(1976)

H2. Haken,H.,Synergetics an Introduction, Springer-Verlag,Berlin Heidelberg New York,1977

H3. Ham,F.S.,Phys.Rev.166,307(1968)

H4. Hirschfeld,A.C.,Phys.Lett.67A,5(1978)

H5. Hirschfeld,A.C.,Leschke,H.,Suzuki,T., in A6

H6. Hori,S.,Progr.Theor.Phys.7,578(1952)

H7. Horsthemke,W.,Bach,A.,Z. Phys.B22,189(1975)

H8. Horvathy,P.A.,Extended Feynman formula for harmonic oscillator, preprint CNRS Marseille (1979)

H9. Hsue,J.Math.Phys.16,2326(1975)

I1. Ito,H.,Progr.Theor.Phys.59,725(1978)

J1. Jansen,H.K.,Z.Phys.B23,377(1976)

J2. Jouvet,B.,Quantum Aspects of Classical and Statistical Fields, Preprint University of Paris VI,1977

J3. Jouvet,B., Phythian,R., Phys.Rev.A19,1350(1979)

J4. Jensen,R.V., J.Stat.Phys.25,183(1981)

K1. Katz,A.,Classical Mechanics, Quantum Mechanics, Field Theory, Academic Press, New York and London 1965

K2. Kerler,W., in A4

K3. Kerner,E.H.,Sutcliffe,W.G.,J.Math.Phys.11,391(1970)

K4. Klauder,J., in A6

K5. Kovalchick,I.M.,Usp.,Math.Nauk.,109,97(1963)

L1. Langouche,F.,Roekaerts,D.,Tirapegui,E., in G4

L2. --, Phys.Lett.68A,418(1978)

L3. --, J.Phys.11A,L263(1978)

L4. --, Physica 95A,252(1979)

L5. --, Physica 97A,195(1979)

L6. --, Nuov.Cim.53B,135(1979)

L7. --, Progr.Theor.Phys.61,1619(1979)

L8. --, Phys.Rev.D20,419 (1979)

L9. --, Phys.Rev.D20,433(1979)

L10. --, in proceedings first chilean symposium on theoretical physics, Seguel,A. et al.(Eds.) Santiago 1978

L11. --, Lett.Nuov.Cim.25,307(1979)

L12. --, Phys.Lett.72A,413(1979)

L13. --, J. Phys.A13,449(1979)

L14. --, Functional Integrals and WKB-Expansions for Propagators in Curved Space, Lecture Notes, Instituto Venezolano de Investigaciones Cientificas(I.V.I.C.) Caracas Venezuela (1979)

L15. --, Phys.Rev.A21,1344(1980)

L16. --, in Dynamical Systems and Microphysics, Blaquière, A.,Fer,F., Marzollo,A.(Eds.)Springer-Verlag,Wien New York,1980

L17. --, Physica 101A,301(1980)

L18. --, Phys.Lett.77A,3(1980)

L19. --, in A6

L20. --, Lett. Nuov. Cim.29,214(1980)

L21. --, in T4

L22. --, Phys.Rev.D23,1290(1980)

L23. --, Physica 108A,221(1981)

L24. --, Phys.Lett. 82A,309(1981)

L25. --, in proceedings second chilean symposium on theoretical physics, Santiago, december 1980; edited by Universidad Tecnica del Estado

L26. --, Nuov.Cim. 64A,357(1981)

L27. Langguth,W., Inomata,A., J. Math.Phys.20,499(1979)

L28. Lavenda,B.H., Riv.Nuov.Cim.7,229(1977)

L29. Lee,T.D.,Yang,C.N.,Phys.Rev.128,885(1962)

L30. Leschke,H.,Hirschfeld,A.C.,Suzuki,T., Phys.Lett.A67,87(1978)

L31. Leschke,H.,Hirschfeld,A.C.,Suzuki,T.,Phys.Rev.D18,2834(1978)

L32. Leschke,H.,Schmutz,M.,Z.Phys.B27,85(1977)

M1. Mc Laughlin,D.,Schulman,L.S., J.Math.Phys.12,2520(1971)

M2. Marinov,M.S.,Fortschr.Phys.28,511(1979)

M3. --, Phys.Rep.60,1(1980)

M4. Marinov,M.S.,Terentyev,M.V.,Sov.J.Nucl.Phys.28,729(1978)

M5. Marshall,J.T.,Pell,J.L.,J.Math.Phys.20,1297(1979)

M6. Martin,P.C.,Siggia,E.D., Rose,H.A.,Phys.Rev.A8,423(1973)

M7. Maslov,V.P.,Theor.Math.Phys.2,21(1970)

M8. --, Théorie des perturbations et méthodes asymptotiques, Dunod (1972)

M9. Mayes,I.W., Dowker,J.S.,Proc.Roy.Soc.(London)A327,131(1972)

M10. --, J. Math.Phys.14,437(1973)

M11. Mehta,C.L.,J.Math.Phys.5,677(1964)

M12. Milnor,J.,Morse Theory, Annals of Mathematics Studies 51 (Princeton University Press,1969)

M13. Mizrahi,M.M.,Ph.D. Thesis,University of Texas (1975)

M14. --, J. Math.Phys.16,2201(1975)

M15. --, J. Math.Phys.19,298(1978)

M16. --, in A4

M17. --, J. Math. Phys.20,844(1979)

M18. --, Phys. Rev. D22,2096(1980)

M19. --, J. Math.Phys. 22,102(1981)

M20. Moreau,M., Physica 90A,410(1978)

M21. Morette,C., Phys.Rev.81,848(1951)

M22. Mortensen,R.E.,J. Stat.Phys.1,271(1969)

N1. Nicolis,G., Prigogine,I., Self-Organization in Nonequilibrium Systems, Wiley and sons, New York 1977

N2. Nelson,E., J.Math.Phys.5,332(1964)

O1. Onsager,L.,Machlup,S., Phys.Rev.91,1505,1512(1953)

P1. Papadopoulos,G.J.,Devreese,J.T., Path Integrals and their applications in Quantum, Statistical and Solid State Physics, Plenum Press,New York,London 1978

P2. Parker,L., Phys.Rev.D19,438(1979)

P3. Pauli,W., in E2

P4. Peak,D., Inomata,A., J. Math. Phys.10,1422(1969)

P5. Phythian,P.,J. Phys.A8,1423(1975)

P6. Phythian,P.,J. Phys.A10,777(1977)

R1. Ringwood,G.A., J. Phys.A9,1253(1976)

R2. Rozen,G., Formulations of Classical and Quantum Dynamical Theory, Academic Press,New York and London,1969

S1. Samuel,S., J. Math. Phys.21,2806(1980)

S2. Schenzle,A., Brand,H.,Phys.Rev.A20,1628(1979)

S3. Schulman,L.S., Phys.Rev.176,1558(1968)

S4. Schulman,L.S., Techniques and Applications of Path Integration, Wiley Interscience,1981

S5. Schwinger,J.(Ed.), Selected Papers on Quantum Electrodynamics, Dover Publications, New York 1958.

S6. Stratonovich,R.L., Topics in the Theory of Random Noise,Vol.1,Gordon and Breach,New York,1958

S7. Stratonovich,R.L., Sel. Transl.Math.Stat.Prob.10,273(1971)

S8. Sudarshan,E.C.G., Phys.Rev.Lett.10,277(1963)

S9. Suzuki,T.,Hattori,C., Progr.Theor.Phys.57,1722(1972)

S10. --,J.Math.Phys.21,918(1980)

S11. --, Functional Integrals and General Correspondence Rules, preprint DPKU-7905,Kanazawa University, Japan,1979

S12. Suzuki,T., Hirshfeld,A.C.,Leschke,H.,Progr.Theor.Phys.63,287(1980)

T1. Takahashi,Y.,Watanabe,S., The probability functionals (Onsager-Machlup functions) of diffusion processes, Proc. LMS Symp.on stochastic integrals at Durham, 1980

T2. Teitelboim,C., in proceedings first chilean symposium on theoretical physics, Santiago 1978,Seguel,A. et al. (Eds.)

T3. Testa,F.J., J. Math. Phys.12,1471(1971)

T4. Tirapegui,E.(Ed.), Field Theory, Quantization and Statistical Physics, Reidel Publ. Co., Dordrecht (1981)

T5. Tisza,L., Manning,I., Phys.Rev.105,1995(1957)

T6. Truman,A., J.Math.Phys.17,1852(1976)

T7. --, J. Math.Phys.18,1499(1977)

T8. --, J. Math.Phys.18,2308(1977)

T9. --, J. Math.Phys.19,1742(1978)

T10. --, in A4

U1. Um,G.S., J. Math.Phys.15,220(1974)

V1. Van Kampen,N.G., Physica 96A,433(1979)

V2. --, J. Stat. Phys.24,175(1981)

V3. Veblen,O., Invariants of Quadratic Forms, Cambridge University Press, Cambridge, 1952

W1. Wall,H.S., Analytic Theory of Continued Fractions, Van Nostrand, Princeton 1948

W2. Weatherburn,C.E., Riemannian Geometry and the Tensor Calculus, Cambridge University Press, Cambridge, 1966

W3. Weiss,U.Z., Z. Phys.B30,429(1978)

W4. Wiegel,F.W., Physica 33,734(1967)

W5. Wiegel,F.W., Physica 37,105(1967)

W6. --, Phys.Rep.16,57(1975)

W7. Wissel,C., Z. Phys.B35,185(1979)

W8. Wong,E., Zakai,M., Ann. Math.Stat.36,1560(1965)

Y1. Yahata,H., Progr.Theor.Phys.52,871(1974)

Y2. Yasue,K., in A6

Z1. Zinn-Justin,J., in Rollnik,J. Dietz,K.(Eds.) Trends in Elementary Particle Theory, Lecture Notes in Physics 37,Springer-Verlag, Berlin Heidelberg New York , 1975

AUTHOR AND SUBJECT INDEX

# Mathematics and Its Applications



1. Willem Kuyk, *Complementarity in Mathematics, A First Introduction to the Foundations of Mathematics and Its History*. 1977.
2. Peter H. Sellers, *Combinatorial Complexes, A Mathematical Theory of Algorithms*. 1979.
3. Jacques Chaillou, *Hyperbolic Differential Polynomials and Their Singular Perturbations*. 1979.
4. Svtopluk Fučik, *Solvability of Nonlinear Equations and Boundary Value Problems*. 1980.
5. Willard L. Miranker, *Numerical Methods for Stiff Equations and Singular Perturbation Problems*. 1980.
6. P. M. Cohn, *Universal Algebra*. 1981.
7. Vasile I. Istrăţescu, *Fixed Point Theory, An Introduction*. 1981.
8. Norman E. Hurt, *Geometric Quantization in Action*. 1982. (forthcoming)
9. Peter M. Alberti and Armin Uhlmann, *Stochasticity and Partial Order. Doubly Stochastic Maps and Unitary Mixing*. 1982.

FSC
www.fsc.org
MIX
Papier | Fördert gute Waldnutzung
FSC® C083411

Zeitfracht Medien GmbH
Ferdinand-Jühlke-Straße 7
99095 Erfurt, Deutschland
produktsicherheit@kolibri360.de